Science at EPA

Information in the Regulatory Process

by
Mark R. Powell

Routledge
Taylor & Francis Group
New York London

First published 1999 by Resources for the Future

Published 2017 by Routledge
711 Third Avenue, New York, NY 10017, USA
2 Park Square, Milton Park, Abingdon, Oxon OX14 4RN

First issued in hardback 2017

Routledge is an imprint of the Taylor & Francis Group, an informa business

Library of Congress Cataloging-in-Publication Data

Powell, Mark R.
    Science at EPA : information in the regulatory process / by Mark R. Powell.
    p. cm.
    Includes bibliographical references and index.
    ISBN 1–891853–00–7 (alk. paper)

    1. United States. Environmental Protection Agency—Decision making. 2. Environmental policy—United States—Decision making. 3. Environmental sciences—Research—United States. I. Title.
GE180.P68 1999
363.7'056'0973—dc21
                  99–20153
                  CIP

ISBN 13: 978-1-8918-5300-5 (pbk)
ISBN 13: 978-1-1384-6570-1 (hbk)

About
Resources for the Future

Resources for the Future is an independent, nonprofit organization engaged in research and public education with issues concerning natural resources and the environment. Established in 1952, RFF provides knowledge that will help people to make better decisions about the conservation and use of such resources and the preservation of environmental quality.

RFF has pioneered the extension and sharpening of methods of economic analysis to meet the special needs of the fields of natural resources and the environment. Its scholars analyze issues involving forests, water, energy, minerals, transportation, sustainable development, and air pollution. They also examine, from the perspectives of economics and other disciplines, such topics as government regulation, risk, ecosystems and biodiversity, climate, Superfund, technology, and outer space.

Through the work of its scholars, RFF provides independent analysis to decisionmakers and the public. It publishes the findings of their research as books and in other formats, and communicates their work through conferences, seminars, workshops, and briefings. In serving as a source of new ideas and as an honest broker on matters of policy and governance, RFF is committed to elevating the public debate about natural resources and the environment.

Contents

Foreword

The book deals with a subject on which many people have strong opinions. The strong feelings stem from the high economic and political stakes of the decisions involved and the importance of "science" as both a means to truth and a political weapon.

Previous works on the subject have focused primarily on the Environmental Protection Agency's Science Advisory Board and other visible, "tip of the iceberg" aspects of science in the regulatory process. This study probes beneath the surface to describe the basic inner workings of how scientific information is processed, used, and occasionally misused in the development of pollution control regulations. We believe it is the first attempt to try to comprehensively describe this process.

The process that Dr. Powell describes does not look like something from a public administration textbook. It is frequently messy, the actors are often motivated by dedication to the public interest (as they perceive it) but also succumb to less elevated motives of pride and ambition. The usual way to describe technical information and its use is to describe an objective, de-humanized, somewhat mechanical set of interactions. That is not what this study has found. Dr. Powell finds a very human world, strongly affected by individual perspectives and personalities. The writing is designed to reflect that.

The process of producing this book also has been a bit messy, in the sense that the original manuscript was almost twice as long as the published version. The change in length reflects a decision to focus on EPA and to delete material that related to the important role of other agencies and institutions.

EPA accounts for only a small portion (less than 15%) of federal environmental research. Agencies like the National Aeronautics and Space Administration (NASA), the Department of Energy, and the Department of Defense invest far more in environmental research than does EPA. And, of course, the federal government as a whole accounts for only a portion of all environmental research. State governments, universities, and private companies are significant contributors to the research effort.

The generation and use of research within EPA also is heavily influenced by other institutions and agencies. Congress influences the general framework for research through legislation and it impacts the day-to-day research work through oversight and informal contact. The courts significantly influence the use of research through their decisions about EPA actions. Universities, the private sector, news media, and other federal agencies often are significant influences on how EPA does and uses research.

Dr. Powell's book does not ignore these outside-EPA influences and, particularly in the case studies, the effects of these influences are well documented. However, a detailed analysis of environmental science in these other institutions and their influence on EPA must await a future volume. There is more then enough material to ponder and to stimulate thinking in this book.

J. CLARENCE DAVIES
Director
Center for Risk Management
Resources for the Future

Preface

Support for this project was provided by the U.S. Environmental Protection Agency (EPA) under cooperative agreement CR821574 and by general support funds made possible by contributors to Resources for the Future (RFF). I wish to acknowledge the guidance and contributions of Terry Davies, Director of the Center for Risk Management, as well as the helpful comments and support from others at RFF, in particular Shelagh Grimshaw, Nicole Darnall, Marilyn Voigt, John Mankin, Jim Wilson, Kate Probst, Winston Harrington, and John Anderson. I would also like to thank the members of the project's informal advisory group: Karim Ahmed, Baruch Boxer, Jeff Foran, Vic Kimm, Marc Landy, Raymond Loehr, John Moore, Monroe Newman, and Terry Yosie. Finally, I would like to thank Peter Preuss, Director of EPA's National Center for Environmental Research and Quality Assurance, for supporting the project, while making it clear that neither he nor the project's advisors are responsible for the accuracy of the information or objectivity of the views expressed herein. Any remaining errors of omission or commission are, of course, my sole responsibility.

EPA has encountered numerous difficulties in carrying out its mission. Some of these obstacles arise from within the agency, but many are from external sources. There remain significant problems to overcome and progress to be made, but it is important to give credit where it is due. Since its inception in 1970, EPA has achieved remarkable progress in controlling environmental pollution. For this, the public servants that make up the agency deserve our thanks.

This study analyzes the numerous factors and players that constrain, promote, and condition EPA's acquisition and use of science for regulatory decisionmaking. This is admittedly a broadly scoped effort. As such, it runs the risk of sacrificing some precision at the level of fine details in some instances. The decision to conduct such a broad study was based on the judgment that a comprehensive assessment of science at EPA was missing from the existing literature. In attacking this broad problem, the study combined bottom-up and top-down analytic approaches by conducting case studies in parallel with the overall assessment.

I have made considerable efforts to ensure that this study represents the best obtainable version of the facts and faithfully reports the observations of more than 100 respondents ranging from bench scientists to former EPA Administrators. Numerous commenters on previous drafts have helped identify and correct factual errors and gaps or flaws in the analysis. However, the study's methodology (case studies and interviews) and the analysis that flows from it are inherently subjective. Readers will undoubtedly draw different interpretations and arrive at different conclusions from the raw material that I present. I hope, however, that the analysis is relevant, thorough, and informed.

I have also tried to present a candid and balanced discussion of the means by which science is gathered and used in environmental regulatory decisionmaking. Some readers may find the tone sensational, but no one should be shocked to find that the environmental regulatory community is politically charged, or that within an organization like EPA there are competing agendas, interests, and views. Thus, the study discusses bureaucratic jousting, congressional parochialism, and the like because they influence the acquisition and use of science. Some readers might also object to the impolitic comments made by some interviewees that I have chosen to incorporate. These comments are included to convey how respondents actually think and feel about the issues that are within the scope of this study. Often, these sentiments are strong, but incorporating science in environmental regulation is correctly understood as a thoroughly human and value-laden process. Because science and policy depend on judgment as well as facts, this book should provide a useful antidote to those singing the unqualified refrain for environmental regulation based solely on "sound science."

Special thanks are reserved for my wife Linda and my sons, D.J. and Tyler, whose love sustains me.

MARK R. POWELL

Science at EPA

1

Introduction

IMPORTANCE OF THE PROBLEM

This study is broadly concerned with the acquisition and use of scientific information by the U.S. Environmental Protection Agency (EPA) in regulatory decisionmaking. The study focuses chiefly on national rulemaking (for example, setting national ambient air quality standards and banning pesticides or toxic substances), as opposed to site-specific decisionmaking (for example, Superfund remedy selection). It aims to help policymakers and others better understand the factors and processes that influence EPA's acquisition and use of science so that they can better evaluate recommendations for improving environmental regulatory institutions, policies, and practices.

EPA's annual budget during the 1990s has been on the order of $6–7 billion, but this represents a small fraction of national spending on environmental protection. Perhaps the best single indicator of the importance of EPA decisionmaking is the nontrivial figure of 2% of U.S. gross national product expended on compliance with pollution control regulations. EPA (1990) estimated that in 1990 national (including federal, state, local, and private) expenditures for compliance with environmental regulations were approximately $100 billion (1986 dollars).

Continuing controversies surround numerous EPA decisions, the 1997 revisions to the national ambient air quality standards for particulate matter and ozone, for example. There are also widespread perceptions that many EPA decisions do not reflect the best scientific analysis and that the agency lacks adequate safeguards to prevent science from being adjusted to fit policy (U.S. EPA 1992). Rafts of criticisms of EPA's use of science approach the agency from all points on the political compass. (But, obviously, criticism from some quarters can have greater effect than that from others.) For example, over the last decade, the prestigious journal

Science has published numerous editorials alleging regulatory excesses and scientific manipulation by EPA and other regulatory agencies, the mass media, and celebrities to support environmentalist values.[1] Because *Science* is the nation's unofficial science newsweekly, it finds its way into the hands of research administrators, congressional staffers, judges, and other opinion leaders.[2] In the popular media, while post-Watergate reporters tended to portray environmental issues as morality plays featuring industry, government, and environmental advocates, contemporary reports are increasingly skeptical of the scientific claims of environmental scientists and advocates (for example, Budiansky 1993). Popular radio talk-show host Rush Limbaugh purports to debunk environmental science over the airwaves on a regular basis.

Over the past several years, policymakers have increased their scrutiny of the use of science at EPA. Most notably, Congress has debated comprehensive reforms with the stated goal of placing environmental regulation on the firm foundation of "sound science." There has been a blizzard of reports and a cornucopia of conferences focusing on environmental regulatory science. In sum, the economic and political stakes regarding science at EPA are high.

NATURE OF THE PROBLEM

As the title of the 1992 report, *Safeguarding the Future: Credible Science, Credible Decisions*, by the Expert Panel on the Role of Science at EPA suggests, the credibility—and by extension legitimacy—of environmental regulatory decisionmaking rests largely on a broadly shared perception that some form of rational analysis underlies it. Over time, science at EPA has become more institutionalized, decentralized, consistent, rigorous, and comprehensive. By design and custom, however, EPA is not a "science agency" like the National Institutes of Health or the National Science Foundation, which have the conduct or support of scientific research as their primary missions. Typically, EPA's norms, staffing patterns, and incentives subordinate science. Instead, EPA is a regulatory agency dominated by a legalistic culture that often looks for engineering-based solutions to meet statutory obligations.

When EPA was formed by administrative reorganization in 1970, it inherited from other agencies a hodgepodge of programmatic and research offices with distinct (sometimes opposing) bureaucratic cultures and operating procedures. Although roughly one-third of the agency's budget was originally devoted to its Office of Research and Development (ORD), as EPA quickly established itself as a regulatory agency with strong programmatic offices (that is, air, water, land, and toxic chemicals), the proportion of resources allocated to ORD dropped markedly. As a percentage of the agency's total operating budget, ORD's budget now hovers in the single digits. In some cases, program offices assumed the major responsibility of acquiring and evaluating scientific information for their own use (pesticides and toxic substances) or shared the responsibility with ORD (air). In other cases, the science was never a high priority, and no office assumed lead responsibility for developing the scientific information (for example, Superfund). Over time, ORD's resources have diminished and its influence within the agency has waxed and waned. A contributing factor to the research office's lack of consistent prominence within EPA is that the office's appropriate role within the agency—a producer of basic research, a scientific resource for program office clients, or a source of scientific analysis independent from the program offices—has never been fully resolved.

Unlike the leaders of the public health agencies (that is, Centers for Disease Control and Prevention, Food and Drug Administration, and National Institutes of Health) who are traditionally doctors and scientists, EPA policymakers are typically attorneys who lack formal scientific training. Many EPA scientists are skillful at developing useful models and drawing inferences from what limited data are available to address questions relevant to regulatory decisions. The agency, however, does not support the level and type of in-house research and analysis necessary to attract or retain a large cadre of high-caliber scientists. Communications between scientists and policymakers within EPA are often poor or missing, and scientists do not always have "a seat at the table" when regulatory decisions are being hammered out.

On the other hand, the current state of environmental science is such that it often invites decisions to be based on economic, political, administrative, or technological criteria. Due to the lack of data and the inadequacies of existing scientific tools to address questions relevant to regulatory decisions, the scientific uncertainties confronting decisionmakers are generally very large. To a considerable extent, the intractability of environmental science has been responsible for the traditional emphasis placed on technology-based solutions by Congress and EPA. But this may be something of a self-fulfilling prophesy as the inability of science to provide definitive and precise answers immediately undermines arguments for allocating to science the time and resources necessary to get the answers.

Even when science can or could inform regulatory decisions, Congress, the courts, and the public often place time constraints on EPA that prevent the agency from conducting a thorough scientific analysis. Of course, the ability to meet deadlines is a function of workload, resources, and desire, among other things. But, while EPA's operating resources have plateaued and are in decline, the number of decisions it must reach to meet statutory obligations has continued to climb. The Safe Drinking Water Act Amendments of 1986, for example, required EPA to set standards for twenty-five new contaminants every three years. (The Safe Drinking Water Act Amendments of 1996 now require EPA to develop a list of unregulated contaminants and evaluate at least five every five years.) The 1990 Clean Air Act Amendments required EPA to develop standards for a list of 189 hazardous air pollutants. Under the 1996 Food Quality Protection Act, EPA is required to reassess thousands of pesticide residue tolerances within ten years. Reporting requirements issued by numerous congressional committees with jurisdiction over EPA add further to the workload.

Legislative provisions can render science moot in environmental regulatory decisionmaking. Provisions requiring technology-based standards, such as those concerning effluent limits under the Clean Water Act, ensure that engineering and affordability criteria dominate regulatory decisions. In other cases, legislative premises distort the use of science. For example, the Clean Air Act mistakenly presumes that an ambient air pollutant concentration always exists below which the most sensitive members of the population will suffer no adverse health effects. Consequently, the standards would appear to be constrained merely by our technical capability to detect biological changes at ever lower concentrations. At the same time, the Clean Air Act's prohibition of considering costs in setting national ambient air quality standards encourages pseudoscientific debate about precisely what constitutes an adverse health effect because regulatory compliance costs will be sensitive to this subjective definition.

In short, there are a host of interconnected factors emanating from inside and outside EPA that appear to impede the agency's use of scientific information in regulatory decisionmaking. These impediments are discussed in greater detail in the body of the report and are illustrated by case studies in the appendixes. By contrast, relatively few factors appear to facilitate EPA's use of science. Among the most frequently discussed are various forms of external scrutiny, including peer review, which have a reasonably good track record of ensuring some level of quality of regulatory analysis. However, given the scientific uncertainties involved in most environmental regulatory decisions, peer review and official science advisory panels are no panacea—generally, reviewers can reveal uncertainties, but only additional research can resolve them. Relying too heavily on external reviewers and advisers to render judgments that inherently involve policy choices can result in shifting problems rather than solving them and abdicating policy formulation to unaccountable experts.

Another factor that strongly conditions EPA's acquisition and use of science is the role of the agency's political leadership in creating demand for science and in appreciating the potential and limitations of science for informing their decisions. The agency's leadership can set the tone by demanding rigorous scientific analysis using available data. However, as a result of the short tenure of most political appointees, their demand for science tends to be focused on decisions arising only in the short term. Furthermore, the occasional tendency of decisionmakers to seek certainty or political cover from science has caused EPA at times to marshal an uncertain science to defend the exercise of administrative judgment.

USE OF SCIENCE IN ENVIRONMENTAL POLICY

The uses of science in environmental policy are many, including "reality definition," agenda setting, setting the terms of debate, political weaponry, and decisionmaking.

Reality Definition

From the perspective of decisionmakers who can no longer observe gross forms of pollution, such as belching smokestacks and burning lakes, science is a primary lens through which the reality of environmental problems can be perceived. Many environmental problems are cryptic, are multifactorial, or have delayed consequences, making problem identification and assessment complex and uncertain. The absence of a simple, direct, and immediate relationship between a pollutant and a readily observable diagnostic illness or "body count" does not dismiss the possibility that otherwise "important" problems result from complex processes or indirect interactions. One such example would be the depletion of stratospheric ozone by chlorinated fluorocarbons (CFCs).[3] Therefore, science is increasingly the principal source of information about what the remaining and emerging environmental problems are, when they occur, who or what is affected, and how big a problem is.

Agenda Setting

Science and scientists are often key in setting the environmental policy agenda. According to Gregory Wetstone, Legislative Director for the Natural Resources Defense Council and former counsel for the House Subcommittee on Health and the Environment chaired by Rep. Henry Waxman (D-Calif.), the model of regulatory agenda setting under Congresses controlled by the Democratic Party was that activist scientists would raise awareness about an environmental problem through the media and thereby exert pressure on Congress (presentation at the Environmental Law Institute

Public Interest Associates Workshop, Washington, D.C., May 9, 1995). Congress, in turn, could exert pressure on the regulatory agency, earmark appropriations for the agency, or statutorily direct the agency to address the problem. Because the environmental regulatory agenda is crowded with issues competing against one another for resources and attention, another mode for science to set the agenda consists of entrepreneurial staff within EPA marshaling science to elevate an issue on the agency's agenda (the lead in drinking water case study in Appendix A provides an illustration of this). In some cases, new scientific information can cause EPA to revisit previous decisions (as in the PM-10 case study in Appendix C) or move forward on pending decisions (as in the case of ethylene dibromide in Appendix E). In contrast, new science can provide a rationale for postponing action on an agenda item (as in the arsenic in drinking water case study in Appendix B). But in other cases, noisy economic and political considerations drown out the signal of new science, and decisions are pushed down the agency's agenda despite—or perhaps because of—the new evidence (as in the ozone case study in Appendix D).[4] Science may also affect the environmental policy agenda in a less systematic fashion. For example, *Science and Judgment in Risk Assessment* (NRC 1994) suggests that EPA's agenda may often be based on the ease of obtaining data on a particular chemical. Discovery of the occurrence of a contaminant where it is not expected also adds a random component to the agenda-setting equation (as in the dioxin case study in Appendix G).

Setting the Terms of Debate

Kingdon (1984) observes that while problem identification and agenda setting tend to be associated with a visible cluster of political activities and actors, policy *alternatives* tend to be generated and narrowed by relatively hidden, loosely knit communities of specialists. Despite potentially diverse orientations and interests, members of these communities ("policy wonks") share their specialized training and knowledge. Framing an environmental problem as "scientific," rather than as a question of what values are important or how the benefits and costs should be distributed, may narrow the scope of policy conflict, raise the ante for a seat at the bargaining table, and exclude from the process those without scientific credentials or resources. Defining the problem as resolvable only through science may result in narrowing the range of policy alternatives that are generated or considered.

Political Weapon

Science can be mustered to legitimize or undermine policy choices and is a favorite weapon in political battles over environmental policy. For example, a September 1995 hearing before the House Science Energy and Environment Subcommittee, chaired by Rep. Dana Rohrbacher (R-Calif.), gave prominent voice to the minority view of skeptic scientists arguing against current plans to rapidly phase out ozone-depleting CFCs. Although the hearing gave "equal time" to proponents of the phaseout, the presence of the skeptics was disproportionate to the prevailing scientific consensus regarding stratospheric ozone depletion. The hearing came as Rep. John Doolittle (R-Calif.) introduced a bill to postpone the ban on CFCs. It also coincided with the award of the Nobel Prize in chemistry to three scientists for their research on the effects of synthetic chemicals on the ozone layer, marking the first time environmental scientists had received the Nobel. Vice President Gore charged that the hearing was like "the old Stalinist approach to science where you try to silence the people who are real scientists and lift up the fringe element." Rohrbacher responded that "Gore and his far-left allies are environmental bullies who turn world history on its head in their hysterical effort to stifle competing views on environmental issues" and suggested that Gore was using the power of government to "intimidate those who had differing views" (*Risk Policy Report*, October 20, 1995, 21–22). The skeptics alleged that scientists who disagree with the establishment line are denied research funds. When supporters of the ban defended the ozone depletion research on the basis of passing the test of peer review, Doolittle responded, "I'm not going to get involved in peer-review mumbo-jumbo" (Greenberg 1995).[5]

Environmental advocates also use science as a political weapon. The Union of Concerned Scientists (UCS) and the Physicians for Social Responsibility (PSR), groups that have traditionally been in the vanguard of nuclear disarmament, are now spearheading an advocacy campaign focused on protecting the environment (see *Science*, October 23, 1992, 551; November 27, 1992, 1433). While UCS and PSR count many eminent scientists and physicians among their number, their policy prescriptions arguably reflect their environmental values more than their scientific credentials. Their "science-policy" positions can also create a political climate within the scientific community that makes it difficult for scientists to challenge the more "politically correct" views of the prominent scientist-activists. For example, Harvard biologist and UCS supporter Edward O. Wilson's widely publicized claim that up to 50,000 species a year are being driven to extinction may be the best "guestimate" available, but even like-minded scientists are dubious about the basis for the estimate.[6] According to Vernon Heywood, a former chief scientist of the International Union for the Conservation of Nature, scientists "are a lot more skeptical in private. They are very cautious about expressing their views [publicly], because they don't want to be seen rocking the boat. This is the fear it might damage 'the cause'" (quoted in "The Doomsday Myths" [Budiansky 1993]).

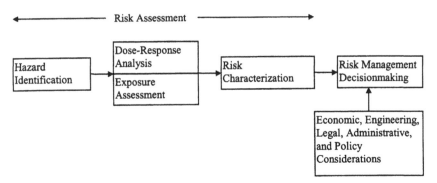

Figure 1.1. The "Risk Paradigm"

Despite concerns over distortions and omissions that occur when science is used as a political weapon, some would argue that science is preferable to other weapons available in the political arsenal (for example, raw emotion, political threats, bribes) because it is more substantive, objective, and transparent than the alternatives.

Decisionmaking

Science does play an important part in environmental regulatory decisionmaking. For the purposes of this study, environmental science refers to information that can be used in assessing risks to human health, welfare, and the environment. It is a useful orientation to think of environmental risk assessment in terms of the four steps described by the National Research Council (NRC 1983) (Figure 1.1). The first step, hazard identification, consists of identifying the adverse effects that may arise from exposure to pollution and assessing the weight of evidence that a causal relationship exists between exposure and adverse effects. The next two steps can occur in tandem. Dose-response analysis seeks to determine the relationship between the level of exposure to a pollutant and the likelihood, extent, or severity of adverse effects. Exposure assessment attempts to ascertain the levels of a pollutant to which individuals or populations have actually been subjected. Finally, risk characterization combines the preceding steps to present decisionmakers with an assessment of the damages that might occur and how likely they are to occur as a result of environmental pollution.

While seemingly straightforward, risk assessment, in practice, is a messy business. Gaps in the data and holes in our understanding of the science routinely cause risk assessors to make assumptions, and certainly much has been made over the plausibility or reasonableness of EPA's risk data, methods, and assumptions. Substantive critiques of EPA's risk assessments differ fundamentally over whether the agency systematically overestimates or underestimates the magnitude of environmental problems, and plausible evidence can be cited to support both sorts of bias (Finkel 1995). Although EPA's risk assessment methods are generally believed to be conservative (that is, result in overestimation of risks), an analytically messy combination of conservative, nonconservative, and missing elements together with underlying scientific uncertainties renders the actual degree of conservatism (if any) hard to nail down in any particular case (NRC 1994).

Another set of factors that makes risk assessment a difficult task is the sheer quantity and diversity of scientific information needed. Unlike biology, chemistry, or physics, risk assessment is not an "organic" scientific discipline. Instead, the field arose to respond to the needs of regulatory decisionmakers who frequently pose policy-relevant questions that cannot—and may never—be definitively answered by science. Scientifically plausible environmental risk estimates can differ by multiple orders of magnitude (that is, factors of ten). Therefore, risk assessment is distinct from scientific disciplines that verifiably test hypotheses. Unlike discipline-bound scientific research, risk assessment draws from a broad array of scientific disciplines. Take, for example, the case of lead in drinking water as discussed in Appendix A. In formulating the health effects assessment for the 1991 lead in drinking water rulemaking, EPA analysts drew from the work of epidemiologists, statisticians, toxicologists, pharmacologists, water chemistry experts, drinking water surveyors, and more. Regarding the quantity of information used in regulatory risk analysis, the 1986 EPA Air Quality Criteria Document for Lead, which served as the *starting* point for the lead in drinking water analysis, cited more than 150 technical references alone.

Role of EPA's Regulatory Responsibilities

To provide a context for understanding the use of environmental science in environmental regulatory decisionmaking, it is useful to consider the nature of EPA's regulatory responsibilities. EPA's regulatory duties can be divided into three broad categories: (1) setting of national environmental standards and approval of standards promulgated by states, (2) contaminated site cleanup decisions, and (3) implementation, enforcement, and other associated activities (for example, reviewing implementation plans and monitoring regulatory compliance). This study focuses primarily on the agency's exercise of its standard-setting duties. These responsibilities differ across agency programs and derive from varying statutory mandates and interpretations of statutory intent by the agency and the courts. A few of EPA's statutes set "numerical" standards for regulation. The numerical standards establish a quantitative "bright line" separating negligible or permissible risks from intolerable risks. For example, under the 1990 Clean Air Act's hazardous air pollution (HAP) section (Section 112), EPA is authorized to delete a category of stationary sources of HAPs if no source in the category poses a risk of premature mortality to the most exposed individual greater than one in one million (10^{-6}). The recently abandoned Delaney Clause of the Federal Food, Drug, and Cosmetic Act of 1958 stipulated that no additive that was found to be carcinogenic could be allowed in the food supply (that is, zero risk).[7]

Most of EPA's statutory provisions provide nonquantitative, or "narrative" direction, and they can be divided into three types, as they relate to human health: health-based provisions, technology-based provisions, and balancing provisions (adapted from *Setting Priorities, Getting Results* [NAPA 1995]). Different types of provisions may appear within the same statute or even within the same section of a statute.

Health-based provisions direct EPA to protect public health with an adequate (or ample) margin of safety. Such provisions provide the agency with discretion to decide what constitutes an adequate (or ample) margin of safety, from what, and for whom. The provisions under the Clean Air Act for setting the national ambient air quality standards (NAAQS) are of this type, and the act specifies that the agency must consider sensitive members of the population and has been interpreted to forbid the agency to consider costs in setting the health-based standards. Such provisions have been called "rights based" because they assert symbolically that all individuals have a *right* to a minimum level of protection from environmental pollution. Health-based provisions provide opportunities for science to play a large—sometimes dominant—role in decisionmaking and cause policy debates to take place in the realm and terms of science. However, controversies regarding health-based standards do not generally originate from the science but emanate from differing values or competing economic and political interests.

Technology-based provisions direct EPA to require polluters to use the "best conventional," "best available," or "maximum achievable" control technology. These provisions permit or direct the agency to consider technical feasibility, as well as the cost or affordability of technologies. Technology-based provisions include the enforceable maximum contaminant levels under the Safe Drinking Water Act, the enforceable maximum effluent limits under the Clean Water Act, and the first stage of regulation of HAPs under the Clean Air Act.[8] The means of achieving the "health-based" NAAQS are intended to be technology based. Technology-based provisions subordinate science to engineering and economic considerations.

Balancing provisions direct the agency to weigh the costs and benefits of a rulemaking intended to protect people from an "unreasonable risk." The Federal Insecticide, Fungicide, and Rodenticide Act (FIFRA) and the Toxic Substances Control Act (TSCA) require balancing. Under balancing provisions, science provides input into administrative decisionmaking in terms of the health and environmental benefits that may derive from regulatory action. Therefore, science cannot dominate decisionmaking but is required to support decisionmaking.

Statutory provisions also determine the standard for judicial review of EPA decisions, and this standard differs among statutes. Under most of its statutes, EPA is held to the "arbitrary and capricious" standard. Under this standard of review, the agency must demonstrate that its rulemaking is supported by a credible record of decision (which includes EPA's scientific analysis) and has been conducted in compliance with procedures required by the enabling statute. However, under TSCA, EPA is judged by the more demanding "substantial evidence" criterion. As a result, when EPA seeks to regulate substances under TSCA, it assumes a greater burden of proof that its scientific analysis supports the finding of an "unreasonable risk." Under the substantial evidence standard, there is greater potential for executive branch decisions to be overturned by the judiciary.

DESCRIPTION OF THIS STUDY

As mentioned earlier, for the purposes of this study, scientific information is considered to be that information that is useful in assessing environmental risks. Although economic and engineering information are also important in EPA rulemaking, they are not a chief focus of this study. (*Economic Analyses at EPA: Assessing Regulatory Impact* [Morgenstern 1997], another recent publication by Resources for the Future, focuses on the development and use of economic analyses at EPA and contains a dozen case studies.) This study responds to recommendations made by two recent panels of experts that addressed EPA's use of science. To support the implementation of an agency decisionmaking process whereby scientific advice is provided early and often, the Expert Panel on the Role of Science at EPA recommended that

the Agency should analyze how it used science in developing one or more major regulations. The goal of this analysis would be to determine the type of scientific and technical information needed to ensure scientifically credible decisions, as well as the points in the regulatory process at which scientific input is most effective. The analysis should take into account the varying needs and decisionmaking processes of the different EPA program offices. (U.S. EPA 1992, 6)

A similar recommendation was elaborated three years later in a report by the National Research Council's Committee on Research and Peer Review in EPA:

The committee also recommends that EPA examine several completed rulemakings to determine the types, degree, and timing of scientific interactions between regulatory offices and ORD staff, and the effects of such interactions (or lack of them) in shaping agency decisions. From there, and taking into account the varying needs and constraints of the regulatory processes, more concrete and effective guidance for future interactions could emerge. (NRC 1995, 3)

In addition to the reports evaluating federal environmental research and decisionmaking alluded to previously, there is a considerable body of regulatory science literature. Many of the case studies in this arena provide a substantive critique of whether regulatory agencies "got the science right" in a particular case, as well as some discussion of the generally high level of scientific uncertainty in assessing particular risks.[9] In the scholarly regulatory policy literature, Weinberg (1972) coined the term "trans-science" to draw attention to the considerable gray zone between science and policy, defined by questions posed by regulatory decisionmakers "which can be asked of science and yet which cannot be answered by science." McGarity (1979) characterized as "science policy" a class of regulatory issues that included questions posed in scientific terms that were unresolvable by science alone, due either to the practical limits of science or to the need to make value judgments. Jasanoff (1987) discussed the often fruitless competition among scientists, public officials, and interest groups to delineate the boundary between science and regulatory policy. More recently, scholars investigating the use of science in policymaking have concentrated on the well-documented, official science advisory process (for example, Jasanoff [1990]; Smith [1992]).

This study is designed to build on and complement these sources by focusing on the processes by which science is generated, transmitted, and used in environmental regulatory decisionmaking. A key objective of the study is to identify (1) the critical impediments (both within and beyond EPA's control) to the agency's use of scientific information in regulatory development that may respond to practical measures and (2) the factors that have facilitated the effective use of science in particular cases. Including case studies of decisions made (or avoided) provides a means for evaluating decisionmaking processes to illuminate lessons learned and assess informational, analytical, and procedural needs.

The eight case studies are contained in the Appendixes:

*   1991 Lead/Copper Drinking Water Rule (Appendix A)
*   1995 Decision Not To Revise the Arsenic in Drinking Water Rule (Appendix B)
*   1987 Revision of the National Ambient Air Quality Standard for Particulate Matter (Appendix C)
*   1993 Decision Not To Revise the National Ambient Air Quality Standard for Ozone (Appendix D)
*   1983–84 Suspensions of Ethylene Dibromide under the Federal Insecticide, Fungicide and Rodenticide Act (Appendix E)
*   1989 Asbestos Ban and Phaseout Rule under the Toxic Substances Control Act (Appendix F)
*   Control of Dioxins (and Other Organochlorines) from the Pulp and Paper Industry under the Clean Water Act (as part of the combined air/water "cluster rule" proposed in 1993 and finalized in 1997) (Appendix G)
*   Lead in Soil at Superfund Mining Sites (Appendix H)

The case studies were selected in consultation with informal advisers to the project and are not intended as a random or representative sample of EPA regulatory decisions. There is some variability among the case studies in the political and economic stakes involved and in the level of development of the underlying science. As a group, however, the cases tend toward the "high-profile" end of the distribution of EPA decisions; none of the case studies could be fairly characterized as pedestrian. Therefore, some of the case study findings probably do not generalize to routine decisionmaking at EPA. On the other hand, many routine decisions flow directly from high-profile decisions, such as those regarding default procedures for conducting risk assessments. The cases also reflect the traditional centrality of human health risks at EPA and contain little discussion of ecological risks.

The cases selected involve each of the "national" environmental regulatory statutes (Clean Air Act; Safe Drinking Water Act; Toxic Substances Control Act; Federal Insecticide, Fungicide, and Rodenticide Act; and Clean Water Act). Two cases involve decisions to maintain the status quo (ozone and arsenic), as opposed to the remainder of the cases, which involve decisions to change from the status quo. The Superfund case study (Appendix H) is intended primarily as a means of comparing the role of science in site-specific decisionmaking with that in the national rulemaking process.

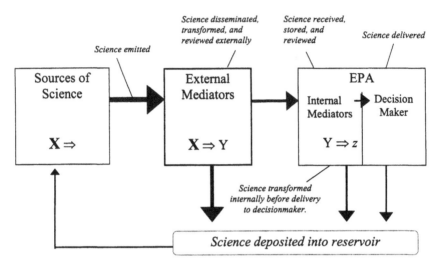

Figure 1.2. Fate and Transport of Science in Environmental Regulation

The interviews conducted as the major focus of the case study research are also not intended to be a survey from which to test statistical inferences. A prominent feature of the case studies consists of an effort to map the origins, flow, and effect of scientific information relating to a particular decision. To accomplish this, the project uses an extended analogy to fate and transport modeling. As used in risk assessment, this modeling procedure predicts the movement and transformation of pollutants from their point of origin to their ultimate destination. Thus, to extend the analogy, one can imagine universities and research institutes "emitting" scientific findings, which are disseminated and "transformed" by the media and consultants outside the agency. (An alternative pattern is the generation of scientific findings within EPA by agency scientists.) Science can enter EPA through multiple "exposure routes," which assimilate information differently. Once inside the agency, information is "metabolized" prior to its "delivery" to the "target organ" (the decisionmaker). This fate and transport terminology is adopted because it is part of the vernacular of many of those providing the information and of many of the ultimate users of the study results. Figure 1.2 presents a simplified model of the fate and transport of science in environmental regulation for illustrative purposes. Note that as scientific information moves from its source of origin toward the regulatory decisionmaker, it decreases in quantity and changes in quality ($X \Rightarrow Y \Rightarrow z$). More detail is added to the model in Chapter 2.

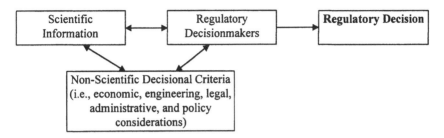

Figure 1.3. The Regulatory Decisionmaking Process

Figure 1.3 combines Figures 1.1 and 1.2 to present a model of the regulatory decisionmaking process, including scientific and other decisional criteria. As Figure 1.3 suggests, there are interactions among the scientific decisional criteria, nonscientific decisional criteria, and the regulatory decisionmaker(s). For example, scientific information is an input for estimating environmental regulatory costs and benefits. The economic and political stakes of a decision influence the volume of scientific information available for decisionmaking, and the operative regulatory statute influences the type of scientific information that is provided (for example, the lowest level at which adverse effects are observed in specific subpopulations or estimated excess cancer cases in the overall population). As well, the type and quantity of science provided for decisionmaking are a function of the regulatory decisionmakers' demands for scientific information.

Making use of these conceptual models, this study attempts to address questions specifically about the *scientific information* in each of the case studies, such as What are the sources and their relative contributions? Where are the points of entry? Who are the gatekeepers? What is the internal transport mechanism? How is the information transformed as it flows through the agency? What does and doesn't get communicated to the decisionmaker? and Where and how is the information ultimately applied?

In addition to the case studies and literature review, interviews were conducted to elicit the observations and judgments of current and former policymakers, senior scientists, specialists in regulatory science issues, and others selected on the basis of their general knowledge of the role of science at EPA. In all, 114 interviews were conducted. (See Appendix I for a listing of interviewees.) Seventy-one interviews focused on one or more of the case studies. The number of interviewees per case study varied roughly from a half dozen to a dozen. Thirty-three interviews were conducted with respondents who could provide an overview of the role of science at EPA (for example, current and former policymakers). An additional ten interviews were conducted to solicit information and comments from individuals knowledgeable about specific issues (for example, program-specific information or the formal elicitation of expert judgment).

Over 90% (104) of the interviews were conducted using one of two structured-question interview guides. One guide was developed for case-specific interviews and the other for interviews regarding the general role of science at EPA. The final interview guides benefited considerably from comments from the project's informal advisory group on early drafts and were further revised after conducting dry-run interviews with two midlevel EPA officials. (Neither of these officials were directly involved in any of the case studies, and their responses to the draft questions were not included in the final analysis.) Appendix J contains copies of the final case-specific and general interview guides. Nine of the case-specific interviewees also agreed to answer a supplementary series of questions, which were a subset of the questions posed in the general interview guide. In addition to responding to questions about a specific regulatory decision, many of the case study interviewees also supplied general observations and perceptions about the acquisition and use of science by the pertinent EPA programs. While the thirty-three "generalist" respondents provided many important insights and substantive comments regarding the case studies (in some instances the respondent was a key decisionmaker in one or more of the cases), the supplementary, case-specific portion of this set of interviews generally followed the intended structured format only loosely, if at all.

The selection of interviewees considered that individuals from the bench scientist through the agency staff analyst to the politically appointed decisionmaker, as well as advocates from outside the agency, would provide informative perspectives. The wide range of interviewees included five of six former EPA administrators, four current or former deputy administrators, and five current or former assistant administrators; four current or former congressional staff; several current and former EPA Science Advisory Board members; various representatives of industry and environmental advocacy groups; environmental journalists; and academics from the diverse fields of biology, public health, economics, political science, psychology, and philosophy. But to better understand the processes occurring *within* the agency, interviewees were disproportionately selected from among current and former EPA officials.

Although there was an effort to ensure balance in the group of respondents, because of the nonrandom nature of the selection process and the limited number of respondents, *extreme* caution should be taken in interpreting the numerical response summaries that are reported. Interviewees were given the option of speaking for attribution or off the record and were informed that all respondents would be identified in an appendix to the report. Almost all interviewees elected to speak off the record.

Allowing interviewees to respond off the record may have the undesirable effect of introducing motivational biases (that is, triggering selective and self-serving memories of some protagonists in regulatory science melodramas). The heavy reliance on interviews can also make it impossible to separate professional judgments from personal or organizational interests. I believe, however, that the use of journalistic methodology is justified because using unattributed sources yields key information about science at EPA that is otherwise unobtainable. Although the official record of decision and other primary source materials carry the patina of legitimacy, they may be unreliable. This is especially true with regard to the information that is missing from primary sources. The language of "authoritative" committee reports is frequently so brokered as to be meaningless. Admittedly, the interview material may sometimes leave the reader with an unsatisfying point–counterpoint dialectic. It is frequently the case, however, that the "hard science" available for EPA decisionmaking is similarly inconclusive and open to competing interpretations.

The major questions addressed in the overall study include the following:

- What are the sources of the science informing agency decisions?
- Who is responsible for assessing risks?
- How does this information move to the decisionmaker (that is, who transmits the information to whom, in what form, and at what stage in the process)?
- What distortions or omissions can occur along the way?
- What steps are taken to evaluate and ensure the quality of the science?
- How important are scientific issues in the decisionmaking process?
- What are the factors that either impede or facilitate the acquisition and use of science?
- How does the role of science in site-specific decisions (for example, Comprehensive Environmental Response, Compensation and Liability Act and Resources Conservation and Recovery Act corrective actions) differ from that in national rulemaking?
- What, if anything, should be done to reform the process by which EPA acquires and uses scientific information?

The remainder of the report is organized as follows. Chapter 2 discusses the sources of science information used in regulatory decisionmaking and the web of communications leading to EPA policymakers. Chapter 3 describes EPA's science resources and the Office of Research and Development's role from the agency's inception. Chapter 4 covers the four program offices and their assessment and use of science. The book concludes with an evaluation (Chapter 5) and policy proposals (Chapter 6). In keeping with the focus of this study, the recommendations do not concern the substance of the science. For example, the study does not critique risk assessment data or methodologies. The focus of this study is on the processes by which the science is generated, transmitted, and used in regulatory decisionmaking.

ENDNOTES

[1] Former *Science* Editor-in-Chief Philip Abelson is responsible for many of these editorials. He suggests that the accumulating history of environmental "doomsayers reveals their lack of judgment, respect for facts, and honesty" (Abelson 1993).

[2] For example, in researching *Breaking the Vicious Circle* (written prior to his appointment to the Supreme Court), Justice Stephen Breyer relied heavily on *Science* (Finkel 1995).

[3] Because this book is chock-full of acronyms, Appendix K contains a list of acronyms for the reader's convenience.

[4] The new findings in the ozone case study suggest biological effects at levels lower than the current standard, and revising the standard downward would entail significant economic and political costs. Therefore, there is a "not on my watch" disincentive for taking action in any administration. The Browner administration proposed new ozone standards shortly after the 1996 presidential election in response to a court-ordered deadline.

[5] For more discussion, albeit from a partisan perspective, on the use of environmental science as a political weapon by the Republican majority of the 104th Congress, see the recent article by Rep. George E. Brown, Jr., ranking Democratic member of the House Committee on Science (Brown 1997).

[6] Wilson was a prominent signatory to the 1995 "Warning to Humanity" issued by UCS.

[7] The Food Quality Protection Act of 1996 replaces the Delaney Clause with a standard intended to provide "reasonable certainty of no harm."

[8] If the residual risk to the most exposed individual from a HAP source after adopting maximum achievable control technology exceeds 10^{-6} (a bright line provision), EPA is required to set second-stage standards that protect public health with an ample margin of safety and is authorized to presume that a 1-in-10,000 (10^{-4}) mortality risk for the most exposed individual is acceptable and that the margin of safety should reduce the risk for the greatest number of persons to less than 10^{-6}.

⁹ See, for example, Alexis de Tocqueville Institution (1994); Fumento (1993); Latin (1988); Lave (1982); Uman (1993); and Whelan (1985, 1993).

REFERENCES

Abelson, P. 1993. Toxic Terror; Phantom Risks. *Science* 261 (July 23): 407.

Alexis de Tocqueville Institution. 1994. *Science, Economics and Environmental Policy: A Critical Examination.* Arlington, Virginia: Alexis de Tocqueville Institution.

Brown, G. 1997. Environmental Science under Siege in the U.S. Congress. *Environment* 39(2): 12–20, 29–31.

Budiansky, S. 1993. The Doomsday Myths. *U.S. News and World Report* (December 13): 81–91.

Finkel, A. 1995. An Environmental Misdiagnosis. *New York University Environmental Law Journal* 3(2): 295–381.

Fumento, M. 1993. *Science under Siege: Balancing Technology and the Environment.* New York: William Morrow and Co.

Greenberg, D. 1995. House of Ignorance. *Washington Post*, October 27, 1995, A25.

Jasanoff, S. 1987. Contested Boundaries in Policy-Relevant Science. *Social Studies of Science* 17: 195–230.

———. 1990. *The Fifth Branch: Science Advisors as Policymakers.* Cambridge, Mass.: Harvard University Press.

Kingdon, J. 1984. *Agendas, Alternatives, and Public Policies.* Boston: Little, Brown & Company.

Latin, H. 1988. Good Science, Bad Regulation, and Toxic Risk Assessment. *Yale Journal of Regulation* 5 (Winter): 89–148.

Lave, L. 1982. *Quantitative Risk Assessment in Regulation.* Washington, D.C.: Brookings Institution.

McGarity, T. 1979. Substantive and Procedural Discretion in Administrative Resolution of Science Policy Questions: Regulating Carcinogens in EPA and OSHA. *Georgetown Law Journal* 67: 729–810.

Morgenstern, R., ed. 1997. *Economic Analyses at EPA: Assessing Regulatory Impact.* Washington, D.C.: Resources for the Future.

NAPA (National Academy of Public Administration). 1995. *Setting Priorities, Getting Results: A New Direction for EPA.* A NAPA Report to Congress. Washington, D.C.: NAPA, April.

NRC (National Research Council). 1983. *Risk Assessment in the Federal Government: Managing the Process.* Washington, D.C.: National Academy Press.

———. 1994. *Science and Judgment in Risk Assessment.* Washington, D.C.: National Academy Press.

———. 1995. *Interim Report of the Committee on Research and Peer Review in EPA.* Washington, D.C.: National Academy Press, March.

Smith, B. 1992. *The Advisers: Scientists in the Policy Process.* Washington, D.C.: Brookings Institution.

Uman, M., ed. 1993. *Keeping Pace with Science and Engineering: Case Studies in Environmental Regulation.* Washington, D.C.: National Academy Press.

U.S. EPA (U.S. Environmental Protection Agency). 1990. *Environmental Investments: The Cost of a Clean Environment.* Washington, D.C.: U.S. EPA, Office of Policy, Planning, and Evaluation, November.

———. 1992. *Safeguarding the Future: Credible Science, Credible Decisions.* Report of the Expert Panel on the Role of Science at EPA. Washington, D.C.: U.S. EPA, March.

Weinberg, A. 1972. Science and Trans-Science. *Minerva* 10: 209–222.

Whelan, E. M. 1985. *Toxic Terror.* Ottawa, Ill.: Jameson Books.

———. 1993. *Toxic Terror: The Truth behind the Cancer Scares,* 2nd ed. Buffalo, N.Y.: Prometheus Books.

2

Acquisition and Use of Science at EPA

Science generally originates from multiple sources and passes through filtering and transformation by intermediaries before its ultimate delivery to environmental policymakers. Mapping the flow of science through EPA reveals a complex web of communications.

PATH OF SCIENCE FROM SOURCES TO DECISIONMAKER: FATE AND TRANSPORT ANALOGY

The path of regulatory science can be traced from its sources to regulatory decisionmakers employing an analogy to fate and transport models used in environmental risk assessment. Figure 2.1 adds some additional detail and complexity to the simple model of the fate and transport of regulatory science introduced in Chapter 1 (Figure 1.2). Here, we see that sources of science are both internal (for example, EPA labs) and external to EPA (for example, academia, other agencies, and industry). EPA receives some science from external sources (for example, EPA contractors and pesticide manufacturers) directly, bypassing external agents. EPA decisionmakers receive some of their scientific information directly from external mediators (for example, environmental and science journalists) without it first passing through the filter of internal mediators (for example, EPA scientists and program managers). In general, as scientific information moves from its source of origin toward the regulatory decisionmaker, it decreases in quantity and changes in quality ($X \Rightarrow Y \Rightarrow z$). Note that the ultimate fate of science that drops out along the path of communication is not necessarily irreversible deposition into a sink; scientific information accumulated in the reservoir can become mobile and reenter the regulatory deci-

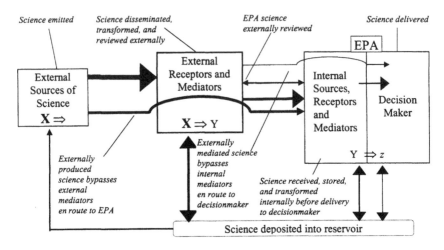

Figure 2.1. Fate and Transport of Science in Environmental Regulation

sionmaking process at various points. Table 2.1 identifies some important sources of science, external mediators, internal receptors and mediators, and decisionmakers.

CASE STUDY EXAMPLES

Examples from the case studies serve to illustrate some of the possible dynamics in the fate and transport of science in environmental regulation.

Lead in Drinking Water

In setting the ambient air quality standard for lead and in phasing out leaded gasoline, the EPA air program relied heavily on an accumulation of academic studies (supported primarily by the National Institutes of Health over the course of many years) regarding neurological impairment in children exposed to lead. The accumulated data formed "an existing body burden" of science in one "body compartment" within the agency. This body burden of data was then supplemented, mobilized, and transported to the EPA drinking water program—a separate body compartment within the agency—by a committed and forceful band of EPA policy entrepreneurs. This group's capacity to produce complex scientific analyses contributed to its clout and gave it a leg up on its challengers within EPA and the Science Advisory Board (SAB) Drinking Water Committee.

Table 2.1. Key Players in the Path of Science from Sources to Decisionmakers

Science Sources	External Mediators	Internal Receptors	Internal Mediators	EPA Decisionmakers
EPA	Official science advisers	ORD	Regulatory development working groups	Assistant administrator
Other federal agencies	Environmental groups	Program office scientists	Assistant administrator	Deputy administrator
State and local agencies	Media	Program managers	Deputy administrator	Administrator
International sources	Trade associations	Decisionmakers		Managers
Academia				
Industry				

Arsenic in Drinking Water

An EPA drinking water program manager received "unfiltered, direct exposure" to an academic study supported by the National Institute of Environmental Health Sciences (NIEHS) Superfund Basic Research Program and reported in the NIEHS journal, *Environmental Health Perspectives*. While previous risk analyses for arsenic in drinking water had focused on skin cancer, this study suggested that internal cancer risks may be substantial. Because internal cancers tend to be more lethal than skin cancers, there was an initial strong reaction to the study at the management level with access to the political decisionmaker. However, an unfounded rumor that the academic researcher had later changed his mind regarding the evidence of internal cancers may have been intended to "repair the damage" caused by unfiltered exposure to the scientific information at this intermediate level. The lack of evidence for internal cancers was cited as the pivotal basis for the drinking water program staff's recommendation that the agency pursue additional research instead of revising the arsenic in drinking water standard.

Particulate Matter National Ambient Air Quality Standard

The economics literature was the source of epidemiological studies of morbidity effects associated with suspended particulate matter. These studies were based on available monitoring and health survey data and enabled researchers to evaluate health effects in economically relevant terms. Within EPA, the Office of Policy Analysis was the internal "receptor and promoter" of these studies, which were incorporated into the regulatory impact analysis (RIA). The Office of Research and Development and the Clean Air Science Advisory Committee, acting in their science gate-keeper role, concluded that too many fundamental questions remained considering the RIA methodology and rendered these studies somewhat "impotent." And, in any event, the Clean Air Act did not permit any official consideration of costs in setting the national ambient air quality standard (NAAQS). Unofficially, the RIA's projection of net benefits at levels below the proposed range for PM-10 helped justify the Thomas administration's decision to select the lowest end of the range for the NAAQS and prevented the Gorsuch administration from relaxing the standard.

Ozone National Ambient Air Quality Standard

Sources inside and outside EPA produced data indicating health effects at ozone concentrations below the existing standard. The data were released into the public domain for several years and subjected to peer review. The information was "transported" to the Administrator's Office and its implications were well understood. However, the elaborate NAAQS procedural requirements had to be satisfied before the science was "available for uptake" and formal consideration by EPA decisionmakers. Furthermore, the evidence of effects at levels below the existing standard was "neutralized" by policy disagreements within the Clean Air Science Advisory Committee.

Ethylene Dibromide Suspensions

Data generated outside EPA on the occurrence of ethylene dibromide (EDB) in groundwater and consumer grain products were released into the public domain. EPA had no control over the data and lost control over interpretation of the data because the agency's reservoir of public trust was "contaminated" by the track record of its outgoing Assistant Administrator for Pesticides and Toxic Substances for manipulating science to provide regulatory relief.

Asbestos Ban and Phaseout Rule

Like nonpest wildlife that is unintentionally exposed to pesticides in the field, the D.C. circuit court—which concluded that EPA had not presented sufficient evidence to justify the asbestos ban and phaseout rule under the Toxic Substances Control Act (TSCA)—was not the intended "target" of scientific information regarding the risks of asbestos. The court received science both indirectly as mediated by EPA and directly through soliciting *amicus curiae* briefs. Despite the court's limited capability to "absorb and metabolize" the science, it substituted its own science policy judgment for that of politically accountable decisionmakers from the more expert administrative agency.

Control of Dioxin (and Other Organochlorines) in Pulp and Paper Mill Effluents

The information provided by the Adsorbable Organic Halide Indicator (AOX), a water quality indicator that measures total organic halides, was first "transported" to EPA from Sweden in the early 1980s via international scientific journals. In 1986, the use of AOX as a regulatory parameter achieved a measure of international legitimacy when the Swedish EPA adopted it. Australia, Austria, Belgium, Finland, and Germany have followed, and AOX was thoroughly assimilated by EPA into the 1993 pulp and paper mill effluent guidelines proposal. However, the "fate" of the information provided by AOX has varied considerably across countries, with only the Swedish standard being more stringent than the final U.S. standard.

Lead in Soil at Mining Sites

The Integrated Exposure Uptake Biokinetic Model for Lead in Children (or IEUBK model) was developed and applied first in the EPA air and water programs and later "assimilated" by the Superfund program. Rodent tests generated by mining site potentially responsible parties challenged the IEUBK model's default assumption regarding the percentage of lead in soil that would be absorbed into the bloodstream of children. A scientist in an EPA regional office designed juvenile swine studies to test whether the IEUBK model's default overstated the bioavailability of lead in soil at mining sites. The results indicated considerable variability among the mining sites tested, with some higher, some lower, and some about the same as the agency's default. It appears, however, that the "fate" of the site-specific data is such that none of the results will increase the liability of the regulated community because EPA deems the cost of removing the contaminated soil from these large sites to be excessive.

FATE AND TRANSPORT ANALOGY LIMITATIONS

Although the fate and transport analogy has some utility for mapping the flow of scientific information from its sources to the decisionmaker in any particular case, there are some critical limitations to the model that preclude too much generalizing. This stems from the fact that unlike an environmental system in which the components and processes are subject to chemical, physical, and biological laws, the regulatory process is not subject to immutable or slowly evolving constraints. When persistent, fat-soluble chemicals are released into the environment, they move inexorably up the food chain. By contrast, in the environmental regulatory development process little, if anything, seems inevitable. The identity and attributes of the components of the regulatory system, as well as the nature of the interacting processes, are subject to change in real time—not over evolutionary epochs. Thus, for example, with a key change in mid-level EPA staff, what once was an impenetrable hide between the science and the decisionmakers can quickly become a permeable, gossamer membrane. A change in administration can result in a dramatic difference in the affinity of regulatory decisionmakers for scientific evidence of an environmental hazard. Statutory amendments can produce fundamental changes in the processes by which the agency acquires and uses science.

Perhaps the most important generalization that *can* be made regarding the fate and transport of regulatory science is that due to the generally short tenure of political appointees, there is a weak or nonexistent feedback loop between decisionmakers and the sources of original research. In at least some EPA administrations, there has been strong feedback from decisionmakers to regulatory analysts working with available data to effect demand for rigorous and transparent scientific analysis. In some cases (such as arsenic in drinking water, Appendix B), there is feedback from decisionmakers to scientists, but it is a diffuse call for further research that provides little guidance about which scientific questions are important to decisionmaking (Figure 2.2).

Some other general observations can be made regarding the fate and transport of regulatory science. One is that regulatory decisionmakers receive very little scientific information directly; the vast majority passes through a web of internal and external intermediaries who filter and transform the information. EPA staff, in particular, occupy key positions in the web of scientific communications to decisionmakers. Internal gatekeepers and intermediaries are critical to the identification, distribution, and packaging of scientific information and its integration within the process of identifying policy options. Despite the establishment of institutions and processes within EPA designed to avoid and resolve scientific disputes by consensus, sometimes there remain competing intermediaries within EPA who send conflicting scientific signals to decisionmakers. Another observation is that due to lack of control over the federal environmental research budget, EPA must rely heavily on external research sources and rarely controls the design of research studies on which it

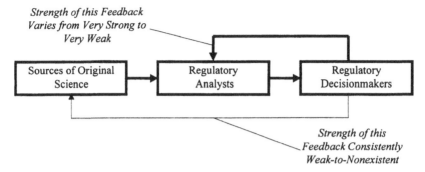

Figure 2.2. Feedback in the Path of Science from Sources to Decisionmakers

depends. As a consequence, regulatory decisions sometimes rely on scientific data not well adapted to the uses to which the agency applies them. A final observation is that as the use of science for regulatory decisionmaking becomes more complex and sophisticated, it becomes more difficult for regulatory decisionmakers to penetrate the increasingly dense black box of environmental risk analysis.

SOURCES OF SCIENTIFIC INFORMATION

The sources of EPA's scientific information used in regulatory decisionmaking vary considerably among recipients, regulatory programs, and types of information and on a case-by-case basis. Internal sources include various offices and laboratories in the Office of Research and Development (ORD), the technical divisions and laboratories in regulatory program offices, and regional environmental services divisions and laboratories. Although EPA's risk assessment guidelines and databases are not sources of original scientific information, they do serve as the source of agency-approved scientific methods, models, and data for regulatory science analysts. External science sources include domestic and international research institutions, other federal agencies, state and local agencies, international agencies, industry and trade associations, environmental groups, and the scientific and mass media.

Some of the research conducted by ORD laboratories has produced seminal work in areas of direct relevance to regulatory decisions—for example, the clinical studies of the effects of criteria air pollutants on human subjects in controlled exposure chambers at the National Health and Environmental Effects Research Laboratory (NHEERL).[1] ORD also conducts substance-specific hazard identification and toxic potency assessments for many regulatory programs, while the pesticides and toxic substances programs generally conduct such assessments on their own. For the most part, the regulatory program and regional offices add to the ORD assessments their own exposure assessments and risk characterizations. Both ORD and the program offices develop guidelines, methods, and models for use in acquiring and assessing scientific information. In some cases, the in-house source of scientific information is unusual or unexpected. The lead in soil at mining sites case study (Appendix H) provides a rare example of a regional office providing applied toxicological research. In the case of dioxin in effluents from pulp and paper mills (Appendix G), preparatory work for an EPA national survey of environmental dioxin levels uncovered an unexpected source of dioxin formation and releases.

Although EPA supports relatively little academic research directly, as the case studies attest, much of the scientific information that the agency uses in regulatory development is academic in origin. The support for the academic work often comes from other federal agencies, such as the National Institute of Environmental Health Sciences. For example, NIEHS was the primary source of support for the studies conducted by New York University researchers that showed that very fine air particles deposit more deeply in the respiratory system than do coarse particles. The results ultimately led to a change in the national ambient air quality standard for particulate matter, which focused control efforts on the particle size of concern to public health. (See Appendix C.) NIEHS was also the principal source of support for academic studies of the cognitive effects of low levels of lead in children. (See Appendix A.) In addition to using academic research findings, EPA makes extensive use of academics as members of or consultants to agency advisory boards, such as SAB or the Federal Insecticide, Fungicide and Rodenticide Act (FIFRA) Scientific Advisory Panel and committees of the National Research Council (NRC).

In addition to supporting basic research, other federal agencies sometimes directly supply scientific information used by EPA. The National Toxicology Program is a primary source of studies on the long-term effects of toxic substances in animals. The National Institute of Occupational Safety and Health (NIOSH), which serves other federal occupational regulatory agencies, also occasionally develops data that are relevant to EPA. In the case of the pesticide ethylene dibromide (EDB), for example, EPA was able to take advantage of a criteria document prepared for NIOSH and a NIOSH mutagenicity study of EDB. (See Appendix E.) EPA also used NIOSH occupational exposure studies in the case of asbestos regulation. (See Appendix F.) Other federal sources of data used in environmental decisionmaking include the U.S. Geological Survey (for example, soils and surface water quality), U.S. Department of Agriculture (food and beverage consumption rates and patterns), U.S. Meteorological Service (weather data), and U.S. Census Bureau (population data).

Industry is the primary source of data for EPA in the pesticides and toxic substances programs. In other program areas, EPA can sometimes cajole industry into providing the data that the agency requires by threatening to use its authority under the Toxic Substances Control Act (see, for example, the dioxin case study). Generally, scientific information that industry produces on its own initiative is intended to screen for hazards that will trigger regulatory scrutiny or expose the firm to potential future liability, to provide information to support a lower estimate of risks to justify less stringent regulation for a given level of environmental protection, or to simply critique EPA's analysis by revealing underlying scientific uncertainties without resolving them. An example of the latter is provided by the arsenic in drinking water case study (Appendix B).

An example of an international research institution developing data used by EPA is provided by the epidemiological studies conducted in Taiwan to assess the relationship between arsenic in drinking water and skin diseases (see Appendix B). The World Health Organization International Agency for Research on Cancer is the source of international scientific consensus on human carcinogenic substance classification. State and local agencies provide a variety of scientific information relevant to regulatory decisionmaking, including public health records and ambient air monitoring data. Environmental groups are not often the source of original scientific data but frequently synthesize and interpret available science.

The scientific and mass media also act as intermediaries but frequently serve as important sources of scientific information for policymakers and opinion leaders. A former EPA Administrator said that he often read about scientific studies for the first time in newspaper articles. In reporting science, journalists exhibit a number of discernible patterns (Stocking 1996). Confronted with scientific uncertainty, journalists tend to downplay caveats and provide little context, so that new findings are not explained in terms of the previous body of research. Generally, disagreements among scientists are not explained. Instead, the affiliation of scientists (that is, industry, academia, advocacy) is reported. At times (for example, when the science or its implications are controversial), journalists tend to play up scientists with fringe or minority viewpoints. An important implication of decisionmakers' receiving their scientific information first through the popular media is that the adage "first impressions are lasting impressions" often holds true.

EPA Risk Assessment Guidelines

The scientific methods and data that are accredited by EPA influence which information gets communicated to decisionmakers and how it is presented. The agency's risk assessment guidelines serve as the source of approved procedures for the acquisition, treatment, and use of scientific information. Thus, the guidelines are intended to establish criteria for the acceptability of scientific data and analysis for use in regulatory decision-making. The guidelines also seek to fill in the gaps in the scientific litera-ture with default assumptions and procedures for use by risk assessment practitioners. Like many of the initiatives to formalize EPA's use of science (including the establishment of the Risk Assessment Forum, the develop-ment of an agencywide toxicity database, and the routinization of ORD representation on regulatory development working groups), a series of episodes during the Gorsuch-Burford administration (1981–1983) gave impetus to the development of a series of formal risk assessment and sci-ence policy guidelines. These episodes (for example, the formaldehyde risk assessment controversy) demonstrated the malleability of environ-mental risk assessment. Since then, the agency has devoted consider-able—though some would argue insufficient—time and resources to developing risk assessment guidelines. The agencywide risk assessment guidelines draw from a variety of scientific sources, including the National Research Council and the Science Advisory Board.

Among the EPA risk assessment guidelines, those for cancer have been most widely used and have arguably had the greatest impact on agency decisions. The use and impact of EPA's noncancer (that is, devel-opmental, reproductive, and neurological toxicity) risk assessment guide-lines are limited by the lack of noncancer toxicity data. According to a for-mer senior EPA official, the pesticides program is the only regulatory area that routinely considers noncancer health effects. This is enabled by the program's strong authority to require pesticide registrants to provide test-ing data. Greater use of the agency's exposure assessment guidelines is limited by the lack of available substance-specific exposure and pharma-cokinetic data. In addition, the use of EPA's new ecological risk assess-ment guidelines is anticipated to be hampered by the lack of data and models for predicting ecological effects at population, community, and ecosystem levels.

The agency's risk assessment guidelines were originally developed to promote a greater level of consistency in the application of science to decisionmaking. Some scientists and stakeholders, however, have criti-cized the guidelines as unduly conservative (that is, risk averse) and their implementation as inflexible. In response, EPA is emphasizing being more faithful to the principles of "sound science" in developing a new generation of guidelines. The underlying science, however, is vague on many policy-relevant questions. As a result, the new guide-lines place increasing burdens on EPA decisionmakers to exercise their science and policy judgment and create new opportunities for the exer-cise of administrative discretion. For example, in EPA's proposed revisions to the cancer risk assessment guidelines, a new narrative approach to cancer hazard identification would place a greater burden on agency decisionmakers to comprehend toxicological information. The new approach for low-dose extrapolation would also require agency decisionmakers to judge explicitly what is an "adequate" margin of safety. The revised guidelines will also generate demands for reassessing the risks of specific toxic substances based on new scientific data. Currently, EPA has no formal agencywide guidelines on the use of epidemiological data or on uncertainty analysis.[2]

In addition to agencywide guidance, regulatory programs and regional offices also develop specialized guidance for the acquisition, treatment, and use of scientific information. Examples of program-specific guidance would include the risk assessment guidance for Superfund and the pesticides and toxic substances testing guidelines. An example of regional guidance would be soil screening values that are intended to identify soil levels of particular contaminants that are of potential concern on contaminated sites. Table 2.2 summarizes the development and status of EPA's agencywide risk assessment guidelines.

EPA Databases and Criteria

According to the General Accounting Office (Guerrero 1995), EPA has developed and maintained hundreds of separate and distinct information systems, each with its own structure and purpose and lacking in standard definitions. As well, many of EPA's scientific data sets have data quality problems and are either incomplete or obsolete. These and other shortcomings make it difficult for the agency and others to conduct scientifically based risk assessments and to evaluate the impact of environmental programs. Many of EPA's data management systems track or manage information for each environmental medium or program. In some cases, the problem has its origins in federal environmental statutes, which direct the agency to collect data and address risks for individual media and pollutants. Another constraint on EPA's ability to acquire more data is the Paperwork Reduction Act of 1980, which requires agencies to submit information-gathering proposals to the Office of Management and Budget (OMB) for review if more than ten people or entities are requested to provide information (5 CFR 1320.3 [c]).

One information system that attempts to integrate and summarize risk analysis data is IRIS (Integrated Risk Information System). IRIS contains summaries of health risk and EPA regulatory information on over 500 specific chemicals. It also contains EPA consensus opinion on potential acute and chronic human health effects related to chemical hazard identification and dose-response assessment.[3] Prior to the institution of

Table 2.2. EPA Risk Assessment Guidelines Development and Status

Guidelines	Development and Status
Carcinogenic/Mutagenic Risk Assessment	Interim guidelines for carcinogenic risk assessment issued in 1976. Revised guidelines for carcinogenicity and mutagenicity issued in 1986. Proposed revisions released in 1996.
Developmental Toxicity Risk	Issued in 1986; revised in 1991.
Reproductive Risk	Proposed in 1988 as separate guidelines for female and male reproductive risks; since merged. Final guidelines issued in 1996.
Neurotoxicity Risk	Proposed in 1995. Finalized in 1998.
Guidelines for Chemical Mixtures	Guidelines issued in 1986. Revised guidelines under development.
Exposure Assessment Guidelines	SAB drafted guidance in 1983. EPA issued Reference Physiological Parameters in Pharmacokinetic Modeling in 1988, Exposure Factors Handbook in 1989, Exposure Assessment Guidelines in 1992, draft revisions of Exposure Factors Handbook in 1995, draft Sociodemographic Data Used for Identifying Potentially Highly Exposed Subpopulations in 1997.
Ecological Risk Assessment Guidelines	In 1989, Superfund program issued interim Environmental Evaluation Manual; ORD issued Field and Laboratory Reference Document for hazardous waste sites. In 1996, EPA issued proposed guidelines, and the Superfund program proposed revisions to its manual.
Uncertainty Analysis	No agencywide guidelines. Region III issued guidance for accepting Monte Carlo simulation in 1994. Policy statement approved in May 1997 permitting use in human health exposure assessments. "Guiding Principles for Monte Carlo Analysis" issued in 1997.
Use of Epidemiological Data	No agencywide guidance.

IRIS, different offices allowed and used different chemical-specific data for risk assessments.

EPA created IRIS in 1986 primarily for use by its program offices in hazard identification and dose-response analysis. IRIS was developed to provide a consistent agency position on the risks of toxic substances and to be accessible to a variety of users. Since the creation of IRIS, the agency's CRAVE (Carcinogen Risk Assessment Verification Endeavor) and RfD/RfC (reference dose/reference concentration, that is, for noncar-cinogen risk assessments) have been standing working groups responsi-ble for developing the unified EPA information in IRIS. Generally, these internal working groups have negotiated the risk summary information entered on IRIS, but occasionally, staff-level disputes have been elevated to senior management and policy levels for resolution (see, for example, the case of arsenic in Appendix B).

Despite the considerable number of chemicals covered in IRIS, it is far from complete. As a result, EPA programs and the regulated community use other databases, including the EPA Health Effects Assessment Sum-mary Tables (HEAST) and Agency for Toxic Substances and Disease Reg-istry toxicological profiles, for information on toxic substances not included in IRIS. In some cases, assessors may ultimately turn to the peer-reviewed or gray scientific literature to find scientific data regarding a substance.

In addition to the lack of coverage, NRC (1994) finds that "the IRIS database has quality problems and is not fully referenced." While the information contained in IRIS is not externally peer reviewed, much of the summary information is derived from the peer-reviewed literature. Critics complain, however, that IRIS oversimplifies complex information and does not adequately convey the uncertainty in its summary informa-tion. For example, IRIS treats some classes of substances or different chemical species of the same substance as toxic equivalents, even though they may have significantly different properties. Also, IRIS values are reported as point estimates without including margins of error. According to an EPA research official, EPA's Office of General Counsel has to this point opposed putting confidence intervals in the IRIS database. One possible explanation is that the agency's lawyers have not wanted to be forthcoming about the uncertainties in the scientific data underlying reg-ulatory decisions.

In response to recommendations for expanding and upgrading IRIS, in 1995, EPA initiated a pilot project intended to streamline the process by which the agency reaches consensus on chemical-specific risk informa-tion that is entered into IRIS. EPA suspended the activity of the RfD/RfC and CRAVE working groups, which were judged cumbersome. The new approach appears to consist of two tracks. Many substance-specific risk assessments will be vetted internally. Based on written comments received within a delimited period, the responsible managers in the pro-gram or regional offices could incorporate the comments, schedule a meeting of reviewers, or convene a conference (*Risk Policy Report*, Sep-tember 22, 1995, 20–21). EPA also formed a new group of agency consen-sus reviewers to conduct an assessment of a handful of chemicals that either are not on IRIS or need major revisions. The new review group is composed of senior-level scientists representing both program and regional offices. The IRIS information will undergo various levels of external review, depending on whether a precedent is set, whether exceptions from defaults are made, and whether unusual information is being considered (*Risk Policy Report*, March 15, 1996, 22).

Health and environmental assessment documents, sometimes generically referred to as criteria documents, provide sources of substance-specific hazard and dose-response information at a much greater level of detail and complexity than does IRIS. Table 2.3 summarizes the health and environmental assessment documents prepared by EPA for the air, water, and hazardous waste programs.

There are a bewildering number of sources of data from inside and outside EPA that can be useful in assessing the magnitude, frequency, duration, and route of human and environmental exposures to pollutants. EPA and the American Industrial Health Council provide handbooks with numerical values for use in modeling exposure, such as body weight and rates of water ingestion, inhalation, and dermal absorption.

Table 2.3. EPA Criteria Documents

Air Quality Criteria Documents (Air CDs) are evaluations of the available scientific literature on the health and environmental effects of criteria pollutants (particulates, ozone, carbon monoxide, sulfur oxides, nitrogen oxides, lead).

Health issue assessments (Tier-1 documents) are the first step in the evaluation process for hazardous air pollutants. The Tier-1 assessment is an initial review of the scientific literature concerning the most important health effects associated with a given chemical or class of substances.

Drinking Water Criteria Documents (Drinking Water CDs) are evaluations of the data available for assessing human health effects of drinking water contaminants.

Ambient Water Quality Criteria Documents (Water CDs) provide an assessment of the potential effects of a pollutant on aquatic life and on human health.

Health and environmental effects documents (HEEDs) are summaries of the literature concerning health hazards associated with environmental exposures to particular chemicals or compounds. They provide health-related limits and goals for emergency and remedial actions under the Comprehensive Environmental Response, Compensation and Liability Act (Superfund). HEEDs contain health effects assessments (HEAs) and reportable quantities (RQs). HEAs are brief, "quick and dirty" quantitative assessments of health effects data. RQs are brief data summaries used to establish levels of chemical substances that must be reported to the National Response Center if a spill occurs.

Health assessment documents evaluate the known health effects data from exposure to particular chemicals or compounds.

The EPA Aerometric Information Retrieval System (AIRS) database contains monitoring data on ambient air pollution levels and air pollutant emissions from point sources. The EPA Toxic Release Inventory includes reports from thousands of companies on their releases of over 600 chemicals into the air, water, and waste streams.

Despite the abundance of sources of exposure data, the data relevant to EPA decisionmaking are sparse and of generally poor quality. Ambient air monitoring data, for example, are widely regarded as superior to the water monitoring data. However, the ambient concentrations typically come from a few, fixed monitors that are not intended to be predictive of actual human exposures. Instead, they are supposed to be located in areas where high air pollution levels are most likely to occur because, given EPA's regulatory mission and budget constraints, the primary purpose of the monitoring system is to evaluate compliance with ambient air quality standards. Air emissions data rely on unvalidated modeled estimates derived from indirect measures, such as production process inputs or outputs. EPA has met with resistance, however, when it has sought to acquire better air quality data. For example, an EPA draft regulation implementing the 1990 Clean Air Act Amendments requiring states to submit more detailed emissions data to AIRS was suspended in 1995 in response to negative comments from states.[4] Similarly, industry has resisted EPA efforts to require continuous air emissions monitoring.

Recognizing that ambient pollution concentration measures may not accurately predict actual human exposures, several federal programs have experimented with population-based monitoring of human blood and tissue samples as an alternative to or supplement for ambient monitoring. The number and type of substances that can be monitored using population-based methods are limited, however, by cost and the availability of analytical procedures. The National Human Adipose Tissue Survey (NHATS), an EPA program, scrutinized mainly pesticide residues from 1967 to 1989, collecting about 12,000 tissue samples, mostly from cadavers. But NRC cited the survey for quality control and other problems, and NHATS is now virtually defunct. The National Health and Nutrition Examination Survey (NHANES) of the National Center for Health Statistics of the Centers for Disease Control and Prevention (CDC) periodically collects blood from a population-based sample of about 20,000 U.S. citizens. EPA used NHANES data in the regulation of lead (Appendix A), but the surveys focus primarily on nutritional data.

A shortcoming of population-based exposure surveys is that if they are not paired with contemporaneous environmental monitoring, there is no way later to predict confidently human exposures based on less costly and intrusive environmental monitoring data. Recently, EPA, the CDC, and other agencies have established a pilot population-based survey, the National Human Exposure Assessment Survey (NHEXAS), to monitor blood, urine, hair, and, in some cases, fingernail samples while also sampling the air, water, soil, food, and dust from inside the houses of subjects.[5]

Risk Analyses

Information on hazards, dose response, exposure, and other relevant topics located on databases, in the scientific literature, or in criteria documents provides the science input for EPA risk analysts. The risk analyses, in turn, are the primary source of scientific information for EPA project and program managers. Briefings and summaries based on risk analyses are a principal source of scientific information for senior managers and EPA policymakers. In some cases, risk analyses are packaged as separate "risk assessment documents." In many cases, however, risk analysis is imbedded in staff papers, rulemaking development documents, and regulatory impact analyses (RIAs). In RIAs, for example, risk analysis serves as the basis for estimating the health and environmental benefits of proposed regulatory action.

According to a 1994 EPA report to Congress (cited in *Setting Priorities, Getting Results* [NAPA 1995, 37–39]), in fiscal year (FY) 1993 the agency completed approximately 7,595 risk analyses. Of these, 6,166 were screening analyses (requiring a few minutes to no more than two person-days of work, once the agency had approved the evaluation method). More than 5,000 of the screens were conducted as TSCA reviews for new and existing chemicals for commercial use. Approximately 1,180 risk assessments were medium level-of-effort projects (requiring more than two person-days and less than four person-weeks). Approximately 249 risk assessments were projects requiring more than four person-weeks. Within this latter group of assessments, the agency is working at any given time on a handful of high-stakes assessments (for example, the dioxin reassessment or the criteria document and staff paper for revision of the ozone national ambient air quality standard) that might take two to four years to complete (excluding review) and cost EPA $1 million or more. Chemical-specific risk assessments are conducted primarily by EPA's Offices of Research and Development, Pesticides and Toxic Substances, and Water. Many, if not most, of EPA's risk assessments are prepared by contractors under agency supervision.

"Hardwired" Science versus Competitive Research Grants

Previous reports (for example, *Safeguarding the Future* [U.S. EPA 1992]) and some respondents in this study suggest that EPA relies too heavily on contractors to perform research and conduct risk assessments. A majority of EPA's research budget supports extramural activities (including contractors, cooperative agreements, research centers, and grants), but the extramural competitive grants program currently accounts for less than 20% of the agency's Science and Technology account.[6] U.S. OTA (1993) notes that even though many contract proposals go through a competitive bid and selection process, the process can lend itself to abuse. The General Accounting Office, for example, criticized the EPA Superfund program for its extensive use and mismanagement of the contracts process (U.S. GAO, 1992).

An academic suggests that "EPA has a strong reputation of developing noncompetitive funding modes that rely a lot on linkages that people develop in the course of their career." Although there are no firm data on what proportion of EPA extramural science is supported by competitive contracts or grants versus noncompetitive funding modes (for example, cooperative agreements), the *perception* that EPA relies on noncompetitive science funding is worth noting. This source points in particular to the agency's practice of supporting research through cooperative agreements rather than through peer-reviewed grant awards. EPA retains more leverage over researchers supported by cooperative agreements than those receiving grants, notes the academic.

According to a former EPA research official, the bureaucratic means of noncompetitive science funding also includes level-of-effort (LOE) contracting, an administrative mechanism by which individual technical projects can be supported if they fall within a class of technical activities that the contractor is well qualified to conduct. The important administrative advantage to LOE contracting is that it can take a year to process a new contract, whereas once established, an LOE contract offers an ongoing vehicle for addressing technical problems that require a quicker response. The potential exists, however, for an office to abuse level-of-effort contracting to ensure that it gets the scientific results it seeks. The Expert Panel on the Role of Science at EPA specifically recommended that program, regional, and policy offices reduce the use of LOE contracts (U.S. EPA 1992).

According to the former EPA research official, because technical capabilities, rather than cost, are the primary criteria for awarding LOE contracts, a contract manager can "hardwire" a contract by specifying the required technical capabilities to match distinctive capabilities of the desired contractor. The upshot is that the contracting office can get the contractor it prefers and may predetermine the results of the work by giving the contract to either a "gun for hire" or an advocate of the preferred policy. It would be virtually impossible, however, to determine how common contract hardwiring is or what the intent of the contracting office was in any particular case. Unfortunately, in many highly specialized areas of environmental science, the number of bona fide experts may be very limited, and any available expert may have strongly held policy views that could color the analysis.

While the academic, peer-reviewed research grant process may also be susceptible to cronyism or other forms of distortion, according to this source, "academics are far less guilty [of seeking to avoid competition for science funding] than the environmental consulting groups—they don't even compare." On the other hand, the peer review process for awarding grants is unable to guarantee quality in science (Jasanoff 1990). In addition, because basic research is acknowledged to be a messy, exploratory process, investigators are not required to adhere to the plans specified in their grant proposals. As a result, the agency compromises its ability to ensure that research will address policy-relevant questions when it awards grants.

There are some valid reasons for EPA to use contracts and cooperative agreements to conduct science activities. Agencies use contracts to support work of a specific, technical nature when methodological development is not a management objective for the task at hand. Contracts allow the agency to ensure that it will get some answer to its questions when it needs the information for decisionmaking. Contracts also permit the agency to stipulate the procedures, data, and methods to be used to ensure that they are consistent with agency practices or comparable to other studies. With cooperative agreements, the agency can support independent analysis of issues that are relevant to itself and the broader public and leverage other resources available to the cooperating institution. Unlike work performed for EPA under contract (which becomes the property of the agency), research supported by a cooperative agreement (or grant) also becomes part of the public domain. Undoubtedly, some fraction of the agency's extramural science resources are diverted by staff to cronies or like-minded investigators outside the agency in ways that avoid competitive pressures and scrutiny. But given the legitimate benefits of contractual mechanisms (including level-of-effort contracting) and cooperative agreements, it is also clear that simply looking at the magnitude of noncompetitive extramural funding overstates the extent of abusive hardwired science.

Official Science Advisers

The agency's official science advisers serve as important sources of scientific information, as a quality assurance mechanism, and as a source of scientific credibility and legitimacy for decisionmaking. EPA has four distinct official science advisory bodies (Table 2.4). The Science Advi-

Table 2.4. EPA Official Science Advisory Bodies

Science Advisory Board (SAB)
   SAB Executive Committee
   Drinking Water Committee
   Environmental Economics Advisory Committee
   Environmental Engineering Committee
   Environmental Futures Committee
   Environmental Health Committee
   Integrated Human Exposure Committee
   Radiation Advisory Committee
   Research Strategies Advisory Committee

Clean Air Act Committees[a]
   Clean Air Scientific Advisory Committee
   Clean Air Act Compliance Analysis Council

FIFRA Scientific Advisory Panel

Board of Scientific Counselors

[a]Administered by the SAB but independent status.

sory Board was created administratively in 1974 to consolidate the fragmentary advisory structure inherited from the component bureaus that formed the EPA. SAB was given statutory footing in 1978 by the Environmental Research, Development and Demonstration Act (ERDDA) and has sought to function as the agency's lead scientific advisory unit. SAB contains an executive committee and eight full standing committees and convenes subcommittees and other panels on an ad hoc basis. The board's small secretariat—headed by the SAB Director—is administratively located in Office of the Administrator. The 1977 Clean Air Act (CAA) Amendments authorized the Clean Air Scientific Advisory Committee (CASAC) to review all criteria documents prior to proposal and promulgation of national ambient air quality standards, and ERDDA required that SAB review the criteria and the NAAQS. Consequently, CASAC is administered by SAB but has a distinct charter and independent status. The 1990 CAA Amendments also established the independent Clean Air Act Compliance Analysis Council to review a cost-benefit analysis of the CAA. The independent status of the CAA committees means that they are not required to clear reports through the SAB Executive Committee. The 1972 FIFRA Amendments established the independent FIFRA Scientific Advisory Panel (SAP). In late 1996, the new Board of Scientific Counselors (BOSC) was administratively established to report directly to the Assistant Administrator for Research and Development.

ERDDA gives SAB fairly broad discretion to review and comment on the scientific and technical basis of proposed criteria documents or regulations relevant to the air, surface and drinking water, toxic substances, and Resource Conservation and Recovery Act programs.[7] Of course, time and resource constraints prohibit SAB from reviewing *all* EPA proposals, and, as a practical matter, the SAB Director negotiates the board's agenda with the regulatory program offices and ORD. In some cases, SAB has initiated studies or exercised its prerogative to conduct review. Most SAB activities, however, are conducted in response to a request by an EPA office or offices, and a charge to the committee is negotiated. The charge to the committee is an important determinant of the kind of review that results. The lead in drinking water case study (Appendix A) provides an example in which reviewers exceeded their charge.

The activities of SAB have increased dramatically over the past twenty years. For example, in 1981, SAB conducted ten reviews, and in 1987, it conducted seventy-seven (Yosie 1988). From 1978 to 1994, SAB and CASAC produced over 450 reports, advisories, letters, commentaries, and consultations (U.S. EPA/SAB 1994). (Advisories, letters, commentaries, and consultations are generally less formal than reports, except that CASAC "closure letters" are the formal means of reporting by CASAC to the Administrator regarding NAAQS criteria documents and staff papers.) In 1989, SAB estimated that 50% of EPA's major activities in one form or another are debated, reviewed, or influenced by SAB (Smith 1992). Currently, SAB and CASAC's ten combined committees include some 100 members augmented by 300 ad hoc consultants and hold approximately fifty meetings and publish about thirty reports each year (*Risk Policy Report*, February 21, 1995). According to the President's FY 1999 budget submission, SAB's budget for FY 1998 was a modest $2.4 million.

Traditionally, SAB comments have tended to follow the form of academic peer review in that science advisers review completed or nearly completed work products. In some cases, SAB comments take the form of "peer input" during earlier stages. Months often elapse between the last public meeting of an SAB panel and the transmittal of a report. Apart from reflecting the normally deliberate academic approach and pace, a contributing factor to this long turnaround time is that SAB panels are frequently given very broad and ambitious scopes of work. Many SAB reviews address a large number of complex, multidisciplinary scientific issues. Furthermore, advisers often work on a virtual pro bono basis. In some cases, EPA administrations have requested that SAB advise them on matters that are more policy than science, such as the relative gravity of terminal versus treatable cancers (see Appendix B).[8]

Under the CAA (Section 109), CASAC was given the following responsibilities (Lippman 1987):

- not later than January 1, 1980, and at five-year intervals thereafter, complete a review of the criteria published under Section 108 of the Clean Air Act and the national primary and secondary ambient air quality standards and recommend to the Administrator any new NAAQS or revision of existing criteria and standards as may be appropriate;

- advise the Administrator of areas in which additional knowledge is required concerning the adequacy and basis of existing, new, or revised NAAQS;

- describe the research efforts necessary to provide the required information;

- advise the Administrator on the relative contribution to air pollution concentrations of natural as well as anthropogenic activity; and

- advise the Administrator of any adverse public health, welfare, social, economic, or energy effects that may result from various strategies for attainment and maintenance of such national ambient air quality standards.

In practice, the NAAQS review process has been messier and more prolonged than the CAA would suggest. CASAC's statutory authority to advise the Administrator on welfare, social, and economic effects gives the committee a standing unlike that of other SAB committees to comment on EPA policy matters.[9] Not surprisingly, CASAC's exercise of its authority to comment on extrascientific matters has been an episodic source of friction between the committee and EPA.[10] CASAC is also unique in that the 1977 CAA Amendments stipulated that the seven members include a physician, a representative of the National Academy of Sciences, and a representative of a state air pollution control agency.

The 1986 Safe Drinking Water Act (SDWA) Amendments require EPA to solicit comments from SAB "prior to proposal of a maximum contaminant level goal and national primary drinking water regulation."[11] However, the law allows SAB to respond "as it deems appropriate" and makes clear that SAB review must be conducted within the statutory timetable for promulgation of drinking water standards. In practice, the SAB Drinking Water Committee (formerly a subcommittee of the Environmental Health Committee) has selectively commented on the more controversial or high-stakes proposed drinking water standards and on overarching scientific issues regarding drinking water. The 1996 SDWA Amendments further require EPA to submit to SAB a list of unregulated contaminants that the agency is considering for standard setting.

The 1972 FIFRA Amendments created the Scientific Advisory Panel. Initially, FIFRA required EPA to consult SAP primarily for decisions to cancel pesticide registrations, but the 1978 FIFRA Amendments directed EPA to consult with SAP on guidelines for scientific analyses. Later, the 1980 FIFRA Amendments required SAP to comment on any EPA decision to suspend a pesticide to prevent an imminent hazard. The 1980 amendments also required SAP review of major scientific studies carried out by or for EPA under FIFRA. FIFRA requires that the National Institutes of Health and the National Science Foundation each nominate six candidates for SAP membership and allows the Administrator to select seven panelists from the list of candidates. Candidates to be selected are to include expertise in toxicology, pathology, environmental biology, and other sciences relevant to assessing the health or ecological risks of pesticides (Jasanoff 1990). In some previous cases, SAP has not had sufficient breadth of expertise to adequately review complex scientific issues, according to a trade association representative. The Food Quality Protection Act of 1996 authorizes the appointment of a scientific review board of sixty scientists to be made available to SAP to assist the panel in its reviews. Thus, SAP is now authorized to adopt the Science Advisory Board practice of employing ad hoc consultants to round out review panels.

SAP activities differ in some important respects from those of SAB. SAP, as a rule, takes on more narrowly scoped topics and reviews than does SAB, none of the SAP projects are self-initiated, and SAP makes a point of issuing its reports within fifteen days. Over time, the scope of SAP's activities has broadened somewhat, and SAP's reviews have become less routine. According to a trade association representative, in the 1980s, SAP was involved in routine, product-specific hazard characterizations (according to EPA's alphanumeric carcinogen classification scheme) and all pesticide special reviews. Now SAP reviews only "the dicey" hazard characterization reviews and focuses on FIFRA testing procedures, risk assessment guidelines, and precedent-setting product-specific reviews. Currently, SAP and SAB are jointly reviewing the development of screens and tests for endocrine disrupters.

There have long been tensions over the number and cost of EPA's advisory committees. According to Smith (1992, 89), when ERDDA was being debated, EPA "was engaged in an intermittent cold war with OMB over the number of its advisory committees … and put intense pressure on Dowd [EPA Administrator Douglas Costle's science adviser and SAB staff director] to abolish or consolidate committees. The 1977 and 1978 congressional actions removed the need for SAB to battle further." More recently, the Clinton administration's "reinventing government" initiative has sought to reduce the number and cost of federal advisory committees (Brown 1995). Despite the calls for expanding peer review of EPA's science and administration requests for SAB to conduct more ambitious projects, the board's resources are limited. The recent establishment of the Board of Scientific Counselors may be indicative of the extent to which growing demand for independent scientific review has outpaced SAB's resources.[12]

Many of those interviewed for this report gave SAB and CASAC credit for improving EPA's acquisition and use of science and for legitimizing the agency's decisionmaking process and outcomes. However, many respondents also expressed concerns about the influence of official science advisers on policy, the potential for conflicts of interest, the "value added" by reviews, and the cost, effort, and time involved in reviews. These issues will be discussed in Chapter 5. Table 2.5 contains a thumbnail sketch of the key role of official science advisory groups in each of the case studies (Appendixes A–H).

THE WEB OF COMMUNICATIONS

The previous section provided a discussion of the various sources of scientific information used by EPA in regulatory decisionmaking. Next, we investigate the web of communications leading to EPA policymakers. It is often an oversimplification to think of a simple chain or easily traceable path of communication linking the sources of science to decisionmakers. Instead, an intricate web of scientific communications often contains multiple intermediaries between the sources of science and EPA decisionmakers.

Research Recipients within the Agency

In most cases, the initial research recipients within EPA are scientists in the research offices and labs or in the technical divisions of the relevant program offices. In some cases, however, research is received initially in the context of one EPA program before being transferred to other affected program areas. Occasionally, EPA managers and policymakers receive scientific information directly from the source (that is, they are briefed by scientists or read science journals) or from an external mediator (for example, the media or interested and affected parties), but more often, they receive scientific information through internal intermediaries. Sometimes staff within EPA receive research directly (that is, unmediated by external agents) and prior to broad disclosure. In other cases, EPA receives research findings mediated by external sources at the same time as does the general public. Sometimes, research recipients within EPA pose the question and know the answer before the results are measured. Other times, they uncover something they did not expect to find.

Table 2.6 summarizes the recipients within EPA of some of the key research findings in each of the case studies in Appendixes A–H. The case studies illustrate that the initial research recipient within EPA is some-

Table 2.5. Role of Official Science Advisory Groups in Case Studies

Drinking Water Standard for Lead (Appendix A)
The SAB Drinking Water Committee is disregarded by a band of EPA policy entrepreneurs and prevailed over by the SAB Environmental Health Committee and CASAC on the health effects of low-level lead exposures in children.

Drinking Water Standard for Arsenic (Appendix B)
The SAB allies with EPA Drinking Water scientific staff to promote arsenic as a precedent-setting departure from the agency's default linear, no-threshold cancer model.

Ambient Air Standard for Particulate Matter (Appendix C)
CASAC suggests that EPA adopt the lower end of the proposed range of the standard and even consider going below the proposed range to provide a greater margin of safety.

Ambient Air Standard for Ozone (Appendix D)
The chair of CASAC is influential in getting new scientific information onto the regulatory agenda, but policy disagreements among CASAC members provide the Administrator with "an out" to justify not revising the standard on the basis of "scientific uncertainty." Despite the broad acceptance of scientific evidence of effects occurring at ozone levels below the current standard, the information is "unavailable" to the Administrator without a full-blown CASAC review.

Suspension of Ethylene Dibromide (Appendix E)
The SAP cautions EPA about the unknown risks of substitute pesticides for EDB. Currently, a primary substitute (methyl bromide) is scheduled for phaseout due to its contribution to stratospheric ozone depletion.

Asbestos Ban (Appendix F)
The SAB reviews EPA's asbestos health effects assessment but is frustrated by its lack of a continued role in reviewing EPA's regulatory analysis.

Control of Dioxins in Pulp and Paper Mill Wastewater Discharges (Appendix G)
EPA uses the results of its previous SAB-reviewed dioxin health effects assessment during an ongoing reassessment. The SAB does not endorse the agency's ecological assessment of the effects of disposing of pulp and paper sludge on land.

Lead in Soil at Mining Sites (Appendix H)
The SAB gives weak support to the use of EPA's Integrated Exposure Uptake Biokinetic model for predicting the concentration of lead in the bloodstream of children exposed to lead at contaminated sites and recommends that the agency develop guidance to prevent misuse of the model at the local level by remediation project managers.

Table 2.6. Research Recipients within EPA for Case Studies

Case Study	Research Findings	Research Recipient	Comments
A. Lead in Drinking Water	Association between lead levels in bloodstream and cognitive performance in children	Health Effects Research Lab in Research Triangle Park, N.C.	Data first used for lead-in-gasoline phaseout
B. Arsenic in Drinking Water	Association between arsenic in drinking water and skin cancer	Pesticides Review Division	Review of arsenical pesticides begun in 1978
C. Ambient Air Particulate Matter	Fine particles penetrate more deeply than coarse suspended particles	Scientists and analysts in air program and research offices in Research Triangle Park, N.C.	Ten years elapse between reception and final regulatory decision
D. Ambient Air Ozone	Clinical studies measuring effects at ozone concentrations below the current standard	Scientists and analysts in air program and research offices in Research Triangle Park, N.C.	The studies were conducted by EPA's Human Studies Division in Chapel Hill, N.C.
E. Ethylene Dibromide	Groundwater contamination by EDB	Pesticides Review Division	Unexpected findings made by affected states
F. Asbestos	Association between asbestos and lung disease	Predates EPA	Asbestosis recognized in 1907; link to lung cancer reported in 1955
G. Dioxins in Effluents from Pulp and Paper Mills	Detection of elevated dioxin levels in fish downstream of pulp and paper mills	EPA National Dioxin Survey	Streams were intended to serve as baseline for uncontaminated areas
H. Lead in Soil at Mining Sites	Absorption of lead in soil by juvenile swine varies about the EPA default assumption	Toxicologist in EPA Regional Office	Juvenile swine replaces rodent as animal model for absorption of lead in soil by children

times far removed—both in terms of time and location within the agency—from the context of the regulatory decision. This finding underscores the importance of EPA's having a strong agencywide science information management system.

Intermediaries between Scientists and Nonscientists

Because EPA policymakers tend to be nonscientists, there are typically intermediaries between them and scientists. Sometimes, the intermediaries play multiple, overlapping roles. Intermediaries occupy positions inside and outside the agency, and, in some cases, there are numerous layers of intermediaries. But, some intermediaries undoubtedly play a greater role than do others. Because they are charged with overall program management and the program scientists report to them, EPA office directors are key intermediaries between scientists and nonscientific decisionmakers within EPA. According to an EPA research official, questions about the science are not generally carried beyond the office directors who shape the information received by assistant administrators. Charged with "making the trains run on time," EPA office directors and program managers are understandably bottom line oriented. Intense time and budgetary pressures can understandably lead to a desire for simplicity. An EPA research official perceives that program managers "don't want to know caveats, subtleties, complexities. Scientists don't feel free to divulge this information to managers. Every [EPA] scientist has had the experience of being criticized by a manager for not giving a simple, clear enough answer. The managers don't want uncertainty."

It should be noted that many EPA career officials are extremely knowledgeable about the various aspects, including the science, of the regulatory programs they manage. In some cases, however, program managers with full command of the scientific uncertainties may seek not to raise them to policymakers, either because, in the manager's judgment, the uncertainties will not affect or usefully inform the policymakers' decision or because the uncertainties might undermine the manager's policy recommendations.

Intermediaries can serve constructively to bridge the gap between the science and policy domains and can legitimately use their interpretations of the scientific information to support policy recommendations. They can also distort, reduce, or block the signal of science received by nonscientific decisionmakers. Intermediaries sometimes use science to buttress or camouflage their policy positions. Because specialized expertise (which includes, but is not limited to, scientific expertise) contributes to one's clout in environmental policymaking, competing intermediaries will sometimes use science-based arguments in attempts to sway nonscientists in policy debates that, in truth, cannot be resolved by science. Intermediaries may also seek or be compelled to satisfy the demands of nonscientific decisionmakers for definitive answers when none are available or for a scientific rationale for a policy choice.

Intermediaries often function as gatekeepers who make judgments about which type of evidence to permit into the regulatory process and what constitutes adequate evidence for regulatory action. The disciplinary training and biases of intermediaries can have important effects. For example, the science culture inside and surrounding EPA has traditionally been dominated by those with toxicological training. As a result, evidence and judgment from other fields of science (for example, epidemiology, physiology, ecology, or statistics) have fewer receptors within the agency and sometimes face greater scrutiny and a higher hurdle of acceptance.

The majority of EPA's scientific and technical staff are not themselves actively engaged in the production of science and can be considered intermediaries at some point along the continuum between scientists and nonscientists. EPA generally forms interoffice working groups to develop or revise regulatory rules. The working group, led by the relevant program office, is responsible for using the available science to formulate regulatory policy options and for briefing senior management and policymakers. Since the mid-1980s, the Office of Research and Development (ORD) has been represented on the working groups regularly. ORD representation was intended to provide independent internal evaluation of the use of science in regulatory development by the program offices. According to a former senior EPA official, however, ORD has had minimal influence in the working groups because it does not have much clout within the agency. In contrast, according to EPA officials, agency lawyers from the Office of General Counsel have traditionally played a strong role in regulatory development working groups. As part of a recent reorganization, the unit within ORD responsible for managing the office's participation in working groups was shifted into a division that traditionally operates on an interoffice consensus basis. According to a senior EPA official, the working group system is biased toward agreement. By implication, this means that regulatory program staff function as the principal mediators of science on specific regulatory decisions.

EPA has developed a number of internal institutions, including the IRIS working groups, the Risk Assessment Forum, and the Science Policy Council, which are intended to provide nonscientific managers and policymakers with agency consensus positions on general scientific matters. Although these institutions have made constructive contributions, the arsenic in drinking water case study (Appendix B) illustrates the limitations of this science-by-consensus approach when broad science policy issues (for example, What constitutes adequate epidemiological data? Should cancer risks be weighted according to the expected lethality of specific cancers?) have direct implications for specific, high-stakes regulatory decisions.

Important intermediaries outside EPA include advocates for industry and the environment, the media, and academics or think-tank policy analysts. The biases of advocates are frequently self-evident. Although they may be viewed as disinterested, objective parties, the other external mediators can also have distorting effects on the translation of science for nonscientists. Scientists from academia and industry often play multiple, overlapping roles as producers of part of the database and as interpreters of all the relevant data for nonscientists. Independent peer review panel members sometimes embody the interface between scientists and nonscientists. For example, then CASAC Chairman Morton Lippman reportedly met with the EPA Administrator to recommend that the agency revisit the ozone NAAQS in light of new data (see Appendix D).

The case studies provide some examples of the importance of key individuals who can constructively bridge the gap between science and policy. For example, several respondents remarked that John Bachmann, an EPA official in the Office of Air Quality Planning and Standards and the lead author of the "Particulate Matter Staff Paper," served effectively as the bridge between the EPA research and air program offices, as liaison to the Clean Air Science Advisory Committee, and as the communicator responsible for briefing policymakers. To serve this bridging function, individuals must be familiar with the science and policy domains, able to grasp the policy relevance of scientific findings, and capable of communicating science to nonscientists in simple terms without jargon.

The case studies also illustrate the potential for key intermediaries to distort or omit scientific information that is conveyed to decisionmakers. For example, the management of the EPA drinking water program was initially alarmed by a scientific report estimating a substantial incidence of internal cancers resulting from arsenic in drinking water. Until that point, the drinking water program management had been willing to wait for additional research before revising the arsenic in drinking water standard because prior cancer risk estimates had been based on less fatal skin cancers. The new internal cancer risk estimate, however, caused some program managers to consider more immediate regulatory action. Although it remains unclear who is responsible for initiating it, a rumor got started that the scientist who had reported the internal cancer estimate, Allan Smith of the University of California, later reversed his position regarding the evidence.

According to a former drinking water official, EPA water program scientific staff told drinking water program managers that Smith "had changed his mind." An EPA official suggests that a miscommunication resulted when the scientific staff reported that Smith had made a 1994 presentation of preliminary data from a new study and that the preliminary data had failed to detect an association between arsenic levels in drinking water and internal cancers. In fact, however, the abstract of Smith's 1994 presentation concludes by saying "these findings provide new support for the evidence" of internal cancers. Furthermore, Smith denies having reversed himself, adding "if anything, the evidence [of internal cancers] has gotten stronger."

An example of the potential for intermediaries to withhold scientific information from decisionmakers is provided by the 1991 lead in drinking water rule. In this case, an EPA official alleges that agency staff advocating more stringent lead regulation (the band of EPA policy entrepreneurs) embargoed a reanalysis of an unpublished study used in the final rulemaking. EPA used an unpublished Centers for Disease Control and Prevention (CDC) study to estimate the relationship between lead in drinking water and lead in the bloodstream of children older than six months. However, there is a discrepancy between the version of the study reported in the final rulemaking and that found in the drinking water docket, with the latter indicating a contribution to blood lead from lead in drinking water lower by a factor of three (for lead levels above fifteen parts per billion, the action level under the 1991 rule). According to an EPA official, the later version of the study was completed prior to the final rulemaking but was not made available to the agency staff responsible for developing the final rule. This official suggests that members of the band of EPA policy entrepreneurs who were reviewers of the CDC paper withheld the revised analysis because it would have substantially reduced the estimated benefits of the regulation. The paper, which the notice of final rulemaking describes as "submitted for publication," was never published. A member of the band of EPA policy entrepreneurs recalled reviewing the CDC report in draft but claimed to be unaware of any discrepancy between the version of the report used in the final rulemaking and the one that is in the drinking water docket.[13]

Whether the allegation against members of the band of EPA policy entrepreneurs is true or false, it is important to acknowledge the *potential* for key intermediaries within EPA to withhold scientific information from decisionmakers. In this case, the potential existed due to the multiple, overlapping roles of intermediaries as producers of scientific information, peer reviewers, regulatory development working group members, and policy entrepreneurs.

Who Does and Doesn't Communicate with the Decisionmaker

Invariably, regulatory decisionmakers have legal, economic, and political counselors. EPA observers have repeatedly voiced concerns about the lack of communication between scientists and agency decisionmakers, and a number of institutional arrangements have been attempted to secure lines of communication. The official science advisory boards, for example, routinely issue letters or memorandums to the Administrator upon completion of reports. The agency has institutionalized ORD's role in the regulatory development working group process. In response to a recommendation of the Expert Panel on the Role of Science at EPA (U.S. EPA 1992), Administrator Reilly appointed a science adviser. In addition to these institutional arrangements, the case studies and interviews suggest that in previous administrations, scientists and science advisers have had at least occasional access to senior EPA officials and to the Administrator.

Whether scientists are involved in the regulatory development process or get an opportunity to communicate with decisionmakers may be less important than *when* they are involved and whether their role is substantive or pro forma. ORD's representation in the regulatory development working group process, for example, may not prevent other members of the working group from excluding scientists from decisionmaking. A former EPA political appointee argues that agency scientists have to be involved *late* in the rulemaking process when decisions are being finalized.

A senior EPA official says that "ORD is usually at the table when major rules are discussed." However, some EPA officials seem to feel that agency scientists have less access to decisionmakers in this administration than in prior administrations. Administrator Browner's decision not to appoint a science adviser has been interpreted by others as being symptomatic of an administration that is disinclined to include scientists in its inner circle. On the other hand, it seems possible that former Assistant Administrator for Pesticides and Toxics Lynn Goldman unofficially or intermittently served the science advisory function in Browner's inner circle.

What Does and Doesn't Get Communicated to the Decisionmaker

By necessity, EPA decisionmakers receive a very digested version of the information that goes into formulating environmental regulations. In addition to scientific information, this package must cover economic, engineering, and administrative topics. As a result, many of the scientific details and complexities are not communicated to decisionmakers. Nevertheless, some of the case studies suggest that when there are high stakes (economic, political, environmental, or a combination thereof), decisionmakers often demand and receive extensive and rigorous briefings on scientific questions that are believed to be central to the decision. Of course, decisionmakers are more likely to demand rigorous scientific analysis if they have substantive inclinations or an affinity for science, and former Deputy Administrators Alm and Habicht were both noted in this regard. Habicht was particularly focused on seeing that the scientific assumptions and uncertainties were aired before EPA decisionmakers. Deputy Administrator Hansen has also assumed substantial responsibilities in improving the quality and characterization of science used for regulatory decisionmaking. In some cases, a policymaker with substantive inclinations may bypass the usual intermediaries and go directly to scientists for information, and this could lead to a more complete communication of science to the decisionmaker.

On the other hand, sometimes important scientific assumptions, uncertainties, and debates are not communicated to policymakers. According to EPA officials, internal debates over science at the agency are often not thoroughly addressed at the level of Assistant Administrator or Administrator. For example, an EPA official notes that a debate between the Policy and Research Offices regarding the adequacy of epidemiological studies on the health effects of inhaled particulate matter (see Appendix C) "didn't get a full airing at the AA [Assistant Administrator] level." Overall, ORD's inability to elevate basic science issues to the top decisionmakers is symptomatic of the office's weak standing within the agency.

The availability of accepted data and methods and the scope of regulatory analyses influence what does and does not get communicated to EPA decisionmakers. The availability of official EPA toxicity values can limit the scope of scientific information communicated to decisionmakers (for example, if IRIS does not contain toxicity data on a particular substance, its potential risk may not be assessed or communicated).[14] Also, most of the available chronic toxicity data is for cancer. Consequently, most of the information on long-term health effects that is communicated to decisionmakers deals with cancer, even though in a particular case, the noncancer effects may be more significant from a public health perspective. The lack of data and agreed-upon methods for quantifying and monetizing a variety of effects, including noncancer and cumulative health effects and ecological and quality of life impacts, also means that these effects are underrepresented in communications to decisionmakers. In some cases, this may mean that decisionmakers do not receive a full accounting of the potential benefits of an environmental regulation.

Conversely, environmental regulatory actions may themselves create spillover risks that are not considered or conveyed to decisionmakers. (Examples would include the effects of shifting pollution from one medium to another, the risks of substitute products, or negative health effects resulting from regulatory costs or economic dislocation.) Graham and Wiener (1995) argue that overlooked adverse effects of regulation are common and neglected benefits rare. Although there are numerous examples of neglected adverse consequences of regulation (for example, see the EDB case study, Appendix E), Graham and Wiener's logic that neglected benefits are rare invokes the joke about the economist who tells her daughter that the $5 bill on the sidewalk does not exist because, if it did, somebody else would have picked it up. By contrast, Mathews (1996) suggests that although cost-benefit studies of environmental regulation may exaggerate particular benefits, their failure to capture other benefits is more than offsetting.[15] Both sides of the debate have a fair amount of anecdotal evidence to buttress their arguments. The bottom line is that there appears to be no firm empirical evidence to support the argument that the information EPA decisionmakers receive is consistently biased in either direction.

The information that EPA decisionmakers receive, however, may sometimes be biased in another way. That is, the range of scientific opinion or uncertainty may be presented as exaggeratedly small or large. For example, the use of point estimates, instead of ranges, in communicating risk information to decisionmakers can create the misperception of certainty and minimizes scientific uncertainties. In contrast, the range of scientific opinion presented to a decisionmaker can be exaggerated, for example, as a consequence of seeking "balance" in a peer review panel. By magnifying the range of scientific uncertainty presented to decisionmakers, staff may invite the decisionmaker to call for further research in lieu of regulatory action (some EPA officials maintain that this transpired in the case of arsenic in drinking water, Appendix B). Alternatively, a large range of uncertainty opens the door for the staff to recommend their own "best estimate," using whatever criteria they choose. Often, says an EPA official, staff present decisionmakers with "the Goldilocks options—too hot, too cold, and just right."

ENDNOTES

[1] The NHEERL facility is located at the University of North Carolina, Chapel Hill. There are a limited number of facilities in the country capable of conducting clinical exposure studies. See Appendix D for a discussion of the role of the chamber studies in EPA's decisionmaking regarding ground-level ozone.

[2] The agency's Science Policy Council recently approved a policy statement on the use of uncertainty analysis techniques that permits their use in the exposure analysis component of risk assessments (*Risk Policy Report*, February 21, 1997, 6).

[3] The core of IRIS is a collection of computer files that contain descriptive and quantitative information in the following categories: oral reference dose and inhalation reference concentrations for chronic noncarcinogenic health effects;

oral slope factors and unit risks for chronic exposure to carcinogens; drinking water health advisories from the Office of Drinking Water; EPA regulatory action summaries; and supplementary data on acute health hazards and physical/chemical properties (*Access EPA*, 1991). The database is maintained by EPA/ORD's Environmental Criteria and Assessment Office in Cincinnati, Ohio.

[4] The draft regulation required data on each emission point within a plant, rather than aggregate data for each facility, and on items related to a factory's process and equipment, such as process rate units, annual process throughput, and typical daily seasonal throughput. The draft regulation also called for annual reports of hazardous air pollution emissions not required by the amendments. States commenting on the draft regulation noted that the reporting requirements exceeded the minimum program needs and noted that additional resources to develop these databases were not available. Office of Air Quality Planning and Standards officials noted that while the reduced level of data would meet minimum program needs, other important data that could contribute to a more effective program would not be collected. Nevertheless, collection of these data would have placed an extra burden on the states (U.S. GAO 1992).

[5] NHEXAS will measure levels of a variety of metals, volatile organics, pesticides, polycyclic aromatic hydrocarbons in about 950 subjects (http://ehpnet1. niehs.nih.gov/docs/1996/104(6)/focus.html).

[6] In FY 1997, the Science to Achieve Results initiative accounted for $97 million of the $552 million Science and Technology account.

[7] Specifically, Section 8 of ERDDA requires the Administrator to provide any proposed criteria document, standard, limitation, or regulations to SAB when these proposals are provided to other federal agencies for formal review and comment. The board is authorized to comment, under time constraints specified by the Administrator, on the scientific and technical basis of the proposals and to make available to the Administrator "any pertinent information" in the board's possession.

[8] In this particular case, SAB declined to review the issue. Barnes (1995) suggests that, in the view of some, Administrator Reilly mistakenly invited SAB to provide advice on setting national environmental priorities in requesting the 1990 report, *Reducing Risk*. Those who objected to the 1990 SAB report may have similar complaints about EPA Deputy Administrator Fred Hansen's July 1995 request that SAB update *Reducing Risk*. In this case, however, the administration's request may have been made to preempt an anticipated Senate appropriations bill reporting requirement (*Risk Policy Report*, August 18, 1995, 12).

[9] This notion was reinforced in the D.C. circuit court's decision in the case of *API v. Costle*, regarding the 1979 ozone NAAQS revision. The court held that SAB, which commented on the criteria document, should also have had the opportunity to comment on the standard itself (Landy et al. 1994, 77). While one could argue that the ERDDA provision authorizing SAB to make "any pertinent information" available to the Administrator opens the door for the board to comment on policy matters, the CASAC authority to advise on welfare, social, and economic matters is clearly more explicit.

[10] Barnes (1995) notes, for example, that CASAC's comments on the policy options of a proposed national ambient air quality standard for sulfur dioxide heightened tensions between the committee and the agency.

[11] The SDWA of 1974 had initially given primary advisory responsibility for drinking water standards to the National Academy of Sciences.

[12] The stated mission of the BOSC is to review ORD administration and research plans (*Risk Policy Report*, November 15, 1996, 13). SAB, however, already has the Research Strategies Advisory Committee, and the Clean Air Science Advisory Committee has responsibilities to provide advice on research within its jurisdiction. This raises the question of why EPA created a new board rather than allocate additional resources to SAB, particularly since this runs contrary to the government streamlining initiative. It is noteworthy that one of the first issues on the BOSC agenda was a review of the agency's arsenic research agenda (*Inside EPA*, January 10, 1997, 6–7). (See Appendix B for a discussion of the formulation of multiple arsenic research agendas and the competing views regarding what research is needed to substantially reduce uncertainties in the risk assessment for arsenic in drinking water.)

[13] Some commenters on the lead in drinking water case study have criticized the CDC study design and analysis. This raises the question, however, of why EPA used the earlier version of the study if its design was flawed, particularly since EPA's contractor had identified a peer-reviewed Scottish study as providing "the most useful data set" for estimating the relationship of water lead to blood lead in older children.

[14] For example, of the twenty-six chemicals that were identified by EPA as contaminants of potential concern in its proposed pulp and paper effluent guidelines, only eleven have noncancer reference doses and six have cancer slope factors available using EPA's primarily data sources (IRIS and HEAST) (U.S. EPA 1993). As a result, the risks of many of the contaminants of concern were not calculated or communicated.

[15] The case of corrosion control treatments required by the 1991 lead/copper drinking water rule (Appendix A) illustrates how spillover benefits may not be captured by EPA regulatory analyses. EPA considered the material benefits of corrosion control in the drinking water distribution system. According to an agency research official, the agency also realized that there could be some ancillary health benefits because implementing the corrosion controls would permit drinking water suppliers to use less chlorine to achieve the same level of disinfection and thereby reduce the formation of hazardous disinfection by-products. What EPA failed to recognize at the time, however, was that corrosion control would also lead to reduced microbial formation in the drinking water delivery system, for example, by reducing pitting of the pipes that provides microbes with tiny refuges where they are safe from contact with disinfectants. Another example of a regulatory control with unintended *positive* benefits is the requirement under the 1990 Clean Air Act Amendments to reduce ground-level ozone formation by limiting the aromatic hydrocarbon content of gasoline. The "reformulated gas" requirement has had the unexpected side benefit of cutting down on the forma-

tion of organic aerosol particulate matter (Odum et al. 1997). These organic aerosols are an important component of the fine fraction of suspended particulate matter.

REFERENCES

Barnes, D. 1995. Sound Science, Schlock Science, Peer Review, and the Science/Policy Interface. *Risk Policy Report*, February 21, 1995, 34–36.

Brown, B. 1995. $144 Million Advice Network. *Washington Post*, January 13, 1995, A23.

Graham, J., and J. Wiener, eds. 1995. *Risk vs. Risk: Tradeoffs in Protecting Health and the Environment*. Cambridge, Mass.: Harvard University Press.

Guerrero, P. 1995. *EPA's Problems with Collection and Management of Scientific Data and Its Efforts to Address Them*. Testimony by Peter F. Guerrero, Director of Environmental Protection Issues, U.S. General Accounting Office, before the Subcommittee on VA, HUD and Independent Agencies, Committee on Appropriations, U.S. Senate. GAO/T-RCED-95-174. Washington, D.C.: U.S. GAO.

Jasanoff, S. 1990. *The Fifth Branch: Science Advisors as Policymakers*. Cambridge, Mass.: Harvard University Press.

Landy, M., M. Roberts, and S. Thomas. 1994. *The Environmental Protection Agency: Asking the Wrong Questions*. New York: Oxford University Press.

Lippman, M. 1987. Role of Science Advisory Groups in Establishing Standards for Ambient Air Pollutants. *Aerosol Science and Technology* 6: 93–114.

Mathews, J. 1996. Environmental Success Story. *Washington Post*, June 17, 1996, A17.

NAPA (National Academy of Public Administration). 1995. *Setting Priorities, Getting Results: A New Direction for EPA*. A NAPA Report to Congress. Washington, D.C.: NAPA, April.

NRC (National Research Council). 1994. *Science and Judgment in Risk Assessment*. Washington, D.C.: National Academy Press.

Odum, J., T. Jungkamp, R. Griffin, et al. 1997. The Atmospheric Aerosol-Forming Potential of Whole Gasoline Vapor. *Science* 276: 96–98.

Smith, B. 1992. *The Advisers: Scientists in the Policy Process*. Washington, D.C.: Brookings Institution.

Stocking, S. 1996. How Journalists Deal with Uncertainty in Science. Presentation, Meeting of the American Association for the Advancement of Science, Baltimore, Md., February 10, 1996.

U.S. EPA (U.S. Environmental Protection Agency). 1992. *Safeguarding the Future: Credible Science, Credible Decisions*. Report of the Expert Panel on the Role of Science at EPA. Washington, D.C.: U.S. EPA, March.

————. 1993. *Water Quality Assessment of Proposed Effluent Guidelines for the Pulp, Paper, and Paperboard Industry.* EPA-821-R-93-022. Washington, D.C.: U.S. EPA, Office of Water.

U.S. EPA/SAB (U.S. EPA Science Advisory Board). 1994. SAB Reports, FY 1978–FY 1994. Mimeographed.

U.S. GAO (U.S. General Accounting Office). 1992. *Superfund Program Management.* GAO/HR-93-10. Washington, D.C.: U.S. GAO.

U.S. OTA (U.S. Office of Technology Assessment). 1993. *Researching Health Risks.* OTA-BBS-570. Washington, D.C.: U.S. Government Printing Office, November.

Yosie, T. 1988. The EPA Science Advisory Board: A Ten-Year Retrospective View. *Risk Analysis.* 8(2): 167–168.

3

Science inside EPA: Office of Research and Development

In tracing EPA's acquisition and use of science, it is useful to begin with an analysis of the relevant internal agency structure and processes. Although much of the scientific information used by EPA comes from outside the agency, its fate in the regulatory decisionmaking process is largely determined within EPA.

AGENCY SCIENCE RESOURCES

As a former EPA political appointee observes, "There is a real schizophrenia inside and outside EPA as to whether it's a science agency." But, as revealed by its budget, EPA has, since its creation, been conceived of primarily as a regulatory and enforcement agency, not a science agency. The long-term trend for EPA's science has been downward. In 1973, apart from construction grants disbursed by the agency, the EPA Office of Research and Development (ORD) accounted for approximately one-third of the agency's budget. This was a carryover from the research and technical assistance functions of some of the various federal divisions from which the agency was constituted. However, the 1974 budget submitted by the President called for a large decrease for ORD accompanied by substantial increases in abatement, control, and enforcement (Davies and Davies 1975).

The decline in ORD's budget reached its nadir in the early 1980s during the first Reagan administration. The EPA Science Advisory Board's Research Strategies Advisory Committee observed that the relative proportion of funding in the agency's total budget devoted to ORD, less construction grants, had declined from over 20% in 1980 to less than 7% in 1995 (*Inside EPA*, May 13, 1995, 12). Table 3.1 and Figure

Table 3.1. EPA Research and Development Budget Appropriations Account Compared with EPA Budget Authority, FY 1976–1996 ($ million, unadjusted)

Fiscal Year	R&D Appropriations	R&D Portion of Total EPA Budget Authority	Total EPA Budget Authority
FY 76	265	34.3%	772
TQ	64	33.9%	189
FY 77	261	9.4%	2,764
FY 78	317	5.8%	5,499
FY 79	333	6.2%	5,403
FY 80	233	5.0%	4,669
FY 81	250	8.3%	3,026
FY 82	154	4.2%	3,674
FY 83	121	3.3%	3,688
FY 84	146	3.6%	4,064
FY 85	189	4.3%	4,346
FY 86	214	6.2%	3,446
FY 87	198	3.7%	5,344
FY 88	186	3.7%	4,968
FY 89	203	4.0%	5,081
FY 90	230	4.3%	5,380
FY 91	255	4.2%	6,004
FY 92	318	4.9%	6,461
FY 93	323	4.8%	6,737
FY 94	339	5.3%	6,346
FY 95	350[a]	5.0%	6,994[a]
FY 96	427[b]	5.8%	7,331[b]

[a] Enacted appropriations.

[b] The figure is the administration's proposed appropriations for ORD. The House-Senate conference bill allocated $525 million to an agencywide Science and Technology (S&T) account (*Washington Post*, November 17, 1995, p. A18). The agency's S&T account authorization for FY 1997 was $552 million (U.S. CRS 1997).

Sources: For R&D Appropriations: Budget of the U.S. Government, Appendix (1978–1995). Actual amounts unless indicated. TQ, or transition quarter, is the bridge between fiscal year (FY) accounting periods. In 1977, the FY accounting period changed from July 1–June 30 to October 1–September 30. *For Total EPA Budget Authority:* Budget of the U.S. Government, Historical Tables (1995). Budget authority (BA) "is the authority provided by law for agencies to obligate the Government to spend" (p. 5). BA includes trust funds but not interfund transactions.

3.1 display the rapid decline and subsequent oscillations of the EPA Research and Development (R&D) appropriation account as a proportion of its total budget authority. The proportion devoted to ORD reached its lowest point in 1983 (approximately 3%). In constant dollars, ORD's appropriations had regained roughly their 1980 level by 1992 (Carnegie 1992).

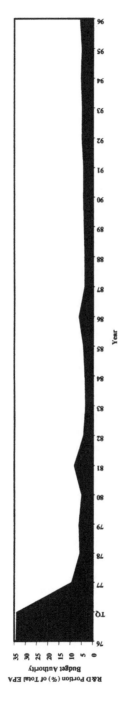

Figure 3.1. Research and Development Account as Percentage of Total EPA Budget Authority, FY 1976–1996

Table 3.2. EPA Estimated Total Environmental Research and Development Budget, FY 1990–1997 ($ million, unadjusted)

FY 90	FY 91	FY 92	FY 93	FY 94	FY 95	FY 96	FY 97	FY 98
424[a]	440[a]	502[a]	—[b]	538[c]	—[b]	525[d]	552[d]	631[d]

[a] Gramp et al. 1992.

[b] No estimate available.

[c] CNIE 1995.

[d] These figures represent EPA's Science and Technology appropriations account, which does not capture all R&D in program offices. EPA's total budget for FY 1997 and FY 1998 was approximately $6.8 billion and $7.4 billion, respectively (*Environment Reporter*, February 14, 1997, p. 2102; U.S. CRS 1997; H.R. 2158).

Although EPA's long-term trend away from R&D rings true to many observers, the foregoing figures should be treated with some caution. The overall level of scientific resources supporting the agency's planning, rulemaking, and enforcement functions is hard to pin down precisely. Current estimates of EPA's annual R&D budget range from $500 million to $1 billion, depending on how R&D is defined (Stone 1994). According to a former senior agency official, the ORD appropriations account may represent only about 50% of EPA's total science budget, with the remainder coming from the air, water, and other program office budgets. Some activities under ORD's budget may also be construed as not genuine R&D.[1] Available EPA science budget estimates are probably underestimates or, at the least, ambiguous.

Table 3.2 provides *available* estimates of EPA's *total* environmental R&D budget, suggesting a nominal increase of nearly 50% over the period 1990–1998. In constant dollars, however, the agency's R&D budget would be relatively flat over the period (assuming the average one-year Treasury rate of 5.4% over 1990–1998). Comparing Tables 3.1 and 3.2 suggests that, as a proportion of the agency's total budget, R&D has hovered at approximately 8% during the 1990s. By comparison, the Food and Drug Administration, which has the authority to acquire much of its data from regulated industry, currently spends approximately 20% of its total budget on research (Lawler and Stone 1995). To place EPA's total R&D budget in context, consider that it is less than 1% of the national cost of environmental regulatory compliance and that EPA controls only about 10% of the total federal environmental R&D budget (see Tables 3.3 and 3.4).

As of 1993, approximately 40% of EPA's roughly 17,000 white-collar employees were trained in science (24%) and engineering (15%) (U.S. OPM 1993). Table 3.5 provides a breakdown of the disciplinary background and educational level of EPA employees with college degrees as

Table 3.3. Estimated Federal Funding for Environmental R&D[a] by Agency ($ million, FY 1992)

NASA	826	USDA	403
DOE	799	ARS	162
DOD	577	CSRS	119
NSF	541	FS	115
DOI	524	Other	7
USGS	367	NOAA	319
FWS	85	AID	45
Other	72	Smithsonian	33
EPA	**502**	TVA	31
DHHS	454	DOT	17
NIEHS	303[b]	**Total**	**5,071**
NIOSH	93		
ATSDR	55[c]		
NCTR	3[d]		

[a] Based on total environmental R&D, including both health and nonhealth categories.

[b] Includes $51 million transferred under Superfund from EPA to NIEHS under the Superfund Basic Research Program.

[c] Funds are transferred to ATSDR from EPA through Superfund.

[d] An estimated 10% of NCTR's $30 million budget is environmentally related R&D.

Notes: AID: Agency for International Development; ARS: Agricultural Research Service; ATSDR: Agency for Toxic Substances and Disease Registry; CSRS: Cooperative State Research Service; DHHS: Department of Health and Human Services; DOD: Department of Defense; DOE: Department of Energy; DOI: Department of the Interior; DOT: Department of Transportation; EPA: Environmental Protection Agency; FS: Forest Service; FWS: Fish and Wildlife Service; NASA: National Aeronautics and Space Administration; NCTR: National Center for Toxicological Research; NIEHS: National Institute of Environmental Health Sciences; NIOSH: National Institute for Occupational Safety and Health; NOAA: National Oceanic and Atmospheric Administration; NSF: National Science Foundation; Smithsonian: Smithsonian Institution; TVA: Tennessee Valley Authority; USDA: U.S. Department of Agriculture; USGS: U.S. Geological Survey.

Source: Carnegie 1992.

of November 1996. Approximately 20% of EPA's employees with college degrees have advanced degrees in agriculture, biological/life sciences, conservation/renewable natural resources, health profession and related sciences, and physical sciences; these scientific disciplines represent less than 40% of EPA's employees with advanced degrees.

EPA has a considerable number of employees with scientific training. But, because much of their time is devoted to contractor management, many of EPA's scientists are unable to practice their craft. Consequently, their technical skills fall into disuse. Since the early 1980s, a large portion of EPA's scientific research, development, and analysis has been per-

Table 3.4. Estimated Federal Funding for Environmental R&D by Agency ($ million, FY 1996)[a]

DOE	1,437
NASA	1,250
DOC (i.e., NOAA)	523
EPA[b]	**457**
DOI	445
NSF	325
USDA	322
DHHS	325
DOT	56
Smithsonian	30
TVA	15
Corps of Engineers	1
Total	**5,186**

[a]Based on Committee on Environment and Natural Resources (CENR) figures for U.S. Federal Environmental R&D. CENR is a committee of the National Science and Technology Council (NSTC), which is a standing executive branch body. Source: Office of Science and Technology Policy mimeograph, August 1996.

[b]New Science and Technology appropriation account ($525 million) includes ORD, funding for several non-ORD labs, Superfund, and environmental technology research.

Notes: DOC: Department of Commerce.

According to NSTC/CENR (1997), $5.05 billion was appropriated to federal environmental R&D in FY 1997. The reductions appear to have been primarily in the area of global change research (see Brown 1997). Note that the Defense Department's sizable environmental R&D budget is not included in Table 3.4.

formed by contractors, many of whom conduct their work "in-house," that is, in EPA labs and offices under the supervision of agency staff.[2] By 1991, EPA extramural resources for R&D accounted for nearly 80% of the agency's total R&D budget (U.S. EPA 1993). Thereafter, extramural R&D resources declined somewhat while the agency's total R&D budget continued to climb. In 1995, the administration's budget request to Congress proposed a broad conversion of EPA resources spent on contractors to operating expenses for hiring additional employees (that is, "direct hires" in federal personnelspeak), and the agency did convert some extramural laboratory and other R&D resources to operating expenses. However, in response to the objections of a number of large environmental consulting firms, the report accompanying the 1996 Senate budget recommendation included language prohibiting further contractor conversion hiring (*Inside EPA*, September 12, 1995).

Despite the agency's limited research budget and the heavy demands of contract management placed on many of its scientists, many are skillful researchers, modelers, or analysts, and EPA has a number of areas of sci-

Table 3.5. Academic Discipline and Educational Level of EPA Employees with College Degrees

Discipline	Doctorate	Master's	College	Total
Agriculture	34	76	86	196
Biological science/life sciences	500	624	801	1,925
Conservation/renewable natural resources	23	154	98	275
Health professions and related sciences	31	157	44	232
Physical sciences	323	613	918	1,854
Engineering	104	992	1,494	2,778[a]
Law and legal studies	50	1,013	42	1,105
Social sciences	58	351	579	988
Other disciplines	125	1,506	1,958	3,589
Total	1,248	5,486	6,020	12,942

[a]188 engineers also have JD or LLB degrees.

Source: Mimeograph from EPA Office of Human Resources Management.

entific strength. For example, a senior EPA official suggests that the agency's atmospheric pollutant transport modeling is world-class. According to an industry scientist, an EPA research laboratory has been responsible for some of the seminal clinical air pollution studies (that is, studies of human subjects in controlled exposure chambers). The agency has also been at the fore in developing integrated exposure models that take into account human exposure to a toxic substance (for example, lead) from multiple exposure pathways (that is, ingestion, inhalation, and dermal absorption). An academic's opinion is that EPA science analysts have become particularly skillful at drawing inferences from limited data and from studies that are not designed primarily for the agency's purposes.

However, EPA's norms and incentives tend to subordinate science, and some respondents feel, for a variety of reasons, that agency staff do not have sufficient training and skill to do a defensible job of scientific analysis. (See Chapter 5 for a more complete discussion of the factors affecting EPA's use of science.) Scientists who excel at basic research turn away from the agency's "fire-fighting" operational mode, which does not lend itself to long-term research, and are drawn to academia and research institutions. Industry and consulting offer expert analysts better financial compensation than does EPA. The agency loses its most up-to-date scientists through the departure of talented young researchers. More senior staff find limited opportunities to remain active researchers in their fields due to the demands of administrative tasks. Perhaps most importantly, the agency's reliance on contractors and other external sources of science limits the development and maintenance of its internal analytic capacity.

Other comments suggested a trade-off between specialization and generalization within EPA's scientific workforce. According to a former EPA political appointee, an undesirable development within the agency's scientific ranks has been increased specialization in experimental sciences, such as toxicology, and fields, such as biochemistry, that focus on phenomena at the cellular and subcellular levels. The loss associated with gaining expertise in effects at the organ-to-cellular levels has been a lack of people trained in fields that consider effects on the whole body, such as physiology and pharmacology. The emphasis on experimental scientific disciplines (for example, toxicology) results in a lack of staff trained in observational sciences, such as epidemiology. Furthermore, biologists traditionally receive "deterministic training" that leaves them ill equipped to analyze the effects of uncertainties in their scientific judgments. Although there appears to be no hard data available on the disciplinary backgrounds of all EPA biological and physical scientists, other observers also perceived a lack of diversity in terms of scientific disciplines among EPA staff. The ambient air particulates case study (Appendix C) discusses the dominance of toxicology within EPA's science culture and its possible origins.

A current senior EPA official, on the other hand, laments the lack of scientific disciplinary rigor of EPA's scientists who have degrees in the relatively new, interdisciplinary field of environmental science. "Their science training is broader but poorer." In a former agency official's opinion, EPA's staff are not highly proficient in their own field, nor in those that are closely related. A core group of scientific "renaissance people" within the agency could bridge the subfields of environmental science, but the former agency official notes that there is a lack of respect in the specialized field of environmental risk assessment for the generalist because the specialists can always pick apart some flaw that is not on the cutting edge in their own subfield.

Apart from EPA technical personnel's level of scientific acumen or breadth, several interviewees commented on the problem of politicized scientists within EPA who, according to a senior careerist, "don't give straight information" to agency decisionmakers. Interviews turned up complaints about particular individuals or groups being either "Republican holdovers" or "environmental activists." Thus, there appears to be some partisan heterogeneity among the EPA scientists and analysts who are perceived as being "politically motivated." The lead in drinking water case study (Appendix A) provides an illustration of activist EPA analysts acting as "policy entrepreneurs" and locking horns with ideologically opposed analysts within the agency.[3]

OFFICE OF RESEARCH AND DEVELOPMENT (ORD)

Since EPA's inception, there has been disagreement both inside and outside the agency regarding the appropriate role of the Office of Research and Development. The principal schism over ORD is whether the office should act primarily as a technical support unit for regulatory programs or as an independent source of scientific research and assessment. EPA researchers, therefore, face two competing standards for science. Internally, the science must be timely and relevant. External validity rests on acceptance by the larger scientific community.

Because the core role of ORD has never been resolved, it carries out a diverse program of regular operations and special initiatives: conducting a small intramural R&D program; administering competitive and noncompetitive extramural R&D; managing some, but not all, of the agency's scientific databases; providing technical support to some program offices, but not to others; conducting major substance-specific risk assessments (for example, the dioxin reassessment) and science programs (for example, the Environmental Monitoring and Assessment Program); and taking the lead in interoffice groups that develop EPA risk assessment guidance and seek to direct agency science policy. Until recently, ORD also carried out independent, internal evaluations of program office regulations for their scientific basis and consistency with agency risk assessment guidance.

The level of various ORD operations has fluctuated over time in response to the interests of Congress and the administrations and as opportunities have presented themselves to internal entrepreneurs. Although former Assistant Administrator Bernard Goldstein initiated the routine inclusion of ORD representatives on internal regulatory development working groups, ORD continues to suffer from a low stature within EPA. Some interviewees regard ORD as "irrelevant" or a "bureaucratic backwater." Others complain that ORD produces "encyclopedic" tomes and other types of information that are not useful for decisionmaking. Staff scientists, on the other hand, bristle that the program offices "treat ORD like contractors," instead of independent analysts.

Following is a more detailed discussion of ORD's organizational makeup and role as a source and user of science in EPA. The total personnel level in ORD has declined from more than 2,000 in 1980 but remained relatively constant during the mid-1980s and early 1990s (U.S. EPA 1993). In 1993, at the request of Congress and in connection with the Vice President's National Performance Review, Administrator Browner called for an evaluation of the agency's laboratories that resulted in a 1995 reorganization of ORD. Prior to the reorganization, ORD had approximately 1,830 staff members—representing less than 30% of EPA's scientists and engi-

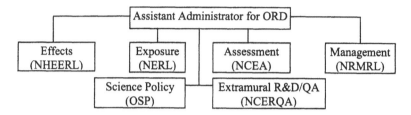

Figure 3.2. ORD Organizational Chart

neers—at twelve geographically dispersed laboratories, three field stations, four assessment centers, and the headquarters office, plus an extramural research budget of $378 million for grants, cooperative and interagency agreements, and contracts (NRC 1995).

The reorganization caused many of the labs and offices to consolidate under organizational units that correspond roughly to elements of the "risk paradigm": hazard identification and dose-response analysis (NHEERL, the National Health and Environmental Effects Research Laboratory); exposure analysis (NERL, the National Exposure Research Laboratory); risk characterization (NCEA, the National Center for Environmental Assessment); and risk management (NRMRL, the National Risk Management Research Laboratory). Two additional ORD divisions are the Office of Science Policy (OSP) and the National Center for Environmental Research and Quality Assurance (NCERQA). Figure 3.2 illustrates ORD's organizational divisions. (This discussion omits the ORD Office of Resources, Management, and Administration.)

To a great extent, the 1995 ORD reorganization focused on consolidating the twelve ORD labs and bringing them back into the fold of EPA. According to a senior EPA official, in response to shrinking ORD budgets during the early 1980s, "The labs started to job-shop, seeking work from other agencies." As a result, the work of the ORD labs diverged from the programmatic needs of EPA, and "twelve independent fiefdoms" were created. "The big loser," according to this source, "was the program offices because they were no longer the labs' major client." The divergence of the labs thus contributed to the overall schism between ORD and the program offices. More recently, however, the ORD labs have received less work from other agencies, justifying a consolidation of the labs and a reduction in their Washington, D.C., management staffs.

National Health and Environmental Effects Research Laboratory

NHEERL, headquartered in Research Triangle Park (RTP), North Carolina, contains the Health Effects Research Lab (HERL) in RTP, as well as the ecology divisions of labs in Gulf Breeze, Florida, Corvallis, Oregon, Duluth, Minnesota, and Narragansett, Rhode Island, which were under the former ORD headquarters Office of Environmental Processes and Effects Research (*Risk Policy Report*, June 16, 1995, 8). The lab's mission is to determine the health and ecological effects of pollution through toxicological, clinical, epidemiological, and ecological research and to develop risk assessment methods (www.epa.gov, May 1995). HERL conducts most of the agency's in-house research on health effects of pollutants (U.S. OTA 1993) and has made noteworthy contributions to the science base for the national ambient air quality standards (see Appendixes C and D). According to NHEERL Director Lawrence Reiter, two of the lab's top priorities are research in the area of endocrine disrupters and the Environmental Monitoring and Assessment Program (EMAP), which is currently being directed from the Corvallis laboratory (*Risk Policy Report*, June 16, 1995, 8; Stone 1995).[4]

In 1988, to exhort EPA to support more long-term research, Congress authorized $10 million for the Research to Improve Health Risk Assessment (RIHRA) program. RIHRA was housed in HERL, and approximately half of the program's resources went to researchers outside EPA (U.S. OTA 1993). According to an independent policy analyst, the RIHRA program and its extramural successors in the Office of Exploratory Research (see discussion under NCERQA later in this chapter) were supported by the office of Sen. Barbara Mikulski (D-Md.), former chair of the EPA appropriations subcommittee. Despite the absence of a special appropriation, a former EPA scientist states that EPA viewed RIHRA as a "congressionally mandated" program, and ORD initially funded the program at $7 million by redirecting funds from other programs. The level of funding for RIHRA declined to $5.1 million in 1993. While some scientists criticized RIHRA for allowing funds to be used for ongoing activities in fulfillment of regulatory needs, other observers argued that the program supported methodological risk assessment research that the agency might not have conducted otherwise (U.S. OTA 1993).[5] A lasting impact of RIHRA, according to a former EPA scientist, has been a trend toward research geared at the uncertainties in risk assessments. Thus, "congressional interference" may have had some beneficial effects on the agency's research program.

National Exposure Research Laboratory

NERL, also headquartered in RTP, North Carolina, contains the Atmospheric Research and Exposure Assessment Laboratory in RTP, the Environmental Sciences Monitoring Laboratory in Las Vegas, Nevada, the Environmental Monitoring Systems Laboratory in Cincinnati, Ohio, and the Environmental Research Laboratory in Athens, Georgia (*Risk Policy Report*, June 16, 1995, 8). NERL's mission is to study and characterize the path and transformations of pollutants from their source to their ultimate receptors, humans or environmental resources (www.epa.gov, May 1995). In a 1995 interview, then NERL Director Gary Foley explained that the lab's priorities include determining mechanisms of tropospheric ozone formation; research on drinking water disinfectant by-products and a comparison of risks posed by various treatment methods; the role of particulate matter on respiratory problems; and research on habitat changes (*Risk Policy Report*, June 16, 1995, 8–9).[6]

National Center for Environmental Assessment

NCEA, headquartered in Washington, D.C., contains the Office of Health and Environmental Assessment (OHEA), including the Human Health Assessment Group and the Exposure Assessment Group in Washington, D.C., and the Environmental Criteria and Assessment Offices (ECAO) in Research Triangle Park and Cincinnati, Ohio (NRC 1995). According to NCEA Director William Farland, NCEA has three primary functions: (1) to write risk assessment reports or develop descriptions of risk information for EPA program offices, (2) to develop agency risk assessment guidance, and 3) to promote research to carry out risk assessments. A priority for NCEA, according to Farland, will be risk assessments "on chemicals that can be used as models by others . . . pushing the state of the science on risk assessment" (*Risk Policy Report*, June 16, 1995, 10).

ECAO in RTP evaluates the state of the science on particular criteria and hazardous air pollutants (see Appendixes C and D for a discussion of decisions regarding two criteria pollutants, particulates and ozone). Its role in the air program is discussed below. ECAO in Cincinnati coordinates the agency working groups that develop EPA's Integrated Risk Information System, the agency's primary official source of chemical-specific health risk information, and maintains the database. The office maintains a secondary health risk database, the Health Effects Assessment Summary Tables for the contaminated site and hazardous waste programs. ECAO in Cincinnati also manages the preparation of some health and environmental criteria and assessment documents for other EPA programs. These documents and databases, and the institutional processes they involve, are discussed in Chapter 2 under "Sources of Scientific Information."

The central office of the NCEA is OHEA, located in Washington, D.C. OHEA is a direct descendant of the former Cancer Assessment Group (CAG), and NCEA Director Farland is a former director of both OHEA and CAG. During the late 1970s to mid-1980s, CAG exercised great influence over EPA's risk assessment practices. A former senior EPA official described CAG during its heyday as "omnipotent" and its mode of operations as cloistered and mysterious. However, during the 1980s, other centers of risk assessment developed within EPA for a number of reasons, according to former EPA Science Advisory Board (SAB) Executive Director Terry Yosie (Yosie 1987). The National Academy of Sciences challenged CAG's data interpretations for blurring the distinction between science and policy. SAB objected to CAG science assessments for not expressing uncertainties in risk estimates. Industry protested what they saw as CAG's overly conservative risk assumptions. Within EPA, program managers were concerned over the ability of CAG staff to meet deadlines and their inability to devote more time to developing risk assessment guidelines and methods. At the same time, the role of risk assessment became more prominent in agency decisionmaking, and many program and regional offices expanded their own scientific staffs and sought to exercise more direct control over the preparation of risk assessments. As a result, many offices became less dependent on the technical skills and leadership centralized in CAG, and the risk assessment process within EPA became one of "bureaucratic free enterprise." The increased competition among ideas and approaches sometimes created distrust among staff or opportunities for stakeholders to intervene either for or against particular risk assessments.

If it ever was, NCEA is no longer a hegemonic risk center. Still, NCEA assumes responsibility within the agency for conducting "blockbuster" pollutant-specific risk assessments. For example, NCEA is primarily responsible for conducting the reassessment of dioxin begun in 1991 (see Appendix G for a discussion of the role of the reassessment in the pulp and paper cluster rulemaking). The dioxin reassessment process departs from the agency's routine assessment procedures by using external scientists to author chapters that summarize and evaluate the science on dioxin disposition and pharmacokinetics, mechanism(s) of action, toxicity, epidemiology and human data, and dose-response modeling. Many EPA observers have been critical, however, of the decision to make NCEA solely responsible for the risk characterization chapter, which synthesizes the science for policymakers. The internal-external division of responsibilities in the dioxin reassessment is consistent with the long-standing practice of the national ambient air quality standards (NAAQS) review process (see discussion of the air program in Chapter 4). However, in the NAAQS process, there is a separation of science assessment from science-based policy recommendations between offices within the agency. With NCEA responsible for both risk assessment and risk characterization, there is little institutional separation of science assessment from science policy. Some of the external criticisms of the NCEA draft dioxin reassessment—

for example, the failure to thoroughly analyze the effects of key uncertainties—also hark back to earlier critiques of CAG assessments.[7]

Office of Science Policy

OSP, located in Washington, D.C., administers the EPA's Science Policy Council (SPC) and Risk Assessment Forum (RAF) and coordinates ORD participation in the EPA regulatory development process through the Office of Research and Science Integration (ORSI). RAF was established in 1984 to reconcile alternative analytical interpretations and scientific assumptions across the agency at the staff level. RAF is also responsible for developing a consensus across the agency on EPA risk assessment guidance developed by NCEA. SPC consists of senior-level managers and scientists from across the agency who oversee RAF and resolve differences when scientists from different offices cannot reach consensus.

Prior to the 1995 reorganization, the office responsible for coordinating ORD participation in the regulatory development process was the Office of Science, Planning, and Regulatory Evaluation, directed by Peter Preuss. As its name implies, the office functioned largely as an independent, internal evaluator of the scientific basis of program office regulations. The ORD regulatory evaluation group was initiated during Assistant Administrator Goldstein's tenure to institute a routine ORD presence on internal regulatory development working groups. Goldstein served during the second Ruckelshaus administration in the wake of Administrator Gorsuch's resignation. The Gorsuch administration was marred by, among other things, controversy surrounding a dubious formaldehyde risk analysis by the Office of Toxic Substances (see Jasanoff 1990).

Predictably, tensions formed over time between Preuss's regulatory science oversight group and the program offices responsible for regulatory development. A program official complained of "late hits from the ORD headquarters oversight people . . . oversight people who wanted to put their own interpretations on things . . . randomly coming into the process wherever they saw an opportunity" to influence a policy decision. In fairness, this source notes that in some cases when ORD's regulatory evaluators elevated a scientific issue late in the regulatory development process, it may have been raised during the staff-level working group process, but program staff "stiffed" them. "That's not out of character with folks" in the program office, allowed the official. "The ORD reviewers also tended to take a broader view than the scientists" in the program office, according to this source. Nevertheless, science can be used as a weapon in the struggle for bureaucratic supremacy. An ORD official, for example, suggests that the regulatory science oversight group's efforts suffered from an inclination toward intellectual one-upsmanship. Although internal review can provide a necessary quality assurance measure, it also can lapse into the time-honored tradition of bureaucratic jousting. According to the program official, some ORD reviewers and program staff "want to practice guerrilla warfare." These sorts of tensions are manageable but perhaps unavoidable consequences of internal quality control processes.

Given that the Risk Assessment Forum and Science Policy Council were designed as consensual institutions, it appears that moving the responsibility for coordinating ORD participation in the regulatory development process under OSP is intended to shift the nature of ORD's participation away from oversight. This organizational change also appears consistent with the Browner administration's de-emphasis of the internal, independent programmatic oversight and review functions previously housed in EPA's policy office.

RAF and SPC are intended to ensure that scientific information is used consistently across the agency. Yosie (1987) states that the goal of RAF and SPC was to form "mediating structures" within EPA to constructively channel internal dissent while fostering a climate of trust within the organization. The reviews are mixed. A former senior EPA official comments, "That's the goal; implementation is spotty." A program office scientist complains that RAF is dominated by ORD scientists. Another EPA official suggested that for those without credentials in conventional environmental risk fields like toxicology, there are barriers to entry onto the Risk Assessment Forum.[8] However, a former EPA scientist remarks that RAF has fostered a sense within the agency of "commonality, you become part of a club." This shared sense of purpose "makes it harder to maintain a lone empire, a separate agenda."

National Center for Environmental Research and Quality Assurance

As currently configured, NCERQA comprises four divisions: the Environmental Engineering Research Division; the Environmental Sciences Research Division; the Quality Assurance Division; and the Peer Review Division. The center was formed by combining the previously separate Offices of Exploratory Research and of Modeling, Monitoring Systems, and Quality Assurance. Its primary responsibilities are managing the agency's expanded research grants and fellowship programs; administering a research centers program; developing quality assurance policy and oversight for all EPA programs involving environmental measurements; coordinating agency activities related to development and validation of environmental monitoring methods; and managing peer reviews for NCEA's activities and for other selected EPA projects (www.epa.gov, May 1995). NCERQA is the latest in a series of EPA responses to calls for more

EPA support for academic research and for greater emphasis on quality assurance. In November 1995, Peter Preuss, former director of OHEA and ORSI, became director of the center.

The House Science Committee provided the impetus for establishing the former Office of Exploratory Research (OER), which managed the agency's extramural research program. The sequence of events leading up to the establishment of the extramural program began in 1974 when concerns surfaced about EPA's management of the Community Health and Environmental Surveillance System (CHESS) and interpretation of the CHESS epidemiological data on sulfur oxides in ambient air. (For further discussion of CHESS, see Appendix C and Jasanoff [1990].) As part of this episode that damaged the credibility of EPA intramural science, Rep. George Brown (D-Calif.), former chair of the Science Committee, held hearings in 1976. In 1977, the National Research Council recommended that the agency initiate a competitive, peer-reviewed, extramural, investigator-initiated research grants program. EPA finally initiated the grants program in 1979. Since then, its budget has fluctuated between $5 million and $25 million per year (NRC 1995). In 1994, a committee of senior EPA staff recommended increasing the entire extramural grants program to $100 million per year within three years (U.S. EPA 1994). The expansion of the extramural grants program has elicited predictable complaints from some at ORD laboratories and EPA regulatory program offices. According to EPA sources, the labs view the extramural grant program as budgetary competition, while the regulatory programs see it as a diversion from the agency's mission.

In 1995, under the STAR (Science to Achieve Results) initiative, ORD doubled the research grants program to $44 million, half being awarded through NCERQA, and the other half being awarded through a process conducted jointly by EPA and the National Science Foundation (NSF).[9] This unprecedented arrangement was prompted in July 1994, when the Senate appropriations subcommittee, expressing its dissatisfaction with the agency's peer-review process, directed EPA to "enter into a partnership" with NSF to review research proposals and called on NSF to "apply its peer-review process" to half of EPA's extramural basic research program (Science, July 22, 1994, 463). Neither EPA nor NSF were prepared for the congressional directive, and the first joint request for proposals (RFP) was a messy operation. EPA assigned some of its own staff to learn how NSF reviews proposals, but there was little opportunity for the EPA research contract managers to learn. The hurriedly prepared RFP was broadly scoped, and, consequently, it was swamped by applications.[10] On the basis of this chaotic experience, EPA geared up its internal grant proposal review capacity, establishing a new division in NCERQA for the 1996 RFPs. Given the research grant staff's incomplete preparation, an

EPA research official felt that the number of complaints about the 1996 tranche of research grant awards was surprisingly low.

However, NSF's research proposal review criteria and its disciplinary focus may not serve as the best model for the EPA research grants program. Because NSF's primary mission is to advance scientific disciplines, the foundation's grant program emphasizes investigator-initiated research, and its review criteria are based on quality first and foremost. EPA, on the other hand, needs to consider the relevance of proposed research to its mission in addition to quality, and the agency's mission creates demand for multidisciplinary problem-oriented research. Although collaboration with NSF provides an opportunity for the agency to leverage its science resources, EPA faces the difficult challenge of striking the appropriate balance between supporting basic academic research and satisfying the science needs of the regulatory programs. This balancing act is reflected in a recent press report that quoted one EPA official who suggested that "with some of these projects, we're leaning towards NSF's framing of the research" (*Risk Policy Report*, January 24, 1997, 24). An EPA regulatory program official fears that by adopting the NSF model, EPA's research program will become less relevant to decisionmaking. "We need fewer, bigger grants for multi-investigator projects rather than a bunch of small, individual investigator grants. [Following the NSF model could produce] a lot of interesting, tiny projects that don't add up to anything." Therefore, a grants program that responds to and anticipates the needs arising from the agency's public health and environmental protection mission may provide a better model for EPA.

In contrast to the EPA extramural research grants program, the agency's research centers program is intended to support long-term, problem-oriented research. Like the grants program, the centers program was initiated in 1979 when the notion of regulatory science was relatively new. A former senior EPA official described the centers as a plan by then Assistant Administrator for ORD Steven Gage to "bribe outside scientists to answer agency questions." Since its inception, the program has been a mixed bag. The initiative's potential has been limited by bureaucratic, scientific, and congressional politics, perhaps predictably so. The program has administered a handful of competed long-term grants to university-based centers, but it has also disbursed support to a generally larger number of congressionally designated, noncompeted grantees. As of 1995, there were nine competed research centers and eighteen noncompeted research centers.[11]

Table 3.6 lists the competed and congressionally designated research centers that received funding through OER during 1984–1996.[12] (EPA labels the congressionally designated centers as "targeted." Many relate to the Superfund program. Most, but not all, are situated at uni-

Table 3.6. Competed and Congressionally Designated NCERQA Research Centers

Competed Centers (with lead institution)
Northeast Hazardous Substance Research Center, New Jersey Institute of Technology
Great Lakes & Mid-Atlantic Hazardous Substance Research Center, University of Michigan
Great Plains & Rocky Mountain Hazardous Substance Research Center, Kansas State University
Western Region Hazardous Substance Research Center, Stanford University
South & Southwest Hazardous Substance Research Center, Louisiana State University
Center for Airborne Toxics, Massachusetts Institute of Technology
Center for Clean Industrial & Treatment Technologies, Michigan Technological University
Multiscale Experimental Ecosystem Research Center, University of Maryland
Center for Ecological Health Research, University of California, Davis

Congressionally Designated Centers (with lead institution)
Center for Environmental Management, Tufts University, Medford, Mass.
Gulf Coast Hazardous Substance Research Center, Lamar University, Beaumont, Tex.
Center for Environmental Policy, Education, & Research, Clark Atlanta University, Atlanta, Ga.
Southwest Center for Environmental Research & Policy, Arizona State University, Tempe, Ariz.
Center for Environmental Resource Management, University of Texas, El Paso, Tex.
National Center for Air Toxic Metals, University of North Dakota, Grand Forks, N. Dak.
Center for Excellence in Environmental Science & Engineering, University of Detroit Mercy, Detroit, Mich.
Urban Waste Management & Research Center, University of New Orleans, New Orleans, La.
Center for Energy & Environmental Studies, Southern University, Baton Rouge, La.
Lower Mississippi River Integrated Cancer Study, Louisiana State University Medical College, New Orleans, La.
Pollution Prevention Research & Development Center, University of Connecticut, Storrs, Conn.
Arkansas Clinical & Developmental Toxicology Program, Arkansas Children's Hospital Research Center, Little Rock, Ark.
California Urban & Environmental Research Center, California State University, Hayward, Calif.
Oil Spill Remediation Research Center, McNeese State University, Lake Charles, La.
National Microscale Chemistry Center, Merrimack College, North Andover, Mass.
Center for Resource Conservation & Waste Management, Oklahoma Alliance for Public Policy, Oklahoma City, Okla.

Source: Undated 1996 mimeograph of EPA response to inquiry from Sen. Christopher Bond (R-Mo.), Chair, Senate Appropriations Subcommittee for VA, HUD, and Independent Agencies.

versities.) For the congressionally designated centers, funding per center varied considerably, ranging from $21 million (Tufts University) to $150,000 (Oklahoma Alliance for Public Policy). The four largest recipients were Tufts ($21 million), Lamar ($16 million), Clark Atlanta ($15 million), and Arizona State ($10 million). By comparison, the largest recipient among the competed centers, a Stanford-led consortium, received $12 million.

The research centers program has met with considerable internal resistance. According to an EPA official, the ORD laboratories object to resources for the ERCs coming from their budgets, while the program offices generally object to EPA's funding academic research that is not directly or immediately useful to regulatory programs. The program also has influential academic critics, some of whom were unsuccessful research center applicants. While scientists have a good basis for arguing that stable funding is needed to conduct long-term research, a hazard of creating centers is that they may become viewed as an entitlement.

The current ERCs constitute the second tranche of the program. The first tranche, which included eight centers, ended in 1989, and EPA reopened the centers program to an externally peer-reviewed competition in 1990. Four centers—Massachusetts Institute of Technology, Michigan Tech, University of Maryland, and University of California, Davis—were awarded from among eighty-seven applicants in 1992, but none of the original centers were re-funded. According to an EPA official, a director of one of the original centers registered a protest with the agency. In response to the appeal, EPA conducted a review of the research center selection process, after which the agency adhered to its decision. Not surprisingly, the controversy did not contribute to the reputation of the environmental research centers program in academic circles. The current centers were scheduled to be externally reviewed prior to their grants expiring in 1996, but no new research centers were solicited in 1997. In line with the administration's accent on children's health issues, the President's FY 1999 budget submission for EPA called for establishing six children's environmental research centers in collaboration with the Department of Health and Human Services, but no monies were designated for the centers. Early in 1998, the agency issued a request for proposals to establish two medical centers of excellence on children's environmental health issues (*Risk Policy Report*, February 20, 1998, 18).

National Risk Management Research Laboratory

NRMRL, headquartered in Cincinnati, Ohio, contains the Risk Reduction Engineering Laboratory in Cincinnati (drinking water, land remediation, and multimedia sector specific); the Air and Energy Engineering Research Laboratory in RTP (air pollution); the Robert S. Kerr Environmental Research Laboratory in Ada, Oklahoma (subsurface pollution); and the Center for Environmental Research Information in Cincinnati (information and technology transfer). In a 1995 interview, then NRMRL Director Timothy Oppelt said that the lab is designed to research technical approaches to reducing risk, including prevention and control, and its priorities include the Browner administration's environmental technology initiative and drinking water research (*Risk Policy Report*, June 16, 1995, 9). NRMRL is also responsible for research related to characterization of pollutant generation and release, remediation of contaminated media, public health protection from indoor pollutants, and occupational protection in industrial environments, agricultural pesticide use, and Superfund site cleanup (www.epa.gov, May 1995).

ENDNOTES

[1] For example, the administration's FY 1996 proposal for EPA R&D included $78 million that corresponded to special White House initiatives for environmental technology, climate change, and a working capital fund (*Inside EPA* , February 10, 1995, 4).

[2] Congressional oversight of EPA during the 1980s focused on improper activities by contractors, including in-house contractors' writing their own scopes of work and drafting proposals on which they would bid. This reinforces the idea that the source of the science informing agency decisions is important because such practices influence not only who does the analysis but also what gets investigated and how. Carnegie (1992) notes, however, that limitations on the number of staff that the agency is permitted to employ have contributed to the disproportionate share of the workforce at ORD laboratories that is on-site contractors.

[3] As used in the public policy literature (for example, Kingdon 1984), the concept of policy entrepreneurs refers to members of the policy community who blend the qualities of the expert analyst and the advocate to effect policy change, particularly through advancing ideas onto the policy agenda.

[4] The geographic scope of EMAP has been reduced from nationwide to three regions—the Pacific Northwest, the Everglades, and the mid-Atlantic coastal watershed (Stone 1995).

[5] According to the former EPA scientist, the RIHRA program has resulted in an added emphasis on dosimetry within EPA. Dosimetry refers to dose measurement and evaluates the relationship between ambient exposure and the delivered dose and involves the field of pharmacokinetics. A recent reorganization of HERL established a new branch that specifically addresses dosimetry.

[6] Foley is currently listed as the Director of EPA's National Risk Management Research Laboratory.

[7] These observations about NCEA reinforce the importance of some of the major questions raised in the introductory chapter: who is responsible for assessing risks and who transmits scientific information and in what form? (In this case, a synthesis of scientific information or a science-based policy recommendation.)

[8] RAF members are nominated by their offices for consideration by the forum.

[9] According to the National Science and Technology Council Committee on Environment and Natural Resources, the EPA STAR program budget totaled $95 million in fiscal year (FY) 1997, and the administration's FY 1998 request was for $115 million (NSTC/CENR 1997). It appears that much of the STAR budget is being programmed through interagency collaborative research programs. NCERQA's FY 1997 extramural research grant solicitation announced approximately $33 million in funding across eight categories. In early 1997, the agency also announced three collaborative research grant programs with four federal agencies with diverse missions. In addition to the ongoing EPA/NSF "Partnership for Environmental Research" ($12 million), the agency established the EPA/Department of Energy/NSF/Office of Naval Research Joint Program on Bioremediation ($5.5 million) and the EPA/National Aeronautics and Space Administration Joint Program on Ecosystem Restoration ($4 million) (*Risk Policy Report*, January 24, 1997, 24).

[10] In contrast to the first joint EPA/NSF request for proposals, the 1996 joint RFP was more narrowly focused and may have provided more opportunities for institutional learning.

[11] The competed centers include four environmental research centers (ERCs) and five hazardous substance research centers (HSRCs). The HSRCs were authorized under the 1986 Superfund amendments and established in 1989. The HSRC program is slated to end after the current grant period.

[12] Many of the research centers are consortia. Table 3.6 lists only lead institutions. The congressionally designated research centers are listed in decreasing order of funding. Table 3.6 is not a comprehensive list of congressionally designated research centers supported by EPA. According to a former EPA/ORD official, such centers are "sprinkled throughout the agency" in program and regional offices, as well as ORD. For example, the Texas Institute for Applied Environmental Research, Tarleton State University, Stephensville, Texas, receives congressionally designated support from the EPA Office of Policy, Planning, and Evaluation, according to an agency official. Congressionally designated centers are also more common than might appear because, in some cases, they are designated in neither committee reports nor appropriations bills. Instead, they are designated informally through back-channel communications.

REFERENCES

Brown, G. 1997. Environmental Science under Siege in the U.S. Congress. *Environment* 39(2): 12–20, 29–31.

Carnegie (Carnegie Commission on Science, Technology, and Government). 1992. *Environmental Research and Development: Strengthening the Federal Infrastructure.* New York: Carnegie Commission, December.

CNIE (Committee for the National Institute for the Environment). 1995. *Federal Funding for Environmental Research and Development—Fiscal Year 1994.* Washington D.C.: CNIE.

Davies, J. C., III, and B. S. Davies. 1975. *The Politics of Pollution,* 2nd ed. Indianapolis, Ind.: Pegasus.

Gramp, K., A. Teich, and S. Nelson. 1992. *Federal Funding for Environmental R&D.* A Special Report of the R&D Budget and Policy Project. Washington, D.C.: American Association for the Advancement of Science.

Jasanoff, S. 1990. *The Fifth Branch: Science Advisors as Policymakers.* Cambridge, Mass.: Harvard University Press.

Kingdon, J. 1984. *Agendas, Alternatives, and Public Policies.* Boston: Little, Brown & Company.

Lawler, A., and R. Stone. 1995. FDA: Congress Mixes Harsh Medicine. *Science* 269: 1038–1041.

NRC (National Research Council). 1995. *Interim Report of the Committee on Research and Peer Review in EPA.* Washington, D.C.: National Academy Press, March.

NSTC/CENR (National Science and Technology Council/Committee on Environment and Natural Resources). 1997. *Program Guide to Federally Funded Environment and Natural Resources R&D.* February. Washington, D.C.: NSTC/CENR.

Stone, R. 1994. Can Carol Browner Reform EPA? *Science* 263: 312–315.

———. 1995. EPA Streamlines Troubled National Ecological Survey. *Science* 268: 1427–1428.

U.S. CRS (U.S. Congressional Research Service). 1997. *Environmental Protection Agency: FY 1997 Budget.* Washington, D.C.: U.S. CRS. Also available at http://www.cnie.org/nle/leg-15.html (accessed March 1999).

U.S. EPA (U.S. Environmental Protection Agency). 1993. *Report to Congress: Fundamental and Applied Research at the Environmental Protection Agency.* EPA/600/R-93/038. Washington, D.C.: U.S. EPA.

———. 1994. *Research, Development, and Technical Services at EPA: A New Beginning.* EPA/600/R-94/122. Washington, D.C.: U.S. EPA.

U.S. OPM (U.S. Office of Personnel Management). 1993. *Federal Civilian Workforce Statistics: Occupations of Federal White Collar and Blue Collar Workers.* OPM-MW56-23. Washington, D.C.: U.S. OPM.

U.S. OTA (U.S. Office of Technology Assessment). 1993. *Researching Health Risks.* OTA-BBS-570. Washington, D.C.: U.S. Government Printing Office, November.

Yosie, T. 1987. EPA's Risk Assessment Culture. *Environmental Science and Technology* 21(6): 526–531.

4

Science inside EPA:
The Regulatory Programs

EPA regulatory programs are organized under the broad program-matic offices of four Assistant Administrators (AAs). The major regulatory programs are listed in Table 4.1 under their respective Assistant Administrator.

Over thirty interviewees selected on the basis of their broad knowledge of and experience in environmental regulatory decisionmaking were asked how their evaluation of EPA's use of scientific information differed among regulatory programs. The general pattern of responses to this open-ended question suggests that most of these respondents regard the pesticides, toxic substances, and criteria air pollutant programs as the most sophisticated users of science among EPA regulatory programs. The surface water and drinking water programs occupy a middle position. Respondents were nearly unanimous in placing the Superfund contaminated site and hazardous waste programs at the bottom of the list. Factors as varied as the geographic proximity and occasional interchange of staff between the air and research programs, the statutory authority of the pesticide and toxic substances programs to define industry data submissions, and the nonregulatory status of the Superfund program all play a role in shaping these programs' use of science in their policy decisions.

Despite the generally low marks respondents gave the Superfund and hazardous waste programs for their use of science, some respondents applaud these programs for their pathbreaking attempts to grapple with the analytically vexing issues posed by multiple pollutants, chemical mixtures, and multiple pathways of exposure to pollutants. By comparison, the other regulatory programs have traditionally tended to consider science on a narrower, more analytically tractable scope. In other words, Superfund's approach to science may not be as clean, but it attempts to address real-world conditions. An agency research official notes that sin-

Table 4.1. EPA's Major Regulatory Programs

AA for Air and Radiation (OAR)
Criteria air pollutants
Hazardous air pollutants

AA for Prevention, Pesticides, and Toxic Substances (OPPTS)
Pesticides
Toxic substances

AA for Solid Waste and Emergency Response (OSWER)
Superfund
Hazardous waste

AA for Water (OW)
Drinking water
Surface water

gle-medium programs (for example, air) are inherently more straightforward and easier to study than are multimedia programs (for example, hazardous waste). It is also easier to defend the more tractable, but narrowly scoped, single-medium, single-contaminant decisions. The comprehensive, multimedia approach represents an important difference between the place-based contaminated site and hazardous waste programs and the national regulatory programs, which are traditionally substance- and medium-specific and frequently ignore site-to-site variation.

Important differences within regulatory program offices also preclude overly broad generalizations about the acquisition and use of science by "AA-ships." An EPA official feels that differences among program managers and political appointees over time are responsible for important differences within the broader offices. In some cases, differences in the use of science within AA-ships may be greater than differences among them. Some differences arise from the maturity of the program. For example, a former senior EPA official suggests that newer regulatory programs show a greater willingness to use new scientific information and, except in cases in which periodic science reviews are required, existing programs are unwilling to revisit old decisions in light of new science. An Office of Research and Development (ORD) official adds that newer programs (for example, hazardous waste) had to develop a science base from scratch, whereas programs that predate EPA (for example, air) are built on decades of research. There is also considerable variation within programs. A senior EPA official, for example, finds that the available scientific information is uneven on an individual decision-by-decision basis, rather than on a program-by-program basis. Another senior EPA official notes that programs vary in the extent to which they consider science on ecological effects in addition to health effects information. A former senior EPA official says that the operative statutory requirements are the main determinants of differences among programs in the use of science. A former EPA Administrator believes that regulatory programs with more statutory flexibility generally do a better job of using scientific information.

A commonality among the regulatory programs, however, has been their reluctance to come to grips with scientific uncertainty in a meaningful way. A former political appointee comments, "There is a lack of clarity about the role of science in EPA. The scientists want to express uncertainties, ranges, and so forth, but the people who write the regulations don't want to. The program people take good science and render it vulnerable to criticism." For an overall evaluation of the factors that affect the use of science in EPA decisionmaking, see Chapter 5.

The following sections introduce the major EPA regulatory programs evaluated in this report and discuss the role of science in each.

OFFICE OF PREVENTION, PESTICIDES, AND TOXIC SUBSTANCES (OPPTS)

An EPA official characterizes OPPTS as sitting atop EPA's "risk culture." According to a former senior EPA official, "OPPTS has a large stable of in-house scientists, including many toxicology experts, and the senior management is continuously involved in risk-based decisionmaking that is often quite sophisticated."

OPPTS churns out more substance-specific risk assessments than any other program office. The overwhelming majority of these assessments, however, are quick, routinized screening analyses for new chemicals or slight modifications of currently registered pesticides. A former senior EPA official suggests that OPPTS is "capable of good work but hopelessly understaffed." For the most part, OPPTS scientific staff develop and apply the rules for industry to submit chemical testing data and analyze those data for substance-specific decisionmaking. There have been inconsistencies in EPA's guidelines for testing chemicals between the pesticides and toxics programs, but OPPTS has undertaken to merge its own guidelines and to harmonize them with international guidelines (*Risk Policy Report*, November 17, 1995, 31–32).

Traditionally, there has been a weak relationship between OPPTS and ORD. In part, this seems to stem from EPA's statutory ability to require or negotiate for data from the chemical and pesticides industries. The pesticides program, in particular, is viewed as being "data rich." Thus, OPPTS has not needed ORD laboratories to conduct toxicity testing. OPPTS also inherited a substantial cadre of toxicologists and health scientists from other agencies when EPA was formed, in particular from the U.S. Department of Agriculture's (USDA) pesticides program. But, while many of these "old-line" science staffers were admired for their technical competency, their bureaucratic culture sometimes clashed with the public health orientation of many EPA scientists and the regulatory culture of the agency overall. Probably as a result of these factors, OPPTS has historically conducted its own chemical-specific environmental and health assessments. What OPPTS has been unable to secure from industry or do for itself, however, is sufficient basic research or the development of needed generic databases, noncancer testing methods, and risk assessment tools and models. Sources from OPPTS argue that the Office of Research and Development should be doing more work specifically tailored to fill these gaps.

The use of science by the pesticides and the toxic substances programs is heavily conditioned by the operative statutes, the Federal Insecticide, Fungicide and Rodenticide Act (FIFRA) and the Toxic Substances Control Act (TSCA). Both statutes call on EPA to balance the costs and benefits of environmental regulation, placing a high analytical burden on the agency. An important difference between the statutes is that whereas FIFRA is a bona fide licensing program that empowers EPA to require the pesticides industry to provide scientific data, the agency must be able to support a finding that a commercial chemical may pose an unreasonable risk or substantial exposure before it can require data from industry under TSCA. Unlike other pollution control statutes, TSCA also subjects EPA to the "substantial evidence" criteria of judicial review instead of the less demanding "arbitrary and capricious" standard.

Pesticides

The EPA Office of Pesticides is responsible under FIFRA for registering new pesticides and reregistering old pesticides to ensure that, when used according to label directions, they will pose no "unreasonable risks" to human health or the environment. FIFRA requires EPA to balance the risks of pesticide exposure to human health and the environment against the socioeconomic benefits of pesticide use. Pesticide registration decisions are based primarily on EPA's evaluation of data provided by the pesticides industry. EPA also sets tolerances for pesticides in most raw and processed agricultural commodities under the Federal Food, Drug and Cosmetic Act (FFDCA).[1]

The pesticides program has well over 300 scientists on its headquarters staff, according to an OPPTS official. The pesticides program also maintains three laboratories. The two major labs are chemistry labs (in Bay St. Louis, Mississippi, and Beltsville, Maryland).[2] The labs' main job is a quality control function: to develop and revise pesticide testing guidelines and to ensure that the procedures (for example, for analyzing pesticide residues on food and in soil) work "as advertised." With its responsibilities under FIFRA and FFDCA, the pesticides program makes thousands of decisions each year. Given this workload, a former EPA Administrator concludes that the pesticides program has inadequate resources to review pesticides data.

Beginning in the late 1970s, EPA began requiring extensive data on pesticides. Currently, the potential data requirements for pesticide registration are extensive, with about 150 types of studies that EPA can require (see 40 CFR 58). The agency can require testing on acute toxic reactions, such as poisoning and skin and eye irritation, as well as possible long-term effects, such as cancer, birth defects, and reproductive system disorders, as well as information on the movement and fate of pesticides in the environment that is useful in exposure assessment. A company may have to spend up to $3 million to support a registration petition (NAPA 1995).

Pesticide data requirements are tiered, however. That is, depending on the results of a first-tier study, a registrant may have to conduct a second, more extensive tier of studies. An industry official believes that EPA's pesticide information requirements got "out of hand" during the 1980s. "Requiring new tests was a way of not making a decision." This source credits pesticide reregistration time lines established in the 1988 FIFRA Amendments and former Assistant Administrator for Pesticides and Toxic Substances Linda Fisher with creating the pressure to overcome the perceived paralysis by analysis. The 1988 FIFRA amendments also required that the Scientific Advisory Panel (SAP) review the pesticide data requirements and testing guidelines proposed by EPA. (For a discussion of the role of SAP, see the section entitled "Official Science Advisers" in Chapter 2). Whether one agrees with the assessment that data requirements were excessive, this evolution of pesticide data requirements suggests the more general observation that requiring new scientific information is not always primarily intended to shed light on a decision; sometimes it is used to avoid or postpone a decision.

EPA processes about 5,000 new pesticide registrations each year (U.S. EPA 1996a, 30). The vast majority of the new registrations are for new manufacturers of an existing pesticide, new formulations using existing active ingredients, and new uses of an existing product. The pesticides program registers about fifteen to twenty-five new active pesticidal ingredients each year. A brand-new active ingredient may need six to nine years to move from development in the laboratory, through full completion of EPA registration requirements, to retail shelves. This time frame includes about two or three years to obtain registration approval from EPA.

FIFRA requires EPA to reregister existing pesticides that were originally registered by USDA before current scientific and regulatory standards were formally established. The reregistration process is intended to ensure that (1) up-to-date data sets are developed for each chemical (or their registrations will be suspended or canceled); (2) modifications are made to registrations, labels, and tolerances as necessary to prevent unreasonable risk to human health and the environment; and (3) special review or other regulatory actions are initiated to deal with any "unreasonable risks." Reregistration has proved to be a massive, time-consuming analytical undertaking that includes an elaborate scientific review of an existing pesticidal "active ingredient," decisions on which additional data will be required for full reregistration, and development of the agency's current regulatory position on the pesticide.[3] Under the Food Quality Protection Act of 1996, EPA is now required to periodically review a pesticide's registration every fifteen years.

Between 1975 and 1985, EPA completed thirty-two special reviews of pesticides, each of which generally took from two to six years and could be followed by administrative hearings taking another two years (Shapiro 1990). As the case study on ethylene dibromide (EDB, Appendix E) indicates, it can take EPA many years to ban a pesticide, and over that period, scientific understanding is constantly evolving as data accretes. Spurred ultimately by public alarm at the detection of pesticide residues in consumer products and groundwater, the final decision on EDB was based to a large extent on information that was not available at the beginning of the reregistration process.

During the 1990s, the agency has reduced its use of the special review process. Instead, EPA has relied more on the reregistration process to characterize the risks of existing pesticides, according to an EPA pesticides official. If there is evidence of a potentially "unreasonable risk," the pesticides office generally enters into negotiations with the manufacturer without formally "bumping it up to special review." Although the special review process is intended to kick in a higher level of scrutiny and does not necessarily lead to a ban, companies are now more willing to negotiate because they "don't like their chemical name and special review to be used in the same breath. They feel it automatically taints their product in the public mind." According to an industry representative, this concern arose from the 1989 "Alar incident" in which public attention was seized by an episode of *60 Minutes* and actress Meryl Streep pleading before a Senate committee to protect children from tainted apples and apple juice.

Regarding pesticide tolerances in food, until recently, FFDCA allowed EPA to balance costs and benefits of pesticide use in the setting of the tolerance for pesticides in *raw* agricultural commodities but not for food *additives* or pesticide residues that concentrate in food processing. (When pesticide residues found on raw commodities [such as tomatoes] are concentrated in the processing of food [to tomato paste], the pesticide is then considered a food additive.) A 1958 FFDCA amendment, the Delaney Clause, allowed no tolerance for carcinogenic food additives. As a consequence, fresh fruits and vegetables were permitted to have more pesticide residue than the same products if they were processed. For many years, the contradiction was often ignored, but a 1992 Supreme Court decision held that EPA must automatically deny any tolerance for a cancer-causing pesticide that is found in processed food.[4] The Food Quality Protection Act (FQPA) of 1996 replaced the Delaney Clause with a uniform standard for food safety requiring that levels of a pesticide in both raw and processed foods pose a "reasonable certainty of no harm."

In what could be regarded as recompense to environmentalists for surrendering the Delaney Clause, FQPA also requires EPA to employ an additional safety factor (of ten) to provide more certainty that children will be protected from noncancer effects of pesticide residues, unless the agency determines that reliable scientific data are available to rebut this presumption.[5] In a February 1998 memorandum to the agency's Assistant Administrators for Pesticides and Research and Development, EPA Administrator Browner and Deputy Administrator Hansen called for a more vigorous application of the rebuttable presumption of the extra safety factor (*Risk Policy Report*, March 20, 1998, 5). In March 1998, Penelope Fenner-Crisp, a senior EPA pesticides official, reported that the agency had applied an extra tenfold safety factor for only about 10% of the 100 pesticides it had reviewed since FQPA was passed. For another 10%, the pesticides program applied an additional safety factor of other than ten (for example, three) (*Food Chemical News*, March 16, 1998, 14; *Risk Policy Report*, March 20, 1998).

The pesticides program received many high marks from respondents on its use of science. The program is particularly noted for not focusing inordinately on cancer risks or exclusively on human health risks. Several respondents cited the technical capability of the "old-line pesticide scientists" as being responsible for the program's scientific reputation. One senior EPA careerist added, however, that many within the agency view the pesticides program as being "captured" by the pesticides industry.[6] Other respondents see clear divisions within the program. When EPA was formed, its pesticides program drew staff from USDA, the Department of Health, Education, and Welfare's Food and Drug Administration, and the Department of the Interior. As a result, there were differences among and within the original units in their approach to regulating pesticides. In the current organizational framework, some observers view the Registration Division (the largest division, which is responsible for new pesticide registrations and was largely inherited from USDA) and the Health Effects Division (which analyzes human health risks) as being relatively more "industry friendly" than the Special Review and Re-registration Division and the Environmental Fate and Effects Division (which evaluates ecological risks from pesticides). Such differences in bureaucratic culture can be important determinants of the amount and type of scientific evidence needed to warrant regulation or justify more or less stringent regulation. It could be that the cautious, methodical nature with which scientists are trained to approach problems contributes to both the technically competent reputation of the pesticides program and the perception of some observers that the pesticides program is reluctant to impose regulatory restrictions.

As the EDB case study suggests (Appendix E), the pesticides program has had a blind spot in evaluating the risks posed by pesticides that are alternatives to pesticides that are under review. Although an EPA policy official suggests that the situation has improved somewhat since the early 1980s, "The risk trade-off analysis tends to be very uneven, depending on the interaction and makeup of a multidisciplinary team and the work group manager." Generally, according to this source, as the number of alternative pesticides lessens, the burden on EPA to evaluate the risk trade-offs grows. An example of EPA's efforts to conduct a comparative assessment of pesticides is the "corn cluster" analysis, which evaluated various insecticides used on cornfields. It is noteworthy that the policy office—not the pesticides program—was the principal institutional driver behind the corn cluster analysis. Whereas the regulatory program has tended to address pesticide risks one at a time, the policy office has urged a comparison of the risks posed by the alternative pesticides that could be applied to any given crop.

According to a pesticides program official, the "lack of sophistication" demonstrated by the program in comparing the risks of alternative pesticides is part of a larger problem. The program has traditionally dealt with both pesticides and exposure pathways in isolation, says this source, rather than look at the composite risk from other pesticides and the various routes by which a person can be exposed to a given pesticide. Faced with a large number of decisions to make regarding individual pesticides, program staff long resisted conducting cumulative assessments on the basis that they are technically intractable, according to this source. While allowing that the cumulative risk assessment "toolbox is only partially full," this respondent protests the program's lack of progress toward addressing the long-standing issue. Also, the pesticides industry, accustomed to being evaluated on the basis of incremental risk, "hates the idea" of basing regulatory decisions on cumulative risks. The program has been spurred into action, however, by a National Research Council report on the risks of pesticides to children (an outgrowth of the Alar incident) and the Food Quality Protection Act of 1996. NRC (1993) not only addressed children as a susceptible subpopulation with particular consumption patterns but also underscored the need to consider cumulative effects of multiple pesticides and multiple exposure routes.

In response, the pesticides program began conducting a composite special review for triazines, a group of herbicides used primarily on corn. Because triazines have a common effect (causing mammary tumors in female rats), the group provides a common reference point that toxicologists can "put their brain around" to make the integrated analysis more tractable. Subsequently, the pesticides program also embarked on a project to address the combined risks of the five organochlorine pesticides remaining on the U.S. market (*Risk Policy Report*, May 17, 1996, 35). In what portends a major shift in the pesticides program (and perhaps throughout the agency), the Food Quality Protection Act of 1996 directed the agency to consider aggregate exposures via multiple pathways (for example, food, drinking water, residential use, and lawn care) and to consider the cumulative effects of pesticides or other substances that have common mechanisms of toxicity (U.S. EPA 1997).

Toxic Substances

The EPA Office of Toxic Substances (OTS) is responsible for administering the Toxic Substances Control Act. TSCA was intended to address nonpesticidal toxic chemicals in commerce and was designed to provide a "life cycle" framework for management of chemical risks, taking into account risks from product manufacturing, use, and disposal and exposures associated with the air, water, and land (Shapiro 1990). TSCA requires premanufacture evaluation of most new chemicals and allows EPA to regulate existing chemical hazards.[7] The risk management tools available to EPA for existing chemicals range from labeling standards to outright bans. As the asbestos case study illustrates (Appendix F), however, the weight of evidence and scope of analysis that TSCA requires—in combination with the court's willingness to substitute its own policy judgment for that of politically accountable Administrators—has bound the feet of EPA's Section 6 regulatory powers. TSCA has legs, but they are feeble by design.

Like FIFRA, TSCA acknowledges the many benefits of commercial chemicals. To enact regulatory controls under TSCA, EPA must have a "reasonable basis" for finding that a toxic substance poses an "unreasonable risk" to health or the environment. The agency must also adopt the "least burdensome" control option that the agency is authorized to require to limit the risk to an acceptable level. This "unreasonable risk" provision has been interpreted as requiring EPA to balance the costs and benefits of proposed regulatory decisions, taking into account the availability of substitutes and other adverse effects (for example, the risks posed by substitutes) that the regulation may have (Shapiro 1990).[8] Unlike FIFRA and other environmental pollution control statutes, however, EPA decisions under TSCA are subject to the "substantial evidence" standard of judicial review. Under other statutes, by comparison, EPA must meet the less demanding "arbitrary and capricious" judicial review standard. Therefore, TSCA requires EPA to provide a considerable weight of evidence and regulatory analysis to document its regulatory decisions and defend them under judicial review.[9]

The administrative decisions subject to the unreasonable risk and substantial review provisions are not limited to those concerning regulatory controls. They also cover decisions for issuing testing requirements. Note the circularity: EPA must provide evidence of unreasonable risk or substantial exposure to request information to evaluate whether there is an unreasonable risk. The U.S. General Accounting Office (GAO) has concluded that EPA's authority to gather data under TSCA is difficult to use and not very effective in supporting the agency's toxic chemical review process (Guerrero 1995).[10]

To require that an existing chemical be tested, EPA must determine that existing data are insufficient, that testing is needed to develop such data, and that the chemical may present an unreasonable risk *or* that there is substantial production *and* potential exposure. To support these findings, EPA must go through a rulemaking process that generally takes at least two years. According to an environmental policy analyst, a series of court decisions has allowed EPA considerable latitude to justify a TSCA rulemaking to require additional information if there is any credible evidence of potential exposure. Unlike FIFRA, TSCA does not require testing of new chemicals, although any test data that may be available are supposed to be provided to EPA in a premanufacture notice (PMN). EPA receives over 2,000 PMNs a year (Dahl 1995). The agency's Office of Research and Development contains a structure activity relationship (SAR) group that supports the TSCA program. (See discussion below concerning SARs.)

EPA has 90 days (extendible to 180 days) to review a new chemical. If the data are insufficient to adequately review the chemical, and if the chemical may present an unreasonable risk or will be produced in substantial volume, the agency can suspend or limit manufacture until the necessary information is provided. A new chemical can be suspended under TSCA if the agency finds a "reasonable basis to conclude that the chemical presents or will present an unreasonable risk of injury to health or the environment."

Recent findings suggest that the structure activity relationships used by EPA in screening premanufacture notices are not as reliable a predictor of chemical risks as some had thought.[11] According to GAO (Guerrero

1994), the accuracy of SARs compared with testing data is too low to adequately characterize chemical risks. As a result, some "unreasonably risky" new chemicals may be passing through the PMN screen. In the past, when new chemicals have failed to pass through the screen, EPA has generally negotiated testing requirements with manufacturers on a case-by-case basis. By comparison, European regulators have a predetermined battery of tests for new commercial chemicals. The European testing scheme is tiered, requiring more testing of chemicals produced at higher volumes.

According to a former EPA political appointee, "Virtually all of the science problems [under TSCA stem from] an incomplete data set on chemicals."[12] This respondent argues that there should be a greater burden on industry to test new chemicals before entering the market. "The agency spends limited taxpayers' money to make its case," says this source. EPA cannot afford to pay for the testing or to routinely go through the elaborate rulemaking to require industry testing, and "the chemical companies and their lawyers . . . know it."

The problem is offset to a limited extent by the TSCA Section 8(e) requirement that industry report information on chemicals or mixtures that reasonably supports the conclusion that existing chemicals present a substantial risk (for example, as a consequence of accidental releases or widespread and previously unsuspected distribution in the environment). From 1977 through 1991, however, industry had submitted only 1,350 Section 8(e) notices to EPA. Suspecting that companies were not complying with the Section 8(e) requirement, the agency established a voluntary audit program that allowed companies to submit studies they should have provided to EPA earlier at reduced penalties. One hundred twenty three companies submitted more than 11,000 toxic effects studies and were assessed over $20 million in penalties (O'Bryan 1997).[13]

Another challenge facing the TSCA program is that because there are so many thousands of chemicals in commerce, it is administratively unfeasible to have agency staff that are expert in all phases of a particular toxic substance. Instead, EPA relies on having flexible staff with specialized training in risk assessment methods. As illustrated in the asbestos case study, OTS must often look outside the agency for substance-specific expertise.

OFFICE OF AIR AND RADIATION (OAR)

Some respondents ranked OAR's use of science "at the top with OPPTS," but others gave OAR a second-tier or middling ranking. Several commented on the diversity within the OAR program. An ORD official, for example, gives the air program office responsible for developing ambient air quality standards very good marks and grades the use of science by the mobile sources air program office as very poor. A senior EPA official feels that the air program considers only human health effects information and shows a lack of consideration for ecological effects information. The following sections are limited to the criteria and hazardous air pollutants, with an emphasis on the former.

Criteria Air Pollutants

Under the Clean Air Act (CAA), EPA sets national ambient air quality standards (NAAQS) for six "criteria" air pollutants: particulate matter, sulfur dioxide, carbon monoxide, nitrogen dioxide, ozone, and lead.[14] These are widespread ambient air contaminants that are predominately associated with noncancer health effects. The most prominent features of the NAAQS process are the statutory requirements for periodic review (every five years) of the standards and for external review—of both science and policy matters—by the Clean Air Science Advisory Committee.

The 1970 CAA required that the NAAQS reflect the latest scientific information and that both the criteria and the standards be periodically reviewed and, if appropriate, revised. Dissatisfied with EPA's progress in reevaluating the original NAAQS, Congress directed the agency in the 1977 CAA Amendments to revise or reauthorize all of the NAAQS by the end of 1980, and every five years thereafter. As the particulates and ozone case studies (Appendixes C and D, respectively) show, the agency has not kept to this schedule, and the requirements for using the latest science and periodic review have provided opportunities for both industry and environmentalists to delay, compel, or challenge administrative decisions.

There are two types of NAAQS under the CAA. Primary NAAQS are intended to protect public health from any reasonably expected adverse health effects, including an allowance for an adequate margin of safety and taking into consideration sensitive members of the population. Secondary NAAQS are intended to protect the public welfare (structures, crops, animals, and so forth) from any known or anticipated adverse effect associated with ambient air pollution. Unlike FIFRA and TSCA, the decisional criteria for NAAQS provide no role for considering costs. (The intent of Congress was for costs to be considered at a later stage in formulating state-by-state implementation plans for NAAQS attainment.) This legislative framing requiring absolute protection implicitly and incorrectly presumes that there exists an identifiable threshold level of pollution below which no adverse effects occur. As the ozone case study illustrates, a continuum of biological effects has been observed down to background levels of ambient air pollution. "This is the problem with ambient air quality standards," concludes a former EPA researcher. "What you're trying to do is ludicrous—set the level below which the most sensitive person in the population will have no adverse health effects." Although the statute does not permit the agency to justify its decision explicitly on the basis of cost-benefit or cost-effectiveness analysis, when susceptible segments of the population are considered, it is virtually impossible to set a scientifically justifiable "zero risk" ambient air quality standard.

The CAA, therefore, has saddled EPA with a scientifically unsound congressional mandate that sometimes drives and sometimes delays the agency's generation and use of scientific information. The misguided presumption of a threshold frames the research question as, what is the lowest level of pollution at which biological effects can be detected? Recognizing that there is no definitively "safe" numerical level for an NAAQS, EPA can add to the margin of safety in the criteria air pollution control scheme only by increasing the stringency of the enforceable plans for attaining the standards. Because scientific advances in the capability to detect biological effects of air pollution inevitably point to more stringent NAAQS, EPA administrations have occasionally tried to postpone the inevitable when the economic and political stakes are high (see the ozone case study).

Scientific advances have also had very constructive effects on the NAAQS. In the cases of both particulate matter and ozone, new scientific information helped refine broad, ill-defined categories of ambient air contaminants (total suspended particulates and photochemical oxidants) and focus pollution control efforts more specifically on pollutants of particular concern to human health (fine particulates and tropospheric, or ground-level, ozone). The particulates case study also provides an example of how the scientific operational definition of a pollutant can have varied effects on different regions and industrial sectors and can mobilize the resources of the interested parties to engage in the regulatory science debate.

The term *criteria air pollutant* derives from the CAA requirement for EPA to prepare criteria documents (CDs) that summarize and interpret the basic research findings relevant to the specific pollutants. The CD process traces its roots to the National Air Pollution Control Administration (NAPCA), which, prior to EPA's establishment, was charged with providing technical assistance to state air quality control agencies (Davies and Davies 1975).[15] The task of developing and revising CDs is managed by the EPA Office of Research and Development Environmental Criteria and Assessment Office (ORD/ECAO), located in Research Triangle Park, North Carolina. The OAR Office of Air Quality Planning and Standards (OAQPS), also located in Research Triangle Park, is responsible for developing a staff paper, a document informed by the criteria document and additional analysis that presents air office program staff recommendations regarding the NAAQS. Much of the additional scientific analysis included in the staff paper involves exposure assessment that permits comparisons of the health and environmental impact of regulatory alternatives.

According to Yosie (1988), OAQPS began the preparation of staff papers in 1979 at the instigation of the EPA Science Advisory Board (SAB). The institutionalization of the staff paper (which other programs have adopted) resulted in "demystifying the process by which EPA uses the results of scientific studies and translates such results into policy options." The Clean Air Science Advisory Committee (CASAC) is statutorily required to review the air quality criteria and, as a distinct committee of the EPA Science Advisory Board, CASAC is authorized to review proposed national ambient air quality standards.[16] (The roles of CASAC and SAB are discussed in more detail in the Chapter 2 section "Official Science Advisers.")

Unlike other EPA regulatory programs, there is an intimate relationship between the air program and ORD. Key basic research in both the particulates and ozone case studies was produced in ORD's National Health and Environmental Effects Research Laboratory. According to an EPA air program official, the colocation of OAQPS and ECAO in Research Triangle Park has forged a special relationship among EPA staff, with a lot of movement back and forth between the offices at the staff level. Whereas staff in other programs (for example, water or hazardous waste) tend to be more mobile, this source notes that "people tend to come into the air program and stay in the air program." An advantage of the close association between OAQPS and ECAO is that staff build personal and professional relationships. Insular thinking, however, is a potential downside.

In the past, OAQPS careerists, with their respected technical ability, expertise in program implementation, contacts, and institutional memory tended not to view the political operatives in EPA air program headquarters as their clients, according to the air program official. In turn, "OAQPS was viewed by [headquarters staff in Washington,] D.C. as hardheaded." Headquarters also complained about the flak that it was getting from Congress and the environmental community as a result of delays in the NAAQS reviews. During the 1980s, OAQPS management got the message to "make sure the trains run on time." (See discussion in Appendix C.) This new orientation apparently changed the dynamic somewhat between OAQPS and ECAO, with air program staff growing increasingly frustrated with what they and some others view as the bottleneck in the NAAQS review process—criteria document development by ECAO.

Principal complaints expressed by interviewees regarding criteria documents include their encyclopedic scope and the extensive use of external authors to write chapters. Rather than supplementing previous criteria documents with critical new information, ECAO produces comprehensive documents that may expound at length on scientific topics that seem peripheral to regulatory decisions. One contributing reason for the level of detail in criteria documents is the use of external scientific authors whose focus is on their narrow areas of expertise and who may lack a broad perspective on the issues important to policymaking. Official scientific review is another factor that tends to expand the scope of the criteria documents. ECAO submits an initial plan for scoping the document to CASAC. In critiquing the scope, "CASAC says the CD should be more focused," comments an environmental lawyer. In reviewing the document, however, CASAC "tells ECAO to expand it. When the scientific information relates to their own disciplinary expertise, each member [of CASAC] feels it needs to be more comprehensive. CASAC is put in the driver's seat of scoping the CD. ECAO's mind-set is that they must be encyclopedic or they'll get beaten up by CASAC." An ORD scientist suggests that part of the reason that a criteria document is "a compendium of information that could reasonably be expected to be useful" is that it can be difficult to anticipate the information needs of air program staff two or more years into the future.

The use of external authors presents a number of problems but is perhaps unavoidable given ECAO staff limitations. In addition to its effects on the scope of criteria documents, the use of external authors can create a lack of continuity in the documents, making integration of the various parts by OAQPS difficult. An environmental lawyer cautions that some academics who have been external contributors to criteria documents "have had very public axes to grind," resulting in unbalanced work and major rewrites. With approximately fifteen scientists, however, ECAO does not have the depth of staff that OAQPS has, and recently, ECAO's workload has increased with additional responsibilities for hazardous air pollutants. Instead of having experts in particular air pollutants, ECAO is staffed along disciplinary lines (for example, toxicology and ecology). Although ECAO staff undoubtedly gain considerable familiarity with specific criteria air pollutants during the course of CD development, there is sometimes so much research relevant to a criteria air pollutant that its staff may be unable to master, or even keep abreast of, all of the literature on, say, ozone, while it is developing a CD on particulate matter. An academic also observes that ECAO scientific staff are primarily toxicologists and that because epidemiological studies are becoming increasingly important in criteria documents, this creates a need for ECAO to supplement its staff with external epidemiologists. As a result, ECAO contracts with external experts (often academics) or seeks detailees from other EPA offices on an as-needed basis. A former EPA scientist believes that the lack of staff in ECAO suggests that developing criteria documents is "not a priority" for ORD management. An ECAO manager, however, estimates that the office would need 100 scientists to "cover all the disciplinary bases and substances" and recognizes that an increase in staff of greater than 500% is "unrealistic."

An OAQPS official also complains that ECAO has assumed too much responsibility for developing and disseminating air pollutant criteria on an international basis. An ECAO official, however, points to the material benefits of the office's international involvement. "Exporting the EPA assessments has helped foster international communication exchange and collaborative research," for example, between EPA and Dutch officials on the effects of ozone. The particulates case study presents a concrete example of the benefits of ECAO's international activities, says this official. Through ECAO contacts in Great Britain, EPA gained access to the raw data relating mortality in London with ambient particulate measurements that permitted an agency scientist to conduct a key reanalysis. (The sharing of raw data is a tremendously useful practice among scientists, but it is all too rare and often comes only as a result of interpersonal contacts and trust.)

As illustrated in the particulates and ozone case studies, the NAAQS review process is elaborate and protracted. Criteria documents and staff papers can go through several iterations over periods spanning several years. Peer review workshops regularly precede release of draft documents for public comment and CASAC review, and CASAC has insisted upon meeting to review the agency's revisions to criteria documents and staff papers before sending a letter of closure to the Administrator. According to an ECAO official, the most recent review of the ozone NAAQS involved 150 reviewers and twenty CASAC members and consultants. Although environmental organizations have sometimes sued EPA to force compliance with the five-year NAAQS periodic review schedule, shortening the process may not always serve their interests. As the ozone case study illustrates, when the agency is forced to make an NAAQS decision under a fixed schedule, it can readily conclude that the currently available scientific evidence is insufficient to warrant changing the standard. Despite the requirement that NAAQS decisions be based on the latest useful scientific information, the ozone case study also demonstrates that a full-blown NAAQS review is required to make science "available" for EPA decisionmakers.

Hazardous Air Pollutants

In contrast to the ubiquitous criteria air pollutants, hazardous air pollutants (HAPs) are often localized pollutants, and the motivating public health concern is often cancer. (HAPs are sometimes called "air toxics.") The 1990 CAA Amendments identified 189 HAPs and required maximum available control technology (MACT) on major emission sources of those chemicals.[17] However, health risk analysis is relevant to HAPs if the technology-based emissions controls are not sufficient to protect public health with an ample margin of safety (defined for major sources as a residual cancer risk of one in one million (10^{-6}) to the individual most exposed to emissions). In this case, EPA must set second-stage health-based standards (NRC 1994).[18]

To obtain data needed to assess the health risks from approximately 40% of the 189 HAPs, the air program has developed a proposed rule using TSCA Section 4 authority, which allows the agency to require chemical manufacturers to provide data. Current data are insufficient to evaluate the cancer hazard of many HAPs, in part because most of EPA's cancer data were generated using oral ingestion studies, not inhalation studies. For more than 50% of the HAPs, data are insufficient to assess noncancer risks (*Risk Policy Report*, November 17, 1995, 30).[19]

In a division of labor similar to the NAAQS process, ECAO has the lead role in producing health assessment documents for hazardous air pollutants, and OAQPS is responsible for formulating the National Emission Standards for Hazardous Air Pollutants. As in the case of the NAAQS, OAQPS assumes responsibility for conducting the exposure assessment component of the risk analysis for HAPs. Prior to 1990, EPA had listed only eight HAPs and had developed standards for seven. Thus, the lion's share of ECAO and OAQPS scientific resources was devoted to NAAQS. An EPA research manager estimates that ECAO now devotes approximately 35% of its resources to HAPs. According to an OAQPS official, science does not play as large a role as it might in the HAP process because the scientists involved are reluctant to extrapolate from the limited information available. The huge uncertainties associated with HAP risk assessments, says this source, leave the decisionmaker wondering, "What is the science really telling me?"

OFFICE OF WATER

In the Office of Water, the Office of Science and Technology (OST) houses scientists and engineers who work on both drinking water and surface water regulatory programs. The Health and Ecological Criteria Division contains science analysts who assess human and ecological risks, while the Standards and Applied Science Division is dominated by engineers and economists who develop technology-based standards. Health and environmental effects information is used in setting water program goals. The most important role of science in the water program rulemaking process, however, is to provide measures of the environmental benefits of alternative pollution control technologies and regulatory schemes. The ORD Environmental Criteria and Assessment Office in Cincinnati also manages the preparation of some health and environmental criteria and assessment documents for the water program. In general, the water office makes extensive use of contractors for providing scientific information and analysis.

Drinking Water

Under the Safe Drinking Water Act (SDWA), maximum contaminant level goals (MCLGs) are nonenforceable health-based goals, while National Primary Drinking Water Regulations (NPDWRs) set enforceable standards. NPDWRs include either numeric maximum contaminant levels (MCLs) or treatment and/or monitoring requirements. SDWA directs the agency to set the MCL for a contaminant as close to the MCLG as is feasible, and, historically, EPA generally—but not always—interpreted this to mean that drinking water standards should be affordable for large systems.[20] The 1996 SDWA Amendments require EPA to analyze a "feasible" standard based on costs to large water systems but authorize the agency to promulgate a different standard if the agency finds that the benefits of the feasible standard do not justify the costs. EPA must also consider effects on susceptible subpopulations, and the agency is required to base its decisions on "the best available, peer-reviewed science."

The 1986 SDWA Amendments listed eighty-three contaminants for which EPA was to simultaneously develop MCLGs and NPDWRs and required the agency to set drinking water standards for twenty-five contaminants every three years. This mandated pace of regulatory development presented a formidable analytical challenge, and the drinking water program did not adhere to the statutory schedule. The 1996 SDWA Amendments overhauled the contaminant selection process by requiring EPA to publish a priority list of unregulated contaminants every five years subject to EPA Science Advisory Board review. EPA is also now required to make a determination of whether to promulgate standards for at least five contaminants from that list every five years. The 1996 SDWA Amendments also address the chronic problem of the lack of data on the occurrence of drinking water contaminants (see discussion in Appendix A) by requiring EPA to establish a national occurrence database.

As the case studies on lead and arsenic in drinking water (Appendixes A and B, respectively) illustrate, the drinking water program has tenuous ties with the Office of Research and Development. In contrast, the drinking water program appears to have a generally cooperative relationship with the American Water Works Association (the de facto trade association for major drinking water suppliers) and its research arm, the American Water Works Association Research Foundation, regarding the supply and analysis of scientific information. The case studies also suggest that the water program scientists seem generally less precautionary than their counterparts in ORD in assessing the health risks of chemicals in drinking water. Proximity to regulatory program implementation, with its attendant costs, may be a partial explanation for the difference. The observation may generalize beyond the case studies as well. According to a former EPA water official, the program's health risk assessors "needed a smoking gun before they saw anything as a contaminant." This source also suggests that the water program health assessors are "pushing for arsenic as a test case" to establish a precedent for a threshold carcinogen, marking a departure from EPA's standard risk assessment procedures. The major force resisting their efforts resides in the ORD National Center for Environmental Assessment (NCEA), which is responsible for developing agency risk assessment guidelines. The threshold issue was particularly salient to the drinking water program because the symbolic congressional goal under SDWA is to eliminate the occurrence of any carcinogen from drinking water supplies (that is, the MCLG for carcinogens has been set at zero). Arsenic is likely to be a test case for the new standard-setting criteria established by the 1996 SDWA Amendments. Drinking water program staff also locked horns with science analysts from ORD and the EPA policy office in the case of lead.

Surface Water

Based on the case studies and comments from respondents, compared with the drinking water program, the surface water program seems more inclined to impose regulations. Since both regulatory programs currently receive scientific and technical support from OST and share the same political leadership, differences at the management level, in the operative statute, and in tradition may be explanatory factors. (Prior to a 1991 reorganization, the drinking water and surface water programs maintained separate analytical staffs.)

The surface water regulatory program is based on the Clean Water Act (CWA, also known as the 1972 Federal Water Pollution Control Act Amendments) and focuses on reducing the concentrations of pollutants in the wastewater discharges from large point sources. EPA is responsible for developing and periodically revising effluent guidelines that indicate the technologies and practices that specific industrial sectors are required to employ to control conventional and toxic water pollutants. An example of the former is organic matter that depletes dissolved oxygen in water. Dioxins are an example of the latter. (See Appendix G for a case study of the proposed guidelines to control dioxins and other organochlorines from the pulp and paper industry.)

The statutory goal of the Clean Water Act is to eliminate entirely discharges of pollutants from point sources (that is, individual discharging facilities) into surface waters. Although eliminating pollutant discharges may be achievable under some circumstances through process changes that prevent pollution formation or recycling, the goal is largely rhetorical. The statutory goal of eliminating discharges influences the use of science because achieving the goal does not require point sources to eliminate all discharges into surface waters. Attainment depends to some extent on which substances are classified as pollutants subject to regulation under the statute. Consequently, the discharge elimination goal has the potential to encourage or distort the use of science in specifying regulated pollutants.

Under CWA (Section 304), EPA is directed to develop water quality criteria for pollutants reflecting the latest science and, "from time-to-time thereafter," to revise them. These criteria documents provide the basis for estimating health and environmental benefits that result from reductions in discharges of specific pollutants (or mixtures) into surface waters. EPA has issued water quality criteria documents for ninety-nine substances, with the criteria for another thirty-two substances in draft form or in development. Since 1980, when the agency published criteria for sixty-four substances and a methodology for deriving the criteria, EPA has reevaluated twelve substances, with no revised criteria being issued since 1987 (U.S. EPA 1996b). A proposal laying out the issues for revisions to the human health criteria methodology and inviting comment was developed after a 1992 stakeholder workshop but was on hold awaiting EPA's revised cancer risk assessment guidelines (*Risk Policy Report*, November 17, 1995, 34).

While the surface water standards are generally regarded as being technology driven, the program uses health and environmental effects information as an input to evaluating the cost-effectiveness of alternative control technologies. CWA does not require, but permits, EPA to consider cost-effectiveness. Despite objections from environmental advocates, in 1997, the agency clearly considered the cost-effectiveness of alternative bleaching technologies in revising effluent guidelines for pulp and paper mills to control discharges of dioxins and other organochlorines (Appendix G).

Because effluent guidelines for an industrial sector generally address multiple pollutants in the wastewater stream, the agency has adopted the convention of standardizing all pollution reductions in terms of "toxic equivalents." The equivalency factors are based on aquatic toxicity tests and standardized such that the toxicity of copper is assigned a value of one. (A pollutant more toxic than copper under the aquatic test conditions has a toxic equivalent factor greater than one.) This practice permits a crude comparison of water pollution control technologies in terms of cost per pound of toxic equivalent.

In addition to setting "technology-based" effluent guidelines for industrial sectors, EPA may also establish ambient water quality standards. These standards are intended to be based on health or ecological criteria and depend on the state-designated use of the receiving water body. Typically, the ambient water quality standards would be "backed out" from a risk-specific pollutant dosage and various exposure assumptions. Due to their administrative complexity (that is, effluent limits derived from water quality standards must be set on a site-specific basis), ambient water quality standards have been secondary to sectoral effluent guidelines in importance in the surface water program.

The mechanics of formulating such an ambient water quality standard are relatively straightforward. The administrative complexity is introduced at the stage of establishing and allocating pollutant discharge limits that will ensure attainment of the ambient standard. Unlike the case of national technology-based standards for particular industrial sectors, basing regulatory controls of wastewater discharges on risk-based standards requires consideration of site-specific characteristics of the receiving water body and (strictly speaking) of the exposed population.[21] Under the current Clean Water Act program, states are required to identify waters that do not meet ambient water quality standards after implementation of technology-based controls and to develop total maximum daily loads (TMDLs), which specify the maximum amount of pollution a water body can receive daily without violating water quality standards. The TMDL is then to be allocated among the various sources contributing to the problem (though only point source dischargers are subject to enforceable limits). If a state fails to fulfill these requirements, EPA is required to do so. A series of recent court decisions citing EPA's failure to complete the tasks after states failed to do so within the statutory time limits could force the agency to make a large number of site-specific determinations under demanding time, data, and resource constraints. (In the state of Idaho alone, for example, a federal district court has required EPA to set TMDLs for over 900 water segments in a five-year time period [*Inside EPA*, October 4, 1996, 4].)

OFFICE OF SOLID WASTE AND EMERGENCY RESPONSE (OSWER)

The principal regulatory programs that OSWER administers deal with contaminated site remediation and hazardous waste management. The Superfund program (administered under the Comprehensive Environmental Response, Compensation and Liability Act, or CERCLA) identifies and cleans up sites contaminated with hazardous substances throughout the United States and assigns the costs of cleanup directly to those parties—called responsible parties—who had something to do with the sites. Although Superfund is not limited solely to closed or abandoned sites, the program has focused primarily on these types of sites. The Resource Conservation and Recovery Act (RCRA) deals broadly with hazardous waste produced at operating industrial sites. Respondents consistently ranked OSWER's use of scientific information at the bottom among EPA programs. An EPA policy official, for example, says that OSWER is generally regarded as the "weak sister" in terms of its use of science.

OSWER's use of science is abridged due to its culture, a desire for consistency despite geographic heterogeneity, and the inherent complexity of the problems it is charged with remedying. An EPA policy official comments that "OSWER's use of risk information is more mechanical than would be considered desirable. [Its approach is,] here's the formula, you plug in the numbers." This succinctly describes the stereotypical engineering approach to science, a technical, action-oriented, problem-solving approach that eschews paralysis by analysis. It also results, however, in a loss of nuance that is inherent in environmental science. "It's a culture that sees things more as black and white than 'let's get a feel for the uncertainties,'" says this official. The OSWER staff also tends to be young and inexperienced, with high turnover, according to an ORD official. Furthermore, "The scientists in the regions working on CERCLA and RCRA feel . . . unrespected by the programs and the public. There is mutual disrespect between managers and scientists, especially in the Superfund program." A Capitol Hill staffer adds that although OSWER is decentralized, with regional EPA and state staff managing program activities, a desire for consistent decisionmaking across sites has led to heavy reliance on formal guidance to govern the use of scientific information. An EPA policy official suggests that the litigious nature of the Superfund and hazardous waste programs "tends to make people have the rules written down" so that local regulatory managers are "less likely to make judgment calls." Unlike the medium- or pollutant-specific programs, Superfund and RCRA also have the disadvantage of dealing routinely with the complex cumulative effects of an admixture of pollutants occurring on a site.

Superfund

Superfund is administered by the OSWER Office of Emergency and Remedial Response. Unlike the air, water, pesticides, and toxic substances programs, Superfund is not a national rulemaking program. Decision-making is largely decentralized and focused on specific sites. The use of science by remediation project managers (RPMs) depends, to a large extent, on administrative guidance provided by EPA Superfund head-quarters, EPA regional offices, or state agencies. Not surprisingly, guidance has not been uniform over time or across regions or states. Nair (1995), for example, pointed out that states have variable site data requirements and that risk assessment exposure models accepted by some states are not accepted by others. On the other hand, Superfund generally adopts information concerning the toxicity of particular pollutants in a uniform fashion from officially recognized databases without making site-specific adjustments. A former EPA Science Advisory Board member complains that EPA regional offices "indiscriminately adopt" values from the agency's Integrated Risk Information System (IRIS). Thus, the use of science by Superfund is subject to a tension between the competing desires for consistency and site specificity. (The case of lead in soil at mining sites provides a rare example of the agency's developing site-specific health effects information. See Appendix H.)

Most of the Superfund technical staff located in the regional offices are reportedly engineers or hydrogeologists.[22] The prominence of hydro-geologists reflects the most common driving concern behind Superfund site cleanups—groundwater protection—while engineers are required to effect technology-based cleanups. In each of the ten EPA regional offices, there is a regional toxics integration coordinator who serves as a point person for Superfund risk assessment issues in the region. However, historically, there have been relatively few regional EPA staff with advanced training in the health aspects of environmental risk assessment. Relatively little staff capability in toxicology and related fields might be required if EPA regional offices are expected to perform risk assessments "by the book" and overlook site-specific conditions. Increasingly, however, environmental agencies are expected to use site-specific health effects information more consistently in the contaminated site remedy selection process. This poses the challenge of having adequate scientific capacity to manage site-specific data acquisition and to critically evaluate the information produced by the regulated community.

Prior to a recent reorganization, the Toxics Integration Branch served as a central risk assessment branch of the Superfund headquarters in Washington, D.C. The reorganization of the Superfund program produced fourteen centers, five of which provide support to the ten EPA regional offices, and risk assessors are distributed among the various centers. A risk network team has been formed to provide a common forum for Superfund headquarters staff with risk assessment responsibilities. In addition, within EPA, ORD/NCEA in Cincinnati provides some technical support for Superfund. (Outside the agency, the Agency for Toxic Substances and Disease Registry [ATSDR] and the National Institute of Environmental Health Sciences Superfund Basic Research Program conduct basic and applied health effects research in support of Superfund.)

For a site to enter the Superfund remedial process, it must be brought to the attention of the federal government. The process is often prompted by complaints from citizens following discovery of abandoned drums of unknown chemicals or the release or seepage of potentially hazardous pollutants. In response, EPA (or the state) conducts a preliminary site assessment that may lead to a site inspection and the collection of soil and water samples from the site. Based on this information, Superfund uses the Hazard Ranking System (HRS) to determine whether a contaminated site belongs on the National Priority List (NPL). (A site that scores over 28.5 on the HRS index is proposed for inclusion on the NPL.)

Once a site is on the NPL, the remedial investigation begins with a more thorough field study and a site-specific baseline risk assessment conducted according to the Risk Assessment Guidelines for Superfund. The risk assessment is performed either by an EPA contractor or by a contractor for a potentially responsible party (PRP) subject to EPA review. In the past several years, Superfund program policy has swung back and forth on the issue of giving PRPs responsibility for conducting site risk assessments. In 1990, the agency ruled that it alone should conduct all risk assessments at Superfund sites. The Chemical Manufacturers Association challenged this decision in court, and the case was settled in 1993 when the agency issued a new policy permitting regional offices to allow PRPs to conduct risk assessments if the PRPs had sufficient capabilities and a good track record for conducting assessments. According to press reports, however, the Superfund program is preparing new guidance that would reverse the burden of proof to allow PRPs to conduct risk assessments unless there is reason to doubt their capabilities (*Inside EPA*, December 22, 1995, 5).

Site-specific exposure assessments of varying scope, depth, and complexity are conducted as part of the risk assessment, but, in many cases, the exposure assessments are formulated as hypothetical exposure scenarios. An EPA regional official notes that arguing for the acquisition of site-specific information on pollutant exposure pathways is sometimes "an uphill battle." Superfund site risk assessments use substance-specific toxicity values, such as slope factors for carcinogenic potency and reference doses (RfDs) and reference concentrations for "safe" levels of noncancer exposure, provided by IRIS, OSWER Health Effects Assessment Summary Tables, and ATSDR toxicological profiles. The risk assessment may be used to formulate a set of preliminary remediation goals (PRGs) for hazardous substances occurring on the site. The PRGs are generally intended to be health based (in some cases, ecological goals may come into play). In addition to the baseline risk assessment, further analysis may be conducted to assess the residual risks after one or more remedial actions.

Under current EPA practices, current or proposed drinking water MCLs or MCLGs are frequently the operative Superfund remedial objectives for groundwater contamination under the applicable or relevant and appropriate requirements (ARARs) provision of the 1986 Superfund Amendments and Reauthorization Act (Walker et al. 1995). RfDs are also commonly used for setting soil cleanup standards. When Superfund managers adopt toxicity values from central databases or standards developed by other regulatory programs, they indirectly use the health effects information that went into developing these values. If the values are adopted without modification, they also carry along with them any presumptions about the conditions of exposure to hazardous substances. Using drinking water standards for groundwater cleanup standards presumes, for example, that the groundwater would be the sole, lifetime source of drinking water for exposed individuals.

Although a great deal of discussion and debate has focused on Superfund site risk assessments, it is often unclear what role or how large a role the assessments and PRGs play in the final remedy selection. RPMs are permitted to consider technical feasibility, cost, and other factors in negotiating the final remedy selection with PRPs and the local community. RPMs are also required to consider ARARs, but there may be room for judgment in determining whether specific requirements (such as drinking water standards or residential soil cleanup levels) are applicable or relevant. The recent analysis by Walker et al. (1995) of 152 Superfund records of decision (RODs) signed in 1991 concluded that of ninety-five sites with stated numerical cleanup goals for groundwater, sixty relied on ARARs alone, another thirty-three relied on ARARs in combination with preliminary remediation goals based on site-specific risk assessments, and two sites set numerical goals based solely on site-specific risk assessments. As suggested by the substantial number of RODs without numerical cleanup goals, the level of cleanup is often defined only implicitly by what the selected remedy will achieve.

Despite the fact that Superfund receives generally low marks in its use of science from many knowledgeable observers interviewed for this study, the program is, in some respects, on the cutting edge of environmental science. It is, for example, one of the few programs that considers the cumulative effects of multiple pollutants and multiple exposure pathways (for example, ingestion, inhalation, dermal absorption, and so forth). NRC (1994) specifically recognized Superfund for conducting multipathway exposure analyses. Regarding the program's use of cumulative effects models, a Superfund official remarks, "We experience things on a day-to-day basis that other people just talk about." Ironically, some have suggested that Superfund may be too far ahead of the science in this area. While there is broad agreement in principle that cumulative effects are appropriate considerations in regulatory decisionmaking, there is no consensus regarding the best practices for doing so, which probably contributes to the controversy surrounding Superfund's use of science.

An EPA regional official suggests that contrary to conventional wisdom, Superfund's scientific data, statistical models, and risk assessments (most of which are conducted by contractors) are, in fact, sophisticated and that the controversy arises because "zealots" in the program refuse to consider economics or place the risks of contaminated sites into any comparative context. Many of the remedies, says this official, "wouldn't pass the laugh test—like a million dollar solution where ten people are exposed to a 10^{-7} [one in ten million] lifetime [cancer] risk if we do nothing." At least in part, the controversy over Superfund's use of science is a proxy for disputes over values.

Resource Conservation and Recovery Act Program

The RCRA program deals broadly with hazardous waste produced at operating industrial sites and is administered by the OSWER Office of Solid Waste (OSW). The principal focus of the use of science in RCRA has been on defining what constitutes "hazardous" waste subject to federal regulation and in developing regulations covering the generation, transportation, treatment, storage, and disposal of hazardous waste. In addition, there have been a few RCRA "corrective actions," or remedial activities on sites of active industrial operations, but much of the debate concerning Superfund's use of science is motivated by the potential liability to the owners of these sites.

The 1997 federal Commission on Risk Assessment and Risk Management found that RCRA rulemaking on assessing the toxicity of waste materials has inadvertently included large volumes of lower-risk materials.[23] The hazardous waste program has generally paid scant attention to assessing science as a result of a combination of factors, including the operative statute. In the Hazardous and Solid Waste Amendments of 1984, for example, Congress promulgated default regulations prohibiting the land disposal of specified types of hazardous waste and prescribed the use of specific technologies for new landfills.[24] Congress has also set tight deadlines for permit issuance and rulemaking. Like the Superfund program, the RCRA program also confronts the complex nature of the questions posed by attempting to deal with a huge number of chemicals under a multitude of variable site conditions. Still, the program bears much of the responsibility for its scientific shortcomings, and the program's culture has been a contributing factor. According to an ORD official, "RCRA weighs economics and politics. They don't put a premium . . . on getting the science. They fear the ambiguity or confusion that debating the science can engender. They fear that they'll lose control of the decision." Another EPA official opines, "Risk assessments don't have a lot of legitimacy in the RCRA program" because program officials "may fear that risk data won't support a rule" that it issues. In 1995, OSW merged its technical assessment and regulatory analysis branches to form the Economics, Methods, and Risk Assessment Division to conduct risk and cost-benefit analysis in regulatory development and implementation and to give a higher profile to risk issues in the program (*Risk Policy Report*, November 17, 1995, 33).

REGULATORY OFFICES

Regional environmental services divisions (ESDs) and their laboratories provide scientific and technical support to each of the ten regional offices. Through inspections, investigations, and laboratory services, the ESDs provide data for environmental and enforcement decisionmaking. In report language attached to the 1996 EPA spending bill, the Senate Appropriations Committee raised concerns that the reorganization of EPA's Office of Research and Development had excluded program office and ESD labs and encouraged the agency to combine or eliminate some of the labs. According to press reports, SAB had also suggested the agency fold program office and ESD labs into ORD during its reorganization (*Inside EPA*, September 15, 1995, 4). Recently, regional offices began augmenting their technical staffs composed of engineers and hydrogeologists with additional staff with qualifications in fields such as toxicology. In the opinion of an EPA regional official, however, the trend may be "too little, too late" to rectify contaminated site remediation decisions that were uninformed by scientific information, such as the variability in the bioavailability of toxic substances among sites. The lead in soil at mining sites case study (Appendix H) provides an uncommon example of applied toxicological research initiated and conducted by a regional office and used in contaminated site remedy selection.

ENDNOTES

[1] EPA regulates the introduction of genetically modified plants with pesticidal properties under FIFRA. Pesticide tolerances are enforced by the Food and Drug Administration. FDA monitors most food products, but USDA's Food Safety and Inspection Services monitors meat, poultry, and some egg products.

[2] A small microbiology lab (in Cincinnati, Ohio) works on efficacy testing for disinfectants.

[3] Not all chemicals present in pesticides are evaluated, just the "active ingredient." Other substances present in the pesticide are not intended to be toxic but are not necessarily inert either.

[4] Following the court case, EPA prepared a list of twenty-seven pesticides for forty-eight different foods that would lose their established tolerance and thus their registration for those uses (U.S. EPA 1993).

[5] Conventionally, maximum allowable pesticidal residue levels based on noncancer effects have been determined by dividing the no-observed-adverse-effect level found in animal studies by a safety factor of 100. One factor of ten is applied to account for interspecies differences, and another factor of ten allows for interindividual variability in susceptibility and response. The 1993 National Research Council (NRC) report *Pesticides in the Diets of Infants and Children* found that children may sometimes be exceptionally sensitive or highly exposed to pesticides in certain foods (for example, apple juice). The NRC committee recommended, therefore, that an additional safety factor be included in setting acceptable levels for pesticides in those foods, when the acceptable level is based on noncancer effects. Dr. Philip Landrigan, chair of the 1993 NRC committee, is currently a special adviser to the EPA Office of Children's Health Protection. NRC (1994) also questioned whether the conventional tenfold safety factor used in noncancer risk assessment to account for variability within the human population is adequate. However, children may not always represent the most sensitive segment of the population for particular noncancer effects. Some of their organs, for example, do not metabolize compounds in the same way as adult organs. Therefore, for particular chemicals, children with immature organ development may be at lower risk than adults.

[6] *Toxic Deception: How the Chemical Industry Manipulates Science, Bends the Law and Endangers Your Health*, a 1997 book written by a pair of environmental journalists, Dan Fagin and Marianne Lavelle, with support from the Center for Public Integrity, raised allegations about the chemical industry having undue influence over EPA policy, particularly in the pesticides and toxic substances program. In response, EPA's Inspector General launched an investigation into accusations that EPA employees had inappropriately accepted travel expenses from industry, improperly sought or negotiated outside employment, or failed to comply with established postemployment restrictions (*Inside EPA*, March 7, 1997, 1).

[7] Food, food additives, drugs, pesticides, alcohol, and tobacco are exempted by TSCA. EPA regulates the introduction of genetically modified microorganisms under TSCA. TSCA may also require manufacturers to notify EPA of a significant

new use of a chemical before the manufacture, importation, or processing of the chemical for that use occurs. A specific rule must be promulgated for the significant new use requirement to be put into effect.

[8] The House report on TSCA specifically states that a quantitative cost-benefit analysis is not required to show unreasonable risk (Shapiro 1990).

[9] Occupational exposure rulemaking by the Occupational Safety and Health Administration is also subject to substantial evidentiary criteria for judicial review.

[10] One reason that the agency's authority to gather data under TSCA is not very effective in supporting the review process is that scientists require toxicological information on substances that are not developed for commerce in order to better predict the toxicity of substances at the premanufacturing review stage. However, EPA has no authority to require testing of substances that will not be produced for commerce.

[11] In the absence of testing data, EPA uses SARs to predict certain characteristics of chemicals on the basis of molecular structure alone. In comparison, the European Community (EC) uses a broader battery of required tests (for example, in vitro mutagenicity tests) before permitting new chemicals to be manufactured. In 1993, EPA and the EC compared EPA's predictions with test results in the European Community. The results indicated that EPA's predictions were highly accurate for some characteristics, such as biodegradation (93% agreement between predictions and tests), but they were often inaccurate for many other characteristics relevant to assessing risks. For example, SARs accurately predicted the vapor pressure of only 63% of the chemicals tested; vapor pressure is an important factor in evaluating potential exposure to a chemical. Recall that under TSCA EPA can require testing if there is substantial production and potential for exposure. Although EPA has deployed an impressive battery of SARs, according to Ashby (1996), SARs for genotoxic carcinogens that have been developed do not alert regulators to nongenotoxic carcinogens, and few SARs have been discovered for endocrine-disrupting properties.

[12] An environmental policy analyst makes the qualifying observation that comprehensive data are not required to make sound regulatory decisions about many, if not most, substances entering commerce. Many such substances, for example, are relatively inert polymers. This source suggests that the real source of concern is the class of substances for which there are virtually no substance-specific data *and* for which the reliability of available SARs is suspect. One constraint that EPA faces in improving the reliability of SARs, according to this source, is the lack of correspondence between substances that represent data gaps in the SARs scheme and those developed for commercial purposes for which EPA has TSCA authority.

[13] O'Bryan (1997) notes that the Section 8(e) submissions for existing chemicals have contributed to the development of structure activity relationships used for evaluating new chemicals.

[14] Originally (December 1970), NAAQS were required for sulfur oxides, particulate matter, carbon monoxide, hydrocarbons, nitrogen oxides, and photochemical oxidants.

[15] The 1963 Clean Air Act directed the Department of Health, Education, and Welfare (HEW) to prepare CDs summarizing scientific information on widespread air pollutants for use by state and local public health agencies. NAPCA resided in the Public Health Service of HEW. The 1967 Air Quality Act specified the role of CDs in the development of state air quality standards.

[16] The CASAC tradition precedes EPA. In 1968, the Surgeon General established a National Air Quality Criteria Advisory Committee to review and advise on NAPCA CDs.

[17] A major source is defined as a stationary source having the potential to emit ten tons per year of a single HAP or twenty-five tons per year of a combination of HAPs.

[18] For setting second-stage standards, EPA is authorized to operate under the presumption that a lifetime excess risk of cancer of approximately 1 in 10,000 (10^{-4}) for the most exposed person would constitute acceptable risk and that the margin of safety should reduce the risk for the greatest possible number of persons to an individual lifetime excess risk no higher than one in one million (10^{-6}).

[19] An environmental policy analyst suggests that this may be overstating the HAPs data problem. Although there may be insufficient data to quantitatively estimate health risks for most of the 189 HAPs, this source believes there are adequate data on most of the HAPs to make a "safe/not-safe" determination regarding the MACT, as stipulated by the 1990 CAA Amendments.

[20] SDWA does not explicitly direct EPA to consider costs, however, on the basis of 1974 House Report language and 1986 statements in the *Congressional Record*, EPA has interpreted feasibility of MCLGs to mean that "the technology is reasonably affordable by regional and large metropolitan public water systems" (*Federal Register* 54 (1989): 22093–22094). While EPA followed this interpretation of feasibility in the case of lead in drinking water despite concerns voiced by some about the costs of regulatory compliance for small systems (see Appendix A), the agency's decisionmaking regarding arsenic in drinking water has been greatly influenced by the projected costs that would be borne by small suppliers if the MCL for arsenic were substantially lowered (see Appendix B).

[21] According to an EPA official, the agency's water program has traditionally considered site-specific characteristics of the receiving water body (that is, in setting total maximum daily loads) but has typically based its exposure assumptions on national averages instead of local data. EPA departed from this pattern in assessing the risks to recreational and subsistence fishers in the case of dioxins from pulp and paper mills. (See Appendix G.)

[22] We were unable to acquire data on the scientific disciplinary background and level of training of EPA staff at the program or regional level.

[23] Source: http://www.riskworld.com/tcnetwrk/riskwrld/nreports/1997/risk-rpt/volume2 (p. 98).

[24] Congress prohibited land disposal of liquid wastes containing cyanide at concentrations greater than 1,000 mg/l, arsenic (500 mg/l), cadmium (100 mg/l),

chromium (500 mg/l), lead (500 mg/l), mercury (20 mg/l), nickel (134 mg/l), sele-
nium (100 mg/l), thallium (130 mg/l), polychlorinated biphenyls (50 parts per million), and solid wastes containing halogenated organic compounds (1,000 mg/kg)
unless the Administrator determined within thirty-two months that the prohibition was not required in order to protect human health and the environment.
Congress specified minimum technological requirements for new landfills that
included a leachate collection system and double liners, the lower of which is to
be at least three feet thick with a permeability of no more than 1×10^{-7} cm/s (42
U.S.C. 6924).

REFERENCES

Ashby, J. 1996. Assessing Chemicals for Estrogenic/Hormone-Disrupting Properties: Lessons from Carcinogenicity Assessment. *Environmental Health Perspectives*, 104(2): 132–133.

Dahl, R. 1995. Can You Keep a Secret? *Environmental Health Perspectives* 103(10):
914–916.

Davies, J. C. III, and B. S. Davies. 1975. *The Politics of Pollution*, 2nd ed. Indianapolis, Ind.: Pegasus.

Guerrero, P. 1994. *Toxic Substances Control Act: EPA's Limited Progress in Regulating
Toxic Chemicals*. Testimony by Peter F. Guerrero, Director of Environmental
Protection Issues, U.S. General Accounting Office, before the Subcommittee
on Toxic Substances, Research and Development, Committee on Environ-
ment and Public Works, U.S. Senate. GAO/T-RCED-94-212. Washington, D.C.:
U.S. GAO.

———. 1995. *EPA's Problems with Collection and Management of Scientific Data and
Its Efforts To Address Them*. Testimony by Peter F. Guerrero, Director of Envi-
ronmental Protection Issues, U.S. General Accounting Office, before the
Subcommittee on VA, HUD and Independent Agencies, Committee on
Appropriations, U.S. Senate. GAO/T-RCED-95-174. Washington, D.C.: U.S.
GAO.

Nair, R. 1995. Presentation by Monsanto Company representative at the State and
Tribal Forum on Risk-Based Decision Making, November 13, 1995, St. Louis, Mo.

NAPA (National Academy of Public Administration). 1995. *Setting Priorities, Getting Results: A New Direction for EPA*. NAPA Report to Congress. Washington,
D.C.: NAPA, April.

NRC (National Research Council). 1993. *Pesticides in the Diets of Infants and Children*. Washington, D.C.: National Academy Press.

———. 1994. *Science and Judgment in Risk Assessment*. Washington, D.C.: National
Academy Press.

O'Bryan, T. 1997. TSCA Section 8(e) Substantial Risk Reporting and the 8(e) CAP.
The Toxic Substances Control Act at Twenty. (http://www.epa.gov/opptintr/cie/
issue04j.htm).

Shapiro, M. 1990. Toxic Substances Policy. In *Public Policies for Environmental Protection*, edited by P. Portney. Washington, D.C.: Resources for the Future, 195–241.

U.S. EPA (U.S. Environmental Protection Agency). 1993. Statement on Pesticide Regulation. *Environmental News*, February 2.

———. 1996a. *Office of Pesticide Programs Annual Report*. Washington, D.C.: U.S. EPA.

———. 1996b. Water Quality Criteria Document. Washington, D.C.: Water Resources Center. Mimeographed.

———. 1997. *Food Quality Protection Act Implementation Plan*. http://www.pestlaw. com/guide/EPA-970300B.html.

Walker, K., M. Sadowitz, and J. Graham. 1995. Confronting Superfund Mythology: The Case of Risk Assessment and Management. In *Analyzing Superfund: Economics, Science and Law*, edited by R. Revesz and R. Stewart. Washington, D.C.: Resources for the Future, 25–54.

Yosie, T. 1988. The EPA Science Advisory Board: A Ten-Year Retrospective View. *Risk Analysis* 8(2): 167–168.

5

An Evaluation

EPA'S SCIENCE AGENDA

EPA's science agenda is alternately squeezed and stretched by competing demands: satisfying the immediate information needs of regulatory decisionmakers; providing the data and analysis needed to rationalize and prioritize items currently on the regulatory agenda; identifying and assessing the magnitude of emerging or ignored environmental problems; and supplying the scientific data required to evaluate and improve the performance of environmental management measures. These demands originate from EPA regulatory programs and research offices, EPA political appointees and careerists, Congress, the Executive Office of the President, industry, the environmental community, academia, and the general public. In sorting through these competing demands, the agency is confronted with the inevitable trade-off between conducting the depth of analysis commensurate with important regulatory decisions (that is, responding to squeezing forces) and covering the gamut of environmental issues of concern to important stakeholders (that is, reacting to stretching forces).

In many cases, EPA has little discretion over its science agenda. For example, the former requirement under the Safe Drinking Water Act that the agency promulgate twenty-five standards every three years dominated the drinking water program's science agenda, and the Clean Air Act's requirement for periodic review of the ambient air quality criteria has largely driven the air program's science agenda. The pesticides program faces statutory timetables for reregistering old pesticides, and both the pesticides and toxic substances programs must react to industry petitions to place new chemicals on the market. EPA does not control the discovery of leaking chemical drums or underground storage tanks. However, in some cases (indoor air quality and climate change, for example) the agency exercises discretion over its science agenda.

The interactions between demand and supply for science at EPA often result in a cyclic pattern of shallow, episodic initiatives. For example, environmental data collection falls in and out of favor over time, resulting in a discontinuous series of broad, shallow efforts. The agency also sporadically launches new science initiatives in an attempt to cover a broader range of environmental issues, generally in response to a congressional mandate or an administrative initiative. President Clinton's 1994 executive order on environmental justice, for example, has spawned agency activities on the cumulative risks associated with exposures to multiple contaminants through multiple exposure pathways.

To cover a growing set of issues under resource constraints, the agency has typically traded off depth of analysis, using few data and simple scientific models for decisionmaking. Consequently, EPA sometimes uses screening-level analyses to support regulatory decisions with important ramifications. (Although the consequences are typically localized, decisions concerning contaminated site cleanups, hazardous air pollutant standards, and total maximum daily loads [for surface water pollutants] frequently seem to fall into this category because they may warrant detailed, site-specific analysis.) When the highly simplified science is applied to important decisions, controversy often ensues, drawing EPA into deeper, more costly research and data gathering in specific problem areas. Without additional resources devoted to science, the agency's tightly stretched science agenda must give somewhere, so other competing problem areas are neglected to deal with immediate controversies. Frequently, the consequences of insufficient science in the neglected areas become manifest with time, resulting in demands to reallocate resources back to them. And so the pattern repeats itself, with new items being added continuously to the EPA science agenda and consequential topics and themes cycling through periods of attention and neglect.

An area that appears to be particularly neglected by EPA's science agenda is research and development in anticipation of regulatory policy decisions that are likely to arise within two to five years. Scientific data collection and analysis for pollution control program evaluation also tends to be neglected, in part because no politically important groups support it (NCE 1993).[1] In contrast, the EPA science program's emphasis on developing methods for risk analysis has helped address an area in the federal science agenda that the Office of Technology Assessment found to be particularly neglected, in large part because developing methods and tools is regarded as too basic by operational programs (which favor data collection) and too applied by science agencies (which stress basic research) (U.S. OTA 1993).

The National Academy of Public Administration finds that EPA uses risk information *and* consideration of public attitudes, statutory requirements, costs, and the potential for risk reduction opportunities to set the agency's risk assessment and research agenda (NAPA 1995). While each of these factors belongs in EPA's science agenda setting, the demand for environmental science and analysis is insatiable. The agency—and the environmental science community at large—have unsuccessfully struggled to develop a stable set of priorities to guide EPA's science agenda or to formulate consistent criteria and a satisfying process for developing such priorities.

Historically, EPA's formal research budget planning process has focused primarily on the Office of Research and Development's (ORD's) budget instead of the overall EPA science agenda. Until recently, the ORD budget planning process has been governed by a medium- or program-specific approach (that is, for air, water, waste, and toxic substances) that extends from within the agency to the Office of Management and Budget and Congress (U.S. EPA 1992). Program- and medium-specific research committees consisted of ORD and program office staff (U.S. OTA 1993). Traditionally, ORD has been predominantly involved in short-term research in response to the needs of EPA's regulatory and regional offices (NRC 1995). Although respondents view EPA's political leaders as key to creating effective demand for science within the agency, the short tenure (generally two to three years) of most political appointees seems to be a principal reason that the agency's science agenda focuses on immediate regulatory decisions and fails to acquire the data and analysis in anticipation of future decisionmaking needs. Political appointees have little personal incentive to devote current resources to providing for the science needs of their successors, and EPA as an institution has been unable or unwilling to commit adequate scientific resources to regulatory decisions that are certain to arise beyond the immediate future.

A former EPA Administrator, for example, "never understood why EPA didn't do a better job of setting up research programs to fill the knowledge gaps on the criteria air pollutants" when they are required to be reviewed every five years. According to a former EPA air scientist, however, research funding for a criteria air pollutant is generally cut after a decision is made. Then, only one or two years prior to revisiting the national ambient air quality standards, research money is reallocated back to the pollutant. But, says the scientist, "the resources arrive too late to conduct long-term . . . studies that may be necessary to address the critical areas of uncertainty about the nature and severity of health effects."

Congress's short planning horizon also helps explain why little of the agency's research and development (R&D) is devoted to stable, long-term research programs. For example, the Senate appropriations committee report on the 1995 appropriations bill cut $4.6 million and $4.4 million from EPA's research budget for ozone and particulate matter, respectively.

Table 5.1. Distribution of ORD Resources for 1992

Applied research: 43%
Basic research: 29%
Development: 19% (development of processes, techniques, methods, and
  equipment)
Technical assistance: 9% (direct technical support of program and regional offices)

Program	ORD Staff (%)	Extramural Funding (%)
Air and Radiation	30	29
Water	21	7
Waste	20	26
Pesticides and Toxics	16	7
Multimedia	13	31

Despite the Clean Air Act's requiring the agency to review the national ambient air quality standards every five years, the committee justified the cuts on the basis that because the ongoing reviews of ozone and particulate matter were close to completion, the research would have little impact on pending decisions (*Inside EPA* July 29, 1994, 17).

Although reliable accounts of the agency's total science budget are unavailable, it is probably safe to assume that the program offices support relatively little long-term research compared with the Office of Research and Development. In 1993, EPA reported that 25–30% of ORD's resources were directed to "fundamental" (that is, basic or long-term) research (U.S. EPA 1993). Therefore, on an agencywide basis, EPA may devote less than 20% of its science resources to long-term research. (See Table 5.1.)

For many years, Democratic congressional appropriators pressured EPA to increase support for academic environmental research. In 1994, an internal EPA steering committee recommended that ORD increase the proportion of its resources allocated to basic research to at least 50% and expand its competitive, extramural, investigator-initiated research program from $20 million to $100 million a year (U.S. EPA 1994). In response, EPA doubled ORD's extramural basic research grant program in 1995. During the 1990s, the agency's overall extramural R&D budget and the percentage of ORD's extramural funding tied to regulatory programs have declined while ORD's extramural budget has increased.[2] According to press reports, several program offices have suggested that ORD should have a constant percentage of its extramural resources dedicated to program support (*Inside EPA*, February 11, 1994, 10–11). Recently, congressional Republicans have sought to shift EPA's research agenda back in the direction of regulatory program applications and have been protective of environmental consulting firms.[3]

For a brief interlude in the early 1990s, the agency experimented with an "issue-based" research planning process organized around a set of thirty-eight environmental problems grouped into twelve broad themes (U.S. EPA 1993). While the issue-based planning process provided a coherent description of the research program, the process was quickly abandoned due to a sense that the participation of ORD's customers within the agency had been unsatisfactory (U.S. EPA 1994). (This could be interpreted to mean that the program offices devoted insufficient attention to ORD's planning process, that ORD paid insufficient attention to program office input, or that some combination of both occurred.) According to the National Research Council, the outcome of the "issue-based" research planning process reflected "the program's historical origins, traditional areas of strength of ORD's laboratories, EPA's regulatory support requirements, directives from Congress, mission ambiguities, and resistance to change" (NRC 1995, 14). An internal EPA steering committee concluded that ORD's "discipline-based structure did not track well with existing media-/problem-based programs and regions or with science based on risk assessment/risk management." As a result, EPA reorganized ORD along the lines of the risk assessment paradigm and embarked on a new strategic planning process for ORD.

The new process is intended to produce five-year strategic research plans and give greater voice to ORD's internal customers and other stakeholders. It is structured by a research coordinating council involving EPA program and regional offices, and a Deputy Assistant Administrator for Science was created within ORD to manage the process (NRC 1995; *Risk Policy Report*, July 21, 1995, 21). The first output of the process was a 1995 draft strategic plan for ORD intended to feed into the 1997 fiscal year budget process and establish near- and long-term research priorities for ORD. The draft plan identified six near-term research priorities, three specific and three general: drinking water pathogens and disinfection by-products; ambient air particulate mater; endocrine disrupters; human health protection; ecosystem protection; and pollution prevention and new technologies (U.S. EPA 1995).

Both the drinking water and particulate matter topics are regulatory driven. Endocrine disrupters, however, is an emerging topic that Congress placed onto EPA's science agenda in 1996. Sen. Alfonse D'Amato (R-N.Y.) inserted language in the Senate-passed Safe Drinking Water Act reauthorization bill that sought to require the agency to develop screening and testing requirements for endocrine disrupters within two years of enactment (*Inside EPA*, May 10, 1996, 10). This provision was added in the midst of the publicity coinciding with the release of the popular environmental science book about endocrine disrupters, *Our Stolen Future* (Colborn et al. 1996), and at a time when moderate Republicans were looking to put on an environmentally friendly face. Eventually, the 1996 Food Quality Protection Act required EPA to develop screening and testing requirements for endocrine disrupters.[4]

Both the Science Advisory Board (SAB) Research Strategies Advisory Committee and a National Research Council committee have questioned whether the broad priority areas in ORD's strategic plan are useful in terms of setting research priorities (*Risk Policy Report*, March 15, 1996, 13). A potentially troublesome aspect of the new strategic planning process is that the "risk-based" criteria used for selecting research projects among those identified by stakeholders do not appear to match the criteria for evaluating research performance—significance, relevance, credibility, and timeliness.

EPA regulatory program officials continue to complain that the ORD strategic planning process is unduly dominated by ORD, unreliable, and nontransparent. Indeed, a senior research official admits that "ORD is often guilty" of not satisfying regulatory program information needs. Part of the frustration of program representatives, however, stems from the pace of basic and applied research and the limited capacity of research to provide definitive answers to the regulatory program's pressing questions. The pesticides program, for example, now has a pressing need for generic methods to test chemicals for their capacity as "endocrine disrupters." However, this field of environmental toxicology is so new that it was only recently that scientists even converged on a name for it. Screening procedures (laboratory tests and relationships that predict chemical activity on the basis of structure or function) for endocrine-disrupting properties are in the earliest stages of development, with a good many candidates (LeBlanc 1995) but few guiding principles for evaluating their sensitivity or reliability (Ashby 1996).[5]

Once EPA identifies a science project as a priority, the agency still faces the daunting task of deciding its resource allocation level. Just because a research project or risk assessment carries a high institutional priority does not necessarily mean that the relevant questions require large resources to answer. NAPA (1995) concludes that EPA has not done a good enough job in deciding how intensive to make individual analyses. Some observers would point to EPA's ongoing dioxin reassessment as a case in point. At the May 5, 1995, SAB review of the EPA dioxin reassessment, William Farland, Director of ORD's National Center for Environmental Assessment, justified the years and millions of dollars spent on the assessment on the basis that it would serve as a "model risk assessment." If the elaborate multimillion dollar dioxin reassessment provides the model, however, the gap between the agency's science supply and demand will grow even wider. It remains difficult to determine the appropriate level of scientific resources to dedicate to individual analyses or particular regulatory decisions.

The new strategic planning process for ORD has been subsumed by a larger effort to address the agency's entire science budget. In 1992, the Expert Panel on the Role of Science at EPA recommended that the agency develop a science planning process that would apply to science throughout the agency (U.S. EPA 1992). In developing the 1996 appropriations for EPA, the Senate departed from the traditional procedure of having a budget line strictly for ORD and pooled much, but not all, of the agency's research and development operating, program, and laboratory funds into a new Science and Technology account. In response, EPA began an agencywide science budget inventory exercise to capture science activities in program and regional offices as well as ORD. So far, however, this exercise has failed to produce an accurate, comprehensive inventory or budget of EPA science efforts. Some EPA projects have been coded as "science" because it is believed that doing so would protect their budgets, while EPA program offices have sometimes avoided coding projects as science to avoid oversight or seizure by ORD, according to an agency official. EPA has also formed a new planning, budgeting, analysis, and accounting office that is intended to give risk and benefit-cost information a more central role in setting EPA's priorities, including those of the agency's science agenda.

In developing an agencywide science agenda, EPA will need to confront the unavoidable trade-offs between basic research and scientific support for pressing regulatory decisions. On the one hand, short-term regulatory support is narrowly focused and disrupts or preempts the long-term research program. Attempting to cover the full scope of EPA's scientific needs prevents the agency from developing core research programs. There is also some risk to the scientific credibility of EPA researchers when they provide technical assistance to regulatory strategies that have been predetermined, or are perceived to be. On the other hand, using EPA scientists for regulatory support can improve the scientific basis of regulations, maintain an experienced scientific core group within the agency, and provide researchers with a knowledge of regulatory and policy issues. Also, when the agency has attempted to stretch out into fundamental research not directly related to its regulatory needs, the results have been mixed. The Environmental Monitoring and Assessment Program (EMAP) provides a case study of a worthy, but flawed, basic research program without direct linkages to regulatory decisions.[6] EPA needs to find a way of balancing these trade-offs in a transparent manner to develop a stable set of priorities to guide the agency's science agenda.

FACTORS THAT AFFECT THE USE OF SCIENCE IN DECISIONMAKING

Over forty interviewees were asked what they consider to be the most important factors that either impede or facilitate EPA's use of science.[7] Their open-ended responses (often including multiple factors) were categorized, tallied, and aggregated under general headings. Table 5.2 summarizes the responses regarding the impediments to EPA's use of science. Overall, based on the responses of interviewees, the leading impediments to EPA's use of science appear to be the agency's regulatory culture; time and budget constraints; large scientific uncertainties; a combination of factors that present resistance to revisiting decisions in light of new scientific information; policymakers who lack an understanding of what science can contribute; a legislative and administrative preference for technology-based environmental standards; and the quality of EPA staff.

Some of the individual factors that impede EPA's use of science are linked and can be organized under broader headings. The first group of responses in Table 5.2 characterizes EPA as a regulatory agency that is by nature and design not science oriented. Because the agency has emphasized the development and enforcement of environmental regulations since its inception, legal considerations have had a strong influence on EPA's use of science. Using the criteria of enforceability, regulations must draw bright lines that unambiguously permit detection of violations. Through the regulatory development working groups and other internal review and clearance processes, the agency's legal culture permeated into its scientific assessments. One manifestation was the agency's treatment of scientific evidence as either admissible or inadmissible, rather than taking a scientific approach to see what could be inferred from all of the available evidence. EPA lawyers have also had a lot to say about how the science was presented; for example, the agency's policy of not acknowledging the scientific uncertainty in estimates of the toxic potency of chemicals.

In general, scientists have little stature and power within EPA. Instead, those with the power to make and influence decisions tend to have an inclination toward law and politics. An underlying reason for EPA's lack of science orientation may be that EPA is not expected to be an even-handed, independent regulatory agency but to represent the interests of the environmental community in federal policy decisions. For example, when Administrator Browner went before the Senate Environment and Public Works Committee for her confirmation hearing, Senator Chafee (R-R.I.) told her, "You are an advocate for the environment." He

Table 5.2. Factors Impeding EPA's Use of Science

Freq.	Responses

By nature and design, EPA is a regulatory agency, not a science agency.

13   Agency norms, staffing patterns, and incentives subordinate science.

9   Policymakers lack an understanding of what science can contribute to decisionmaking. There are poor communications between scientists and policymakers.

8   EPA staff don't have sufficient training and skill to do a defensible job of scientific analysis. EPA lacks scientists with stature. EPA scientists are politicized.

3   ORD's mission within EPA is unclear, and it is not a force within the agency or on regulatory development working groups.

1   EPA's use of upper-bound point estimates of risk is nonscientific.

1   EPA is not a patron of science. The agency relies on contractors over competed research and has no external scientific constituency.

1   EPA lacks strategic planning for science.

Environmental science is an inadequate/inappropriate basis for regulatory decisionmaking.

10   Large scientific uncertainties invite standards to be based on criteria other than estimated risks.

8   Legislation prescribes and EPA prefers technology-based standards over risk-based standards.

5   Existing scientific tools can't address regulatory decisionmakers' needs.

2   Scientific disciplinary boundaries result in narrow, fragmented analyses.

1   Scientists lack an understanding of policymakers' information needs.

Other impediments

11   Time and budget constraints prevent the agency from conducting a thorough review of the science.

9   Organizational and disciplinary inertia, the desire for regulatory stability, and the legacy of past risk perceptions prevent consideration of new scientific findings that might lead to revision of current standards.

6   Economic and political considerations overwhelm science.

3   Some statutory goals are scientifically meaningless, and the statutes are medium specific.

3   The federal government and EPA inadequately fund environmental research.

2   Peer reviewers have vested interests. External scientists are politicized.

2   Bureaucratic turf prevents coordination between EPA offices.

1   EPA fails to consider ecological science because it lacks a clear environmental protection mandate.

1   Rulemaking under the Administrative Procedures Act consumes too much time and money and is too public.

1   The regulatory development process is both too short to permit good science and too long, since science input occurs early in the process and is often lost or diluted by the final stages.

went on to instruct her not to go too far to accommodate industry (*Congressional Quarterly Yearbook*, 1993, 289). "In the spirit of pluralistic government," explains an academic, "EPA was established to ensure that in every administration, someone with environmental interests is represented in policy decisions. EPA is not supposed to be representing the scientific community but the environmental advocacy movement. This wouldn't happen if EPA were buried in the Public Health Service." A dysfunctional element of the pluralistic model, however, is that EPA's primary constituencies tend—with some justification—to view science and analysis as an obstacle to regulatory action. Regulatory action, however, is not synonymous with environmental progress.

EPA's role as an institutional representative of environmental interests is only a partial explanation for science's lack of stature within the agency. The responses in the second group in Table 5.2 combine to suggest that the state of environmental science is often a serious impediment to its use in decisionmaking. In the opinion of a former EPA Administrator, for example, the quantity and quality of data available to decisionmakers represent the greatest impediment to the agency's use of science. Unfortunately, the basic scientific tools of environmental regulation, toxicology, and epidemiology are simply inadequate to provide precise answers regarding the effects of pollutants at low ambient levels. The scientific uncertainties exacerbate the frequently blurry distinction between science and policy, and the failure of environmental science to reduce the range of viable policy alternatives invites environmental policymakers to base their decisions on other criteria, such as engineering feasibility, economic impacts, or what the traffic will bear politically. Formulating the problem as the failure of science to inform decisionmaking, however, is dangerously circular. Although environmental science is a relative newcomer as sciences go, the inability of the science to provide useful answers to regulators is in no small part due to decisions by administrative and congressional policymakers not to allocate the time and human and financial resources necessary to supply better answers.

The capacity of even highly developed science to guide policymaking should not be overstated, however. Yosie (1995), for example, questions "whether scientists, in their eagerness for a place at the policy table, inadvertently oversold risk assessment over a number of years as the elixir for ailing environmental policies." Many environmental regulatory choices have been incorrectly framed as issues that can be resolved through science. It is wholly appropriate that economic and value trade-offs often determine policy choices. The failure of both policymakers and scientists to recognize what science can and cannot contribute to regulatory decisionmaking is a principal factor in impeding the *wise* use of science.

Table 5.3. Factors Facilitating EPA's Use of Science

Freq.	Responses
15	Demand for and understanding of science by EPA leadership. Leaders taking long view of EPA credibility. Leaders recognizing that science confers legitimacy on decisionmaking.
11	Peer review in general and the EPA Science Advisory Board in particular.
9	External pressure and scrutiny, particularly from Congress, the courts, and industry. Increased external scientific sophistication. Forcing decisionmaker to publicly articulate scientific basis for decisions.
8	Adequate research funding, research infrastructure, data, and time.
5	Statutory flexibility that permits consideration of science, statutory requirements for independent review, periodic review, risk-benefit balancing.
4	Communications between scientists and policymakers.
2	Strong ORD–program office linkages.
2	EPA peer review and risk characterization guidelines.
1	Explicit risk characterization with uncertainty analysis.
1	Use of competed science funding mechanisms.
1	Allowing EPA program offices to manage their own science.
1	Current environmental problems are less obvious, requiring more science.
1	Coincidence of available science with public alarm regarding problem.
1	Institutional learning by EPA.
1	Acknowledgment that policy decisions are public choice, not scientific.

Table 5.3 summarizes the responses regarding the factors that facilitate EPA's use of science. Noteworthy is the critical role that the agency's political leaders are perceived to play in creating demand for science. According to an EPA senior careerist, for example, the agency's use of science depends critically on the Administrator's ability to "understand, appreciate, and know how to use science." The perception of EPA political leaders "that science is important," observes a former EPA official, "translates down through the agency." An EPA official stated, "It is the political-level leadership which has been primarily responsible for driving the change toward increased use of science by EPA over time. Senior career people weren't pressured to question the science without the pressure from political leadership. There were people in ORD and elsewhere who were pushing this, but without the pressure from political people, it wouldn't have happened." The leadership factor appears to be at least as important as independent peer review, pressures brought to bear on the

agency by external agents, adequate scientific resources and time, and statutory provisions that enable or require science.

Some factors may not necessarily impede or facilitate the use of science but nevertheless can have important effects. For example, the case studies suggest that high economic and political stakes have a way of focusing the mind of at least some policymakers on substantive matters like science. On the other hand, the high stakes may make it more likely that economic and political considerations overwhelm the science, despite the greater attention that scientific information may receive from decisionmakers. Also, as an EPA Science Advisory Board member points out, the "magnitude of the decision" helps to account for how elaborate EPA's scientific review is. These effects are no doubt related to the relationship between the decisionmaking stakes and the level of external scrutiny.

The economic and political stakes also largely determine the extent to which EPA's use of science is controversial. According to a former EPA Administrator, "To the extent that the decision can be attacked on scientific grounds, the science is very important. To the extent that there is public controversy about the science, the importance of research increases." As an academic observes, controversy regarding the use of science does not generally originate from the science itself. For example, if the cost of reducing levels of arsenic in drinking water were minimal, and if the drinking water standard had no relationship to contaminated site cleanup levels, it would be inconsequential that EPA based its assessment of the skin cancer risks of ingested arsenic on an old Taiwanese epidemiological study and a linear cancer model (see Appendix B).

Another important factor is the degree of centralization in the development and use of science. The pesticides program is very centralized and highly regarded for its use of science. The Superfund program, on the other hand, is very decentralized. Respondents generally hold the program's use of science in low regard due to the "engineering mentality" and lack of scientific expertise in the state and regional offices. On the other hand, whereas the Superfund program may consider site-by-site variability and the integrated effects of co-occurring contaminants, the national programs have routinely ignored differences among sites, as well as the effects of multiple contaminants in order to make their regulatory analyses tractable. Although the national programs may be admired for their analytical sophistication, they achieve sophistication at the expense of limiting the scope of their analysis. In the controversy over Superfund's science, the perceived disadvantages of decentralized scientific analysis may be confounded with a sense that the risks posed by Superfund sites are not commensurate with the level of resources allocated to the program. In other words, perceptions about the quality of Superfund's science may be colored by judgments about whether the program, driven by a legislative preference for permanent treatment, strikes the proper balance between environmental risks and cleanup costs.

The centralization of EPA science in the Office of Research and Development appears to involve trade-offs. On the one hand, centralization of scientific resources in ORD achieves some administrative efficiencies and promotes some institutional independence that decreases the ability of the leadership of the program offices to manipulate or ignore science for political ends. Concentrating scientific resources may also improve the quality of the agency's scientific efforts by creating a critical mass of scientific staff or by preventing scientific staff from being diverted to nonscientific administrative duties. On the other hand, the centralization of science makes it more difficult to establish clear lines of responsibility to ensure that the information needs of regulatory decisionmakers are met. Furthermore, to the extent that regulatory decisions are "science driven," centralizing the agency's science in ORD may simply substitute the policy agenda of the regulatory program leadership with that of ORD's leadership.

Another factor that affects EPA's use of science is a scientific culture within the agency and in the environmental regulatory science community at large that is dominated by the disciplinary norms and training of toxicology. In some respects, this is a natural outgrowth of the fact that to assess the environmental effects of substances *before* they are released into the environment, many EPA regulatory decisions (for example, Toxic Substances Control Act [TSCA] premanufacturing notices) are based on predicted, rather than observed, outcomes. However, as a result of the agency's science culture, most EPA officials feel fully justified in basing public health decisions on the results of studies in which rodents are administered very high—but controlled—doses of chemicals through tubes stuck in their bellies but ironically balk at the thought of regulating on the basis of observed effects in human populations subjected to an uncontrollable mixture of environmental insults (including smoking and diet, for example).

Tables 5.4 and 5.5 display the responses of more than thirty interviewees regarding environmental regulatory decisions in which, in the respondent's opinion, science played a particularly large or small role.[8] It is noteworthy that some decisions (that is, the regulation of lead) were mentioned in both categories. It is also worth noting that the nature of some of the decisions was taken out of EPA's hands (for example, by technology-based statutory provisions). An EPA Science Advisory Board member observes that "the volume of [scientific] involvement doesn't tell you very much about the role of science in the decision." Development of national ambient air quality standards, for example, involves an elaborate scientific review, but large uncertainties may remain, so decisions may be

Table 5.4. EPA Decisions in Which Science Plays a Large Role

Program Area	Decision	Freq.
National Ambient Air Quality Standards	NAAQS in general	12
	Lead NAAQS/lead in gas	4
	Ozone NAAQS	3
Toxic Substances	Decision to reassess dioxins/dioxin reassessment	10
	Lead (multimedia)	3
	Asbestos	3
	TSCA premanufacture screening	2
Pesticides	Many pesticides decisions	11
National Emission Standards for Hazardous Air Pollutants	Decision not to regulate gasoline vapors as carcinogenic	4
	NESHAPs in general	2
	Benzene NESHAPs	2
Drinking Water Standards	Lead in drinking water standard	4
	Some/most drinking water standards	2
	Drinking water MCLGs	2
Regional/Global Air Pollutants	Acid rain program	4
	Stratospheric ozone depleter/CFC phaseout	2
Indoor Air/Radiation	Radon action levels in homes	2
	Secondhand tobacco smoke	2

based on other nonscientific considerations. On the other hand, while the screening of premanufacturing notices under TSCA is a very rapid and routinized process, science may still play a significant role in the decision of whether to request additional information. Thus, the nature of the decision (for example, whether to request additional information, where to set a national standard, what technology to employ, what contaminated site remedy to choose) affects the level of scientific effort but may shed little light on whether science drives decisionmaking.

THE STATE OF ENVIRONMENTAL SCIENCE

Practical men proceeded to preserve foods while the savants continued to dispute spontaneous generation.
—Davis et al. (1967)

As indicated in the previous section, environmental science is frequently regarded as an inadequate or inappropriate basis for regulatory decision-

Table 5.5. EPA Decisions in Which Science Plays a Small Role

Program Area	Decision	Freq
Hazardous Waste	RCRA and Superfund decisions	7
	Ban on ocean disposal of medical waste	3
	Superfund decisions	2
	Hazardous waste listing	2
	Nonhazardous waste determination	2
Toxic Substances	Asbestos removal in schools	5
	Lead in air, drinking water	2
	Times Beach, Mo. (dioxin)	2
	Some asbestos decisions (school removal, small use bans)	2
	PCBs	2
	Worker protection	2
Technology-Based Programs	Hazardous air pollutants under 1990 CAA	6
	Technology-based programs in general	4
	Water effluent guidelines	2
Pesticides	Alar (children)	5
	Food safety Delaney Clause	2
National Ambient Air Quality Standards	MTBE (gas oxygenate)	3
	Initial ozone NAAQS	2
Hazardous Air Pollutants	Municipal solid waste combustion strategy	2
	Delegation of HAPs to states under Reagan administration	2
Drinking Water Standards	Allocation of municipal wastewater treatment grants	2

making. Quantifying the health and environmental effects of substances at low-level exposures with any precision generally lies beyond the current reach of science. Nevertheless, the current state of science is generally sufficient to provide a basis for sound regulatory decisions in routine cases in which the stakes (in terms of compliance costs and health and environmental impacts) are relatively low and in cases in which the "rank" of the alternatives available to the decisionmaker is insensitive to large scientific uncertainties. (An example of the latter would be decisions in which the feasible alternative with the highest estimated net benefits would remain so even if the risk estimate varied by many orders of magnitude.)[9]

The state of environmental science is characterized by both a chronic lack of data and a primitive understanding of many biological, physical, and ecological processes. The quantity and quality of data available vary greatly among chemicals, but for most chemicals, data for toxic mode of

action, metabolism and toxic potency, exposure, and differential susceptibility are lacking. NRC (1984) found *no* toxicity data on more than 80% of
the roughly 50,000 industrial chemicals (a category that excludes pesticides, food additives, cosmetics, and drugs) used in the United States and
concluded that data were incomplete for the remaining 20%. The share of
pesticides—which are expected to be biologically active—without minimal toxicity information was estimated at 64%. For industrial chemicals
produced in amounts exceeding one million pounds a year, NRC (1984)
found that virtually no neurological, reproductive, or developmental toxicity testing had been done. U.S. OTA (1993) reported that "good" toxicity
testing data are available for only 10% of all chemicals. More recently,
EPA's 1998 *Chemical Hazard Data Availability Study* found no basic toxicity
information (that is, neither human health nor environmental toxicity) is
publicly available for 43% of chemicals manufactured in the United States
in amounts over one million pounds per year and that a full set of basic
toxicity information is available for only 7% of these chemicals (*Risk Policy
Report*, May 15, 1998, 9). Human epidemiological data are available for
less than 1% of chemicals in use (Green 1993).[10]

Monitoring data on the sources, releases, fate, transport, and ambient levels of pollutants in air, water, land, and biota are generally
unavailable for most substances. Available data tend to be sparse, of
poor quality, or both. Part of the explanation for this state is that most
environmental monitoring is conducted for the narrow purposes of
evaluating regulatory compliance, instead of the broader needs of
understanding and assessing environmental risks. In addition, data
that relate levels of pollutants in environmental media (air, water, land,
and biota) to actual exposure profiles (which consider levels, duration,
and timing of exposures) are scarce, as are those for specific doses
delivered to target organs on which substances ultimately have biological effects.

Scientists also lack a fundamental mechanistic understanding of how
pollutants cause harm. For example, cancer—the most studied environmental health effect—is currently understood as a multistep process
requiring deregulation of cellular growth (NRC 1993). Fidelity to this
understanding requires the use of biological models that are more complex than that implied by EPA's traditional (linear multistage) cancer
model. However, this advance in basic cancer biology does not necessarily reduce uncertainty in cancer risk assessment. Regarding the more
complex conceptual models of carcinogenesis, an independent toxicologist observes, "Some of the concepts have been out there for a long time"
but are not used due to a lack of basic biological data needed to implement them. NRC (1993) noted that given current toxicological data and
methods, application of even the simplest multistep model of carcinogen-

esis (the two-stage model) would still require numerous scientific assumptions (including mechanism of action, appropriate target cells, time dependence, and how these influence the shape of the dose-response relationship). Furthermore, NRC concluded that "even if an agent's mechanisms of [carcinogenic] action are well understood, it will still be very difficult to determine its dose-response relationship accurately enough to predict doses that correspond to [cancer] risk as low as one in a million" (NRC 1993, 10). Our scientific understanding is even more primitive in the area of noncancer health effects and ecological impacts.

Environmental risk analysis falls under the category of what John Horgan, author of the popular and controversial book *The End of Science: Facing the Limits of Knowledge in the Twilight of the Scientific Age* (Horgan 1996) would call "ironic science." Its "predictions" are not currently, and perhaps may never be, empirically testable. Cothern et al. (1986) may be exaggerating somewhat—but perhaps not by much—when they say that estimating environmental risks to human health "is as if you had no idea whether your wallet contained enough money to pay for coffee or to pay for the national debt, and no way of finding out." Consider, for example, that while a weather or business forecast could be evaluated and refined on a daily or quarterly basis, a cancer prediction would take decades to verify, if it could be verified at all. Many environmental risk assessments can be neither proven or disproved scientifically. Because a typical risk assessment consists of about fifty separate assumptions and extrapolations (NRC 1983), it is often easy to poke apart or deconstruct, if one so desires.

Much of what passes for "environmental science" is, in fact, a negotiated consensus among experts. In many cases, individual studies do not produce conclusive evidence, and analysts attempt to reach consensus by weighing the accumulated evidence. Needless to say, different experts tend to weight data differently, and consensus can be difficult to achieve. However, in the cases in which expert consensus has been achieved, or when some credentialed group deems it so, it is easy to lose sight of the fact that what is perceived as a "scientific fact" remains an unvalidated, sometimes untestable, negotiated scientific judgment.

For example, the International Agency for Research on Cancer (IARC) regularly convenes working groups of experts to consider and evaluate epidemiological and toxicological evidence from animal studies. The working groups represent a formal scientific consensus-building process developed to assess and categorize the strength of evidence on specific agents suspected of causing cancer in humans. However, there remains disagreement among scientists regarding particular IARC cancer classifications. The international agency's recent upgrading of dioxins to the status of a "known human carcinogen," for example, has been controversial (*Food Chemical News*, November 21, 1997, 4). Because there are fewer data on noncancer effects, there is even greater potential for scientific disagreements regarding the strength of evidence of noncancer effects (NRC 1994).

Rules of thumb negotiated by scientists can be helpful benchmarks for decisionmaking, but they remain somewhat arbitrary. One example is the conclusion reached by many in the field of epidemiology that, absent any additional supporting evidence, the observed rate of disease in the exposed study group should be three or four times that in the control group for a study to be taken seriously (Gori 1995). Although this convention is founded in the observation that the probability of a spurious correlation (due to chance or confounding factors) increases as the "relative risk" value decreases, there is no "objective" basis for this particular range of relative risk values. The convention of expressing aquatic toxicity as the concentration that kills half of the test organisms (the LC_{50}) derives in part from an old rule of thumb in fisheries management that assumes (sometimes incorrectly) that annually harvesting 50% of a fish stock is sustainable.[11]

Negotiated science is not restricted to the conference table; it occurs in the lab as well. Pathology, for example, bears some resemblance to the ancient, now-discredited science of reading goat entrails. The terms *lesion, abnormal cellular growth,* and *benign, precancerous tumor,* or *malignant tumor* are misleadingly precise word models for biological damages that are often hard to categorize. While consensual protocols help ensure the replicability and comparability of different experiments, rules developed for coding laboratory observations are nonetheless often subjective. (See Appendix G regarding different interpretations of the pathological evidence on dioxins.)

Future advances in the state of environmental science will tend to reduce current uncertainties as well as reveal our current ignorance. Although there are bound to be exceptions, it seems likely that most currently recognized environmental hazards are not as risky as they appear—at least if they are considered on a one-by-one basis. This stems from the intent of regulatory risk assessment to be plausibly conservative (that is, systematically overstates risks) in the face of scientific uncertainty. The opposite may be true for identifying environmental hazards. As more research reveals our current ignorance, we are more likely to find unforeseen hazards (Goldstein 1991). If, however, future advances in environmental science do tend both to lessen the burden of achieving existing risk-based objectives and to confirm more hypothesized environmental problems than are dispelled, EPA faces a considerable challenge in gaining public trust.

Table 5.6. Trend in EPA's Use of Science over Time

Overall Rating of EPA's Use of Science	Current	10 Years Ago	20 Years Ago
Very good	5	3	1
Good	12	7	1
Fair	16	16	9
Poor	3	6	5
Very poor	0	0	1
Don't know	6	10	25

VARIATION IN EPA'S USE OF SCIENCE OVER TIME

As environmental science and EPA have co-evolved over the past twenty-five years, the extent and nature of EPA's use of science for regulatory decisionmaking has varied. Over forty respondents were asked how they would rate EPA's overall use of scientific information for regulatory decisionmaking currently, ten years ago, and twenty years ago.[12] Their responses are listed in Table 5.6. In each time period, the most frequent rating is "fair," but the overall trend in respondents' ratings appears slightly upward over time. Of those who supplied a rating, 47% of respondents rated EPA's current use of science as good to very good, compared with 31% ten years ago.

Several general trends emerge from the responses. Although the trends are not even across programs or consistent over time and respondents were not in complete agreement, most agreed that science at EPA has become more institutionalized, decentralized, consistent, rigorous, and comprehensive. In particular, the use of risk assessment in decisionmaking has become more formalized, refined, and ingrained at EPA. A former Administrator observes that the use of risk assessment has gained legitimacy within the agency. According to another former Administrator, "There is a greater awareness that EPA's credibility is tied up in its scientific pronouncements, and therefore, it better be careful what it says."

The trend toward decentralized science at EPA has both logical and strategic explanations. According to an agency research official, "In part, science can't be managed centrally—it has to be tailored to specific program needs. In part, it's a control issue. The programs want control of the science." Programmatic control over the agency's science agenda is necessary to ensure that regulatory decisionmakers' information needs are satisfied. Control over the regulatory analysis function promotes scientific support for policy decisions.

The agency has also markedly increased its use of external scientific review and become more willing to acknowledge scientific uncertainties.

According to an academic, the agency's early approach to science was legalistic. A study was judged either admissible or inadmissible. If admissible, the evidence was an adequate basis for decisionmaking. Increasingly, EPA is taking a more scientific approach that weighs all of the available evidence. At the same time, there is a growing awareness at senior levels that many aspects of the development and use of science inevitably present policy choices.

It is important to note, however, that while EPA's use of science has improved gradually, the bar for what passes as adequate regulatory science has been raised substantially. This is due, in part, to advances in environmental science and to the increased scientific sophistication of those outside the agency who scrutinize its decisions. The increasing marginal costs of pollution control and the increased complexity of the second generation of environmental problems (those that remain after the most obvious problems are controlled) also have contributed to raising the standards for sufficient regulatory science. In expanding the scope of the environmental science that it develops and considers, the agency has made bona fide efforts toward a more comprehensive and integrated approach to the use of science in decisionmaking. However, EPA's limited science resources have been stretched increasingly thin as the agency's agenda expands with time. The expectations and demands on the agency have increased faster than the quality and the breadth of the science intended to inform its decisions.

An EPA official notes that the increased sophistication of EPA's science has "also created a black box. Risk assessors get asked a question by their management, they go away for six weeks, and come back with the answer—2.4. The process has gotten less transparent to the managers." Thus, the increased complexity of science makes it easier for undisclosed assumptions to be buried, consciously or unconsciously, in the analysis and, in some cases, shifts decisionmaking power within EPA from managers to scientists.

There has been considerable variation within and among EPA administrations regarding scientific sophistication and the centrality of science to decisionmaking. Many interviewees regard Ruckelshaus's second administration (1983–1984) and the administrations of Thomas (1985–1989) and Reilly (1989–1992) as the ascendancy and peak of science at EPA. A driving force behind the emphasis on science during this period was the effort to regain EPA's credibility after a series of scientifically dubious decisions and political controversies. Previous administrations, however, may have overstated the role that science can play in decisionmaking to legitimize policy choices.

An EPA research official says that over the past several years, "I don't see great improvements or declines. A status quo is setting in." Many observers inside and outside EPA, however, sense a disengagement from science by the Browner administration (for example, the decision to eliminate the position of science adviser) and believe that EPA's current political leadership has demonstrated less of a commitment than its predecessors to *using* science in regulatory decisionmaking. If so, however, then any disinclination toward science on the part of the Browner administration is not reflected in a reversal in the overall trend in EPA's use of science over time indicated by Table 5.6. It may be that risk analysis and other means of using science have gained such institutional momentum at EPA that less attention from the political level is required to sustain the trend. Perhaps the primary distinction is that whereas previous administrations relied more heavily on science to legitimize policy choices, the current administration looks primarily to stakeholder negotiations to serve this function.

The Browner administration has demonstrated support for improving the quality of science EPA uses by accepting a senior staff recommendation for additional long-term research within ORD; encouraging increased use of peer review of science used throughout the agency; and continuing the evolution punctuated by Deputy Administrator Henry Habicht's 1992 internal memorandum (U.S. EPA 1992) calling for a fuller characterization of environmental risks (including uncertainty and variability) and a more complete and explicit description of scientific assumptions, data, and methods. The Browner administration has also devoted additional resources to the agency's extramural, competitive research grants program.

Supporters of the administration's policy of increased extramural basic research argue that the policy will provide a stronger and more credible scientific basis for future regulatory decisions. An alternative interpretation of the agency's increased support for extramural basic research, however, is that it is part of a pattern of constituency building. For many years, the regulated community and vocal members of Congress have criticized EPA's use of science while university scientists have complained about the agency's historically modest support for academic research. A skeptical view of the administration's extramural science initiative is that it seeks to develop a cadre of academics who support EPA and its mission and to provide a political counterweight to industry-affiliated experts. Of course, the multiple, possible objectives of the extramural research initiative are not mutually exclusive.

By and large, the environmental governing philosophy across the Clinton administration has been to eschew policy analysis and to seek consensus by convening stakeholders around the bargaining table.[13] Participatory exercises (for example, the sector-specific Common Sense Initiative and the firm-specific XL) have been the central pillars of the Browner administration's regulatory reform efforts and arguably represent the administration's flagship initiatives. Further evidence of the low priority the administration places on science and analysis at EPA is provided by the decision to eliminate the independent, internal scientific and economic regulatory analysis functions previously conducted by the agency's policy and research offices, and more recently, by the President's FY 1999 budget request, which called for a substantial reduction in EPA's science budget and allocated approximately 6% of the agency's total budget to the Office of Research and Development.[14] If the perception that the Browner administration places a decreased emphasis on the use of science at EPA is indeed accurate, it may reflect some healthy skepticism about what "objective analysis" can contribute to value-laden environmental policy decisions. It would also arguably represent a correction from the overselling of science-driven regulatory decisionmaking by preceding administrations and influential voices in the scientific community.

Nevertheless, as inclusive as the convening processes have appeared, some interests still feel either left out or unplacated. As unsatisfied stakeholders still have access to a variety of veto measures, the administration has been confronted with the limits of environmental policymaking by consensus.[15] Coglianese (1997) found that EPA has not seen its negotiated rules emerge in final form any sooner than rules developed through conventional administrative procedures and that the litigation rate for negotiated rules issued by EPA has actually been *higher* than that for other significant EPA rules. Coglianese (1997) concludes that negotiated rulemaking adds new sources of conflict (for example, over membership on negotiating committees and the consistency of final rules with negotiated agreements) and raises unrealistic expectations about what participants can gain from their participation. We should be skeptical about the potential to forge a meaningful and stable consensus from diverse stakeholders with conflicting or competing values. As Fisher and Ury (1981, 84) concede—

> However well you understand the interests of the other side, however ingeniously you invent ways of reconciling interests, however highly you value an ongoing relationship, you will almost always face the harsh reality of interests that conflict. No talk of "win–win" strategies can conceal that fact.

ADDRESSING UNCERTAINTY

For several years, the agency has been wrestling over the extent to which it should reveal the scientific uncertainties underlying its decisions and how best to articulate uncertainties to decisionmakers, overseers, and the public. At the same time, EPA and the scientific community at large have also puzzled over how best to evaluate and communicate uncertainty to inform decisionmaking.

In 1992, former Deputy Administrator Henry Habicht issued an internal memorandum (U.S. EPA 1992) calling for a fuller characterization of environmental risks (including uncertainty and variability) and a more complete and explicit description of scientific assumptions, data, and methods. Habicht's memo is so frequently cited that it has come to be interpreted by many as marking a fundamental transition in agency science policy. According to a former EPA official, internal support for the Habicht memo came primarily from the science policy branch of the policy office, and there was some support for the policy within the research office. EPA tentatively began to address quantitative uncertainty analysis in its 1992 exposure assessment guidelines and in the 1995 draft revisions of its *Exposure Factors Handbook*. In 1997, the agency issued a "Policy for Use of Probabilistic Risk Assessment" permitting the application of uncertainty analysis to the exposure portion of human health risk assessment but not to dose-response assessment (the area of greatest uncertainty). Stern (1997) notes that EPA's limitation of probabilistic analysis is policy based rather than limited by theoretical considerations.[16] Still, EPA lacks operational guidelines on uncertainty and variability analysis.[17]

Habicht's 1992 memo presents an interesting case study. Almost all respondents agreed that it represented an important advance in EPA's use of science, but it has been almost entirely ignored in practice. Following in the general vein of Habicht's memo, EPA's Science Policy Council (SPC) issued risk characterization guidelines in 1995 (U.S. EPA 1995). The SPC guidance consists of general principles and emphasizes narrative discussion of uncertainty and variability and description of assumptions, data, and methods. However, Assistant Administrators and Regional Administrators were responsible for developing their own office-specific risk characterization procedures, thus leaving them considerable discretion in deciding the extent and formality of addressing scientific uncertainties. The bottom line for now is that the risk characterization guidance appears to require no substantial operational changes in how the agency addresses scientific uncertainty. Perhaps the guidance indicates that Habicht's memo marked a turning point for EPA rather than an advance per se in how the agency uses science for decisionmaking. The risk characterization guidance seems to suggest that the policy of transparency and full disclosure in scientific assessments is gradually gaining broader acceptance within the agency.

Respondents offer a variety of reasons for EPA's traditional reluctance to acknowledge scientific uncertainties underlying its standards or to analyze the effects of uncertainty on risk estimates for decisionmaking. Disclosing uncertainties can be perceived as a legal, political, and bureaucratic liability. EPA lawyers, for example, may resist acknowledging scientific uncertainty for fear that doing so will weaken their case in defending agency decisions from legal challenges. Similarly, regulatory enforcement officials may have concerns regarding the impact of scientific reevaluations on pending enforcement cases. Another explanatory factor may be that agency officials are reluctant to lose the anchor that clearly defines agency positions in negotiations with the regulated community. An EPA policy official, for example, notes that refusing to acknowledge uncertainty is part of a regulatory negotiation strategy because agency officials feel they need to take a firm stand at one end of the spectrum in anticipation that the regulated community will take the opposite position. Regulatory program officials may also be reluctant to admit uncertainty out of a desire not to provide the Office of Management and Budget, other agencies (which represent such constituencies as agriculture, business, the electric utility sector or, like the Departments of Defense and Energy, are themselves part of the regulated community), or Congress with additional ammunition in reviewing proposed rules. More generally, an EPA policy official explains that part of the administrative culture is an expectation that an "expert agency should have the answers" and observes that "the agency's self-perceived clout is tied to its expertise." Also, the general public, the media, and policymakers dread uncertainty. Instead, what they often seem to want is a trustworthy, reassuring answer to the question, "Is it safe or not?" If acknowledging scientific uncertainty results in EPA's departing from precedent, it may also be interpreted as waffling under pressure from stakeholders.

EPA may resist analyzing the effects of scientific uncertainty for regulatory decisionmaking because uncertainty analysis is perceived as serving the interests of the regulated community. Resistance may also occur because uncertainty analysis adds undesired complexity and time to regulatory analysis. Given industry's advocacy for the application of probabilistic risk analysis (for example, using Monte Carlo simulation), the perception that uncertainty analysis suits industry's purposes is undoubtedly accurate in some cases. Using these methods (that is, replacing high-end exposure scenarios and default assumptions with probability distributions) often has the effect of reducing risk estimates relative to current EPA assessment practices. However, using the same approach (for example, to consider the variability in human susceptibility or exposures) can also have the effect of increasing estimated risks under some circumstances. For example, due to their small size and developmental stage, children are more highly exposed to certain dietary pesticide residues than average members of the adult population. On occasion (for example, for assessing the health effects of environmental tobacco smoke, indoor air radon, and drinking water disinfection by-products), EPA has employed meta-analysis, a procedure designed to deal with the uncertainty arising from multiple scientific studies that produce varying estimates. Although meta-analysis has some substantive critics, objections to its use for regulatory decisionmaking often seem to stem primarily from its capability to detect statistically significant health effects at levels lower than any single epidemiological study can show. The bottom line, therefore, is that analyzing scientific uncertainties can cut both ways—it does not always serve the purposes of regulatory relief.

Another impediment to explicitly analyzing scientific uncertainty for environmental decisionmaking is its complexity and cost. Formal, quantitative uncertainty analysis makes great demands for information that is often missing, requiring risk analysts to replace simple assumptions (that is, point estimates) with more complex assumptions (that is, entire distributions). A recent model study commissioned by the federal Commission on Risk Assessment and Risk Management demonstrated just how costly and tricky it is to perform a sound quantitative uncertainty analysis. In its final report, the commission strongly supported using quantitative descriptions of variability in exposure to contaminants but concluded that little value is added to regulatory decisionmaking by routinely conducting formal quantitative analyses of uncertainty.[18] Currently, EPA has virtually no in-house capability to review such analyses, much less to conduct them.[19] The complexity of quantitative uncertainty analysis makes it more difficult to review than nonprobabilistic risk analysis, which uses singular point estimates to characterize random variables and uncertain quantities. There are concerns that formal, quantitative uncertainty analysis would result in information overload for decisionmakers. The complexity of quantitative uncertainty analysis also tends to increase the number of assumptions imbedded in the analysis (for example, regarding the shape of underlying probability distributions).[20]

On the other hand, qualitative uncertainty analysis is often unsatisfying and of questionable value to decisionmaking. An EPA research official notes, for example, that official science advisory review meetings "can become bull sessions" in which scientists qualitatively discuss all of the scientific uncertainties without addressing the "so what question." According to this official, "It's easy to list the uncertainties and the differing viewpoints. It's often hard to convey the essence of that uncertainty and the consequences they'll have on management alternatives. The question is, what does this uncertainty mean? It's more than a communications problem. It's simply hard to propagate the uncertainties through the system to answer the bottom line." Of course, all of this begs the question of how EPA decisionmakers actually address scientific uncertainty, whether expressed qualitatively or quantitatively. When confronted with counterfire between the experts, the decisionmaker may simply tune out both sides and make his or her decision on some other basis.

QUALITY CONTROL

Since the early 1980s, EPA has substantially expanded and institutionalized a variety of quality assurance and quality control mechanisms, including independent peer review. However, the internal scientific regulatory review and quality assurance functions housed in the Office of Research and Development have had a limited effect on the acquisition and use of science by EPA regulatory program offices. This result appears to be largely due to ORD's lack of clout within the agency, the limitations of quality review during the late stages of any process, and the lack of a clear answer to what constitutes the requisite quality of data for regulatory decisionmaking.

The—now defunct—internal economic regulatory review function that formerly resided in EPA's policy office apparently had some clout within EPA due to a natural parallel to the Office of Management and Budget economic review function. In contrast, EPA's internal scientific regulatory review function has never had a close analog within the Executive Office of the President from which to derive power within the agency. The internal science review function has also suffered to some degree from the ORD political leadership's lack of interest or effectiveness in being directly involved with the agency's regulatory programs. The responsible office has also never been delegated substantive authority from the agency's political leadership (for example, remand authority), thus limiting its powers of persuasion. In general, the internal science review office was regarded by regulatory programs as an unhelpful impediment. Internal scientific reviewers are in an awkward position, however. An EPA official commented, "It's hard to avoid being perceived as an intellectual gadfly or snob when you understand your mission to be cajoling program offices into taking science seriously and not playing games with the numbers to prop up a political position."

Perhaps more productive than cajoling would be a value-added approach in which internal scientific reviewers offer an alternative analysis. For this reason, the Office of Research and Science Integration (ORSI), which is responsible for representing ORD in the regulatory development process, would be a good place to begin seriously developing EPA's in-house capacity for conducting rigorous uncertainty analysis. This office would require substantial analytical capabilities to perform the proposed regulatory science review function adequately. It probably would not have the opportunity, however, to acquire much original data nor command a large budget. By concentrating the development of EPA's intramural uncertainty analysis capability in ORSI, uncertainty analysis practices could be put into use sooner, more pragmatically, and with greater effect on major decisions.

EPA's official science advisory panels and public comment under the provisions of the Administrative Procedures Act provide means of independent scientific quality control, with the agency's science advisers widely regarded as playing the more important and often constructive role. The growing influence of the external scientific community through avenues such as the Science Advisory Board represents one of the most significant changes in EPA's use of science over the past ten years. The Environmental Research, Development, and Demonstration Act requires EPA to provide any proposed criteria document, standard, limitation, or regulations to the SAB when these proposals are provided to other federal agencies for formal review and comment.[21] However, only the criteria air pollutants, drinking water, and pesticides programs are explicitly required by statute to use official science advisory review. In addition, EPA has not submitted many of its important scientific databases (for example, the Integrated Risk Information System [IRIS]) and models to external scientific review.[22] Time and resource constraints have limited broader use of independent scientific review by other EPA programs. The agency may sometimes avoid external scientific review, however, because it might undermine policy objectives or narrow policy options. In contrast, requesting independent scientific studies or reviews is sometimes intended primarily to buy time to delay decisionmaking.

With demanding statutory deadlines to issue numerous regulations enforced by court orders, time has often been the factor limiting EPA's use of scientific review. Many months are required to plan and prepare for external peer reviews, and a 1994 SAB internal review found that six months often elapse between the last public meeting of an SAB panel and the transmittal of a report to the administrator (Johnson 1994). Even with excellent staff support, the act of writing a consensus committee report by scientists scattered among different institutions across the country is time consuming. Furthermore, the interests of academic reviewers are not tied to meeting a regulatory deadline, and scientific consensus does not necessarily coalesce in response to a court order. The traditional norms of academia call for scientists to be objective and skeptical, and they may resist making definitive statements without strong empirical support. In contrast, administrative agencies like EPA value the exercise of professional judgment. "The graybeards need to steep for a while to gain a comfort level with how EPA is treating the science," one EPA official comments. As the most recent reviews of the national ambient air quality standards for particulate matter and ozone demonstrated, when there are deadlines requiring a compressed review schedule, the reviewers "chafe at the bit." One factor contributing to the lengthy period currently required to conduct Clean Air Science Advisory Committee reviews has been the committee's insistence on reviewing criteria documents and staff papers twice—first in draft and again after EPA makes revisions. While some argue that this practice delays the review process unnecessarily, it also adds a degree of assurance that EPA will not simply ignore the committee's scientific advice.

In 1994, EPA Administrator Browner issued a policy statement on peer review (U.S. EPA 1994) making the Science Policy Council responsible for overseeing implementation of agencywide peer review policy. Repeating the pattern observed in the agency's risk characterization guidance, Assistant and Regional Administrators remain responsible for developing office-specific peer review procedures; therefore, it remains to be seen whether the agency will actually extend the coverage of independent peer review to include the scientific basis for more of its decisions.

In 1996, the General Accounting Office (GAO) reported that although EPA had made progress in implementing its peer review policy, implementation remains uneven. GAO cited two primary reasons for this unevenness: (1) confusion among agency staff and management about what peer review is, what its significance and benefits are, and how and when it should be conducted and (2) inadequate accountability and oversight mechanisms to ensure that all relevant products are properly peer reviewed. For example, some agency officials told the GAO that the public comments obtained through the rulemaking process would suffice for peer review of their work products, although EPA's peer review procedures state that these are not substitutes for peer review (U.S. GAO 1996).

In response to the concerns raised by the GAO report, EPA Deputy Administrator Fred Hansen signed an internal memorandum late in 1996 calling for the Office of Research and Development to conduct a review of implementation of the peer review procedures. The subsequent internal review of 2,000 agency science products requiring peer review reported that fewer than 4% either failed to be peer reviewed (36) or merited more extensive peer review (38). However, a December 1997 EPA environmental modeling workshop concluded that of thirty models recently under development, only 5% of staff involved were aware of the agency's peer review policy, and only 2% acknowledged using it. In response, Deputy Administrator Hansen endorsed a Science Advisory Board initiative to address the haphazard manner in which agency risk models are developed, validated, and peer reviewed. Another possible outcome, stemming from the GAO report, is the establishment of an ongoing internal auditing program to review whether peer review procedures are being followed and are improving the quality of the agency's scientific work products (*Inside EPA*, December 6, 1996, 8–9; *Risk Policy Report*, January 23, 1998, 4–17).

External peer reviewers and science advisers can serve an important quality control function, and there are strong arguments for EPA to submit the scientific basis for all major regulatory decisions to external, independent review, but peer review is no cure-all and should not be viewed as a means of arbitrating policy decisions. Peer review at the end of the regulatory development process adds some value, as an EPA official says, "to make sure you're not sending out clunkers." The anticipated scrutiny may also cause the agency to be more diligent. In addition, review contributes to an iterative, institutional learning process. However, in many cases, end-of-the-line review cannot repair mistakes or omissions made early in the regulatory development process or fill data gaps. Back-end inspection may be able to identify scientific uncertainties, but rarely can it reduce them. The benefits of regulatory science quality control must also be balanced against the potential for peer reviewers to intrude on the policy domain. If determining whether the data and analysis are adequate for regulatory decisionmaking is the problem, then peer review does not solve the problem. It shifts the problem from decisionmakers to reviewers.

External peer review is also susceptible to problems of quality control and conflict of interest and may result in excessive encroachment by unaccountable experts into policy decisions. As the asbestos case study (Appendix F) demonstrates, "expert" scientific panels are not immune to making technical mistakes. Part of the problem is that environmental regulatory issues generally present very complex problems requiring expert input from a wide variety of scientific disciplines. It is also a matter of getting what you pay for: being a member of a National Academy committee or blue-ribbon panel may be prestigious, but most of the scientists are serving essentially on a pro bono basis and have a very limited amount of time that they can sacrifice from the research grant treadmill.[23] Much of the time served on expert panels is consumed by meetings and travel, thus limiting the time for independent analysis. Supplementing the panels with consultants to conduct new analysis or in-depth reviews can add substantially to the costs of the review.

A number of environmental and public interest groups have complained about the participation of corporate employees and consultants as members and consultants of EPA science advisory committees. On the other hand, a few respondents complained about the participation of scientists from environmental groups and academics with ties to advocacy groups or who receive research support from EPA. While most would agree that affiliated scientists should not be disqualified from serving on science advisory committees, concerns remain about achieving balanced committees, particularly in view of the numerical superiority of industry-affiliated scientists relative to those affiliated with "public interest" groups.

An internal SAB review completed in 1994 reported that EPA staff voiced concern about the diversity of SAB membership, openness in the member selection process, and potential conflict-of-interest problems when board members receive federal research funds, work for the regulated community, or serve on other federal committees (Johnson 1994). EPA's current policy requires SAB members to disclose to the agency any potential conflicts when they participate in a specific SAB activity. Some EPA observers recommend public disclosure of any conflict of interest and argue that affiliated experts be restricted from chairing science advisory committees. However, it can be difficult sometimes to identify qualified, unaffiliated candidates.

Even if SAB members receive only public or foundation research funding, they may have a vested interest in the perception of an environmental problem or in the use of particular scientific methods. An EPA research official, for example, complains that "nobody on CASAC [Clean Air Scientific Advisory Committee] says, 'This is not a problem.' They've been on CASAC a long time and they have a vested interest in the existence of a problem." In some cases, academics may have sunk investments in being able to perform highly specialized methodologies. An industry representative, for example, recounts an instance in which a Federal Insecticide, Fungicide and Rodenticide Act scientific advisory panel member complained in a public meeting that if EPA stopped using small changes in the level of a particular enzyme (acetylcholinesterase) as an indicator of neurological damage, rather than as a biomarker of exposure, "there goes our research." Academia is also a stakeholder in EPA's science program, having a vested interest in greater agency support for extramural research. The interests of researchers may be served by calling for additional research in lieu of making a regulatory decision.

To EPA's credit, it has attempted to broaden the disciplinary representation on the agency's official science advisory panels and generally seeks to achieve some degree of balance among differing viewpoints on the committees. Perhaps the agency could further improve the composition of its scientific review panels, but it is constrained by the available pool of scientific talent willing and able to participate. If scientific consensus is viewed as a prerequisite to regulatory action, then balance can be problematic because a well-balanced committee is less likely to produce a meaningful consensus. Furthermore, policy disagreements among the panelists can be misinterpreted as purely scientific disputes. Seeking balance on a scientific panel to avoid biasing the central tendency of expert opinion can also tend to exaggerate the spread in the range of expert opinion, perhaps reducing the value of the review for decisionmaking. Conversely, narrow panels can be equally uninformative because they may be skewed one way or the other.

Ultimately, an official science advisory panel is not the appropriate forum for providing EPA with external policy advice. There are other forums, including other advisory committees and the public comment process, that are designed for this purpose.[24] While it may be true that scientific review of policy-related issues can add substantial value and credibility to regulatory decisions, this is an argument for scientists to participate in the public comment process and on policy advisory committees—not for official science advisory committees to be invited to participate in the policy process. To do so only encourages the tendency to treat official science advisory review as a forum for negotiating regulatory decisions. Perhaps the best that can be hoped for is that Administrators, Congress, and the courts appreciate the strengths and limitations of external peer review panels and recognize that although they can contribute a necessary measure of quality control to the regulatory decisionmaking process, they must not be asked or allowed to act as the final arbitrators of agency decisions.

ENDNOTES

[1] Environmental monitoring data collection and analysis conducted by or for EPA generally serves to support enforcement of pollution control regulations rather than evaluation of their performance.

[2] While EPA's overall extramural research budget fell from $202 million in fiscal year (FY) 1992 to $166 million in the FY 1995 request, ORD's extramural budget during the same period grew by 19%. The Office of Research and Development's extramural funding for research supporting program office regulatory activities decreased from 62% in FY 1992 to 59% in FY 1993 to an estimated 51% in FY 1994 and 43%, based on the administration's FY 1995 request (*Inside EPA*, February 11, 1994, 10–11). For FY 1996, the administration proposed to double funds for extramural basic research to a level of $85 million (*Science*, March 31, 1995, 1903). However, much of what is labeled extramural research is not basic, competitive research.

[3] Press reports suggest that House Science Subcommittee Chairman Dana Rohrbacher (R-Calif.) supports an environmental research, demonstration, and development reauthorization bill that directs EPA research to regulatory support (*Inside EPA*, March 17, 1995, 4). In addition, expressing concern that EPA would neglect science applicable to regulations, Sen. John Warner (R-Va.) blocked the agency's plans to double funds for ORD extramural research again in FY 1996 by pushing through an amendment to the EPA appropriations bill that requires prior

approval from House and Senate appropriations committees (*Science*, October 6, 1995, 19). A Warner staffer suggested that the applied research was needed to relax overly stringent regulations. In part, however, the issue may have been that resources devoted to additional academic research translated into reduced resources available for applied research and technical assistance supplied by environmental consulting firms. Unlike the environmental contractors that service the states and regions, largely, though not exclusively, in hazardous waste and contaminated site issues, those in the Washington, D.C., area, including many in Virginia, service primarily the regulatory development and policy functions of the headquarters offices.

[4] The 1996 Safe Drinking Water Act Amendments *authorized* EPA to test under the estrogenic screening program substances that may be found in drinking water sources *if* a substantial population may be exposed to such substances.

[5] Although EPA is required to do so by the 1996 Food Quality Protection Act, some representatives of industry argue that it is premature to develop a testing program for regulatory purposes (*Inside EPA*, May 10, 1996, 10). The Office of Prevention, Pesticides, and Toxic Substances organized a workshop in May 1996 that launched an EPA effort to develop an endocrine disrupters screening and testing program (*Risk Policy Report*, May 17, 1996, 3). In a 1996 seminar cosponsored by the International Life Science Institute and the Society for Risk Analysis, National Academy of Sciences member John Giesy asserted that "at this time legislating [endocrine] testing is unbelievably stupid and a waste of resources and will lead to false positives in screening, but more importantly, to false negatives" (*Risk Policy Report*, November 15, 1996, 29–30).

[6] EMAP was launched in 1989 to establish environmental baseline conditions and monitor trends nationally in response to the EPA Science Advisory Board's 1988 *Future Risk* report (U.S. EPA/SAB 1988). The program is an outgrowth of EPA's involvement in the interagency National Acid Precipitation Assessment Program. Despite the fact that the agency has spent over $150 million on EMAP, making it and the acid precipitation assessment among the biggest research projects in the agency's history, the program has been the subject of numerous critical external evaluations. In response, the Browner administration narrowed the scope of the program and recruited an experienced research administrator for the program (see Stone 1995).

[7] Thirty-three interviewees were selected on the basis of their broad knowledge of and/or experience in environmental regulatory decisionmaking. Nine case-specific interviewees also agreed to answer a supplementary series of questions. The forty-two interviewees are not in any sense a random statistical sample.

[8] Tables 5.4 and 5.5 have been simplified by eliminating all decisions that received only one mention.

[9] Because these types of regulatory decisions are often uncontroversial, they may not be the focus of the regulatory science debate. Such decisions may be controversial, however, if the distribution of costs and benefits is sufficiently uneven. The estimated costs of a regulatory decision may seem minimal in the context of

the entire economy, for example, but may seem large if a small set of firms bears the entire costs. Likewise, the health benefits of a regulatory decision may seem small in the context of the entire population but may be large for an at-risk sub-population.

[10] As indicated in the Chapter 4 discussion of the EPA toxic substances program, developing comprehensive toxicity data on all substances is unnecessary to make sound regulatory decisions. Therefore, these figures serve primarily to indicate the general lack of substance-specific data and should not be interpreted to mean that elaborate testing is required for all substances.

[11] In general, the common use of the LC_{50} to characterize lethal toxicity stems from the fact that the LC_{50} can be estimated more precisely than lethal concentrations in the tails of the distribution (for example, LC_{10} or LC_{90}) with a given number of test organisms (due to the steepness of the cumulative distribution function in the central range of lethal concentrations). This practice too, however, reflects a scientific convention rather than a consideration of what information may be required under particular circumstances.

[12] Thirty-three interviewees were selected on the basis of their broad knowledge of and/or experience in environmental regulatory decisionmaking. In addition, nine case study respondents were willing to answer questions regarding the general role of science at EPA.

[13] Examples of the administration's stakeholder orientation include the "Forest Conference" convened to resolve conflicting interests in the Pacific Northwest, the Habitat Conservation Plan initiative designed to broker regional, multispecies management plans without resorting to the Endangered Species Act, and the 1994 executive order on environmental justice requiring community participation in rulemaking and enforcement processes.

[14] The reduction from the FY 1998 appropriation appears to be on the order of 9%, but accounting changes make it difficult to assess the exact level of the decreased request for EPA science funding.

[15] In January 1997, for example, the automobile and petroleum industries announced their defection from the Common Sense Initiative (*Inside EPA*, January 31, 1997, 1). Environmentalists and local government representatives abandoned the National Environmental Dialogue on Pork Production (*Inside EPA*, November 21, 1997, 16–17).

[16] It is worth noting that the 1997 policy statement endorses the use of uncertainty analysis in all aspects of ecological risk assessment.

[17] In May 1996, EPA's Risk Assessment Forum sponsored a workshop on Monte Carlo analysis intended to kick off the agency's efforts to develop guidance on the use of uncertainty and variability analysis in exposure assessment (*Risk Policy Report*, June 14, 1996, 30). The Risk Assessment Forum held another workshop on probabilistic risk analysis in April 1998 and announced that informal draft guidance was forthcoming (*Risk Policy Report*, June 19, 1998, 28–29). EPA sources report that there are deep divisions among the regional offices over the use of uncertainty analysis, and this may be a contributing factor to the pace of developing administrative policy and guidelines.

[18] Source: http://www.riskworld.com/tcnetwrk/riskwrld/nreports/1997/risk-rpt/volume2. Although the language seems highly brokered, the commission appears to be recommending for nonroutine decisions probabilistic analysis of variability and parametric sensitivity analysis of model uncertainty, but rather less of the latter. The interested reader should refer to Morgan and Henrion (1990).

[19] Currently, EPA's modest uncertainty analysis capabilities are located primarily in ORD's National Center for Environmental Assessment and in the air program's Office of Air Quality Planning and Standards.

[20] Seiler and Alvarez (1996) discuss the informational requirements for selection of distributions for stochastic variables in probabilistic risk analysis. Seiler and Alvarez note that specifying distributions for risk analysis variables requires more information than does point estimation (for example, specifying the mean or ninety-fifth percentile of the distribution) and often requires more information than may be available to the analyst.

[21] As a nonregulatory program, Superfund is exempt from this provision.

[22] There are currently plans to open IRIS and EPA's panoply of environmental models used for regulatory purposes to some degree of routinized peer review, but it remains unclear what shape this will take.

[23] SAB members are reimbursed for travel and per diem and receive special government employee pay for hours worked, but this is significantly less than the experts could earn as private consultants.

[24] For example, EPA established the Clean Air Act Advisory Committee under the Federal Advisory Committee Act (FACA) in 1990 to provide independent advice and counsel to the agency on policy issues associated with the implementation of the Clean Air Act of 1990. The agency established the Pesticide Program Dialogue Committee under FACA in 1996 to provide EPA with advice and guidance regarding regulatory, policy, and implementation issues. The Safe Drinking Water Act (Section 1446) establishes a National Drinking Water Advisory Council to advise, consult, and make recommendations to the Administrator on activities and policies derived from the act.

REFERENCES

Ashby, J. 1996. Assessing Chemicals for Estrogenic/Hormone-Disrupting Properties: Lessons from Carcinogenicity Assessment. *Environmental Health Perspectives* 104 (2): 132–133.

Coglianese, C. 1997. Assessing Consensus: The Promise and Performance of Negotiated Rulemaking. *Duke Law Journal* 46: 1255–1349.

Colborn, T., D. Dumanoski, and J. Peterson Myers. 1996. *Our Stolen Future: Are We Threatening Our Fertility, Intelligence, and Survival?* New York: Dutton.

Cothern, R., et al. 1986. Estimating Risk to Human Health. *Environmental Science and Technology* 20(2): 111–116.

Davis, B., R. Dulbecco, H. Eisen, H. Ginsberg, and W. Wood. 1967. *Microbiology*. New York: Harper & Row.

Fisher, R., and W. Ury. 1981. *Getting to Yes: Negotiating Agreement Without Giving In*. New York: Penguin Books.

Goldstein, B. 1991. The Shift to Exposure Data. *Regulating Risk: The Science and Politics of Risk, A Conference Summary*. Washington, D.C.: National Safety Council and the International Life Sciences Institute, June.

Gori, G. 1995. Letters. *Science* 269: 1328–1329.

Green, M. 1993. Letters. *Science* 262: 637–638.

Horgan, J. 1996. *The End of Science: Facing the Limits of Knowledge in the Twilight of the Scientific Age*. Reading, Mass.: Perseus Press.

Johnson, J. 1994. "Reinventing" the EPA Science Advisory Board. *Environmental Science and Technology* 28(11): 464.

LeBlanc, G. 1995. Are Environmental Sentinels Signaling? *Environmental Health Perspectives* 103(10): 888–890.

Morgan, M. G., and M. Henrion. 1990. *Uncertainty: A Guide to Dealing with Uncertainty in Quantitative Risk and Policy Analysis*. New York: Cambridge University Press.

NAPA (National Academy of Public Administration). 1995. *Setting Priorities, Getting Results: A New Direction for EPA*. NAPA Report to Congress. Washington, D.C.: NAPA, April.

NCE (National Commission on the Environment). 1993. *Choosing a Sustainable Future*. Washington, D.C.: Island Press.

NRC (National Research Council). 1983. *Risk Assessment in the Federal Government: Managing the Process*. Washington, D.C.: National Academy Press.

———. 1984. *Toxicity Testing: Strategies to Determine Needs and Priorities*. Washington, D.C.: National Academy Press.

———. 1993. *Issues in Risk Assessment*. Washington, D.C.: National Academy Press.

———. 1994. *Science and Judgment in Risk Assessment*. Washington, D.C.: National Academy Press.

———. 1995. *Interim Report of the Committee on Research and Peer Review in EPA*. Washington, D.C.: National Academy Press, March.

Seiler, F., and J. Alvarez. 1996. On the Selection of Distributions for Stochastic Variables. *Risk Analysis* 16(1): 5–18.

Stern, A. 1997. The Future of Monte Carlo Analysis in Human Health Risk Assessment. *Risk Policy Report* (November 21): 37–40.

Stone, R. 1995. EPA Streamlines Troubled National Ecological Survey. *Science*. 268: 1427–1428.

U.S. EPA (U.S. Environmental Protection Agency). 1992. Memorandum from F. Henry Habicht, Deputy Administrator. Washington, D.C.: U.S. EPA, February 26. Mimeographed.

———. 1993. *Report to Congress: Fundamental and Applied Research at the Environmental Protection Agency.* EPA/600/R-93/038. Washington, D.C.: U.S. EPA.

———. 1994. *Peer Review and Peer Review at the U.S. Environmental Protection Agency.* Policy Statement. Washington, D.C.: U.S. EPA, June 7.

———. 1995. *Guidance for Risk Characterization.* Washington, D.C.: U.S. EPA, Science Policy Council. February.

U.S. EPA/SAB (EPA Science Advisory Board). 1988. *Future Risk: Research Strategies for the 1990s.* EPA-SAB-EC-88-040. Washington, D.C.: U.S. EPA/SAB.

U.S. GAO (U.S. General Accounting Office). 1996. *Peer Review: EPA's Implementation Remains Uneven.* GAO/RCED-96-236. Washington, D.C.: U.S. GAO.

U.S. OTA (U.S. Office of Technology Assessment). 1993. *Researching Health Risks.* OTA-BBS-570. Washington, D.C.: U.S. Government Printing Office, November.

Yosie, T. 1995. Regulatory Reform: A Preliminary Look Back and Look Ahead. *Risk Policy Report* (August 18): 28–30.

6

Policy Proposals

This study concludes with a set of policy proposals concerning political leadership, the EPA science budget, and peer review. The policy proposals are targeted at three factors over which EPA exerts at least some control and that were identified in Chapter 5 as having a major influence on the agency's acquisition and use of science.

POLITICAL LEADERSHIP

More EPA policymakers need to come better prepared to understand the application of science to policy decisions and to distinguish between and explain the scientific and policy bases for their decisions. With a few noteworthy exceptions, the vast majority of EPA's political and senior career leaders have limited scientific training. Typically, political appointees are lawyers. Indeed, an agency political appointee expressed concern that "senior managers aren't educated about risk assessment and about science." Every administration, it seems, talks about the need for training its nontechnical decisionmakers in the essentials of regulatory science, but little comes of it. The agency's leaders are typically overwhelmed by the day-to-day challenge of meeting regulatory deadlines, budgeting, putting out brushfires, and dealing with the White House, Congress, and stakeholders. Once policymakers are in position, there are too many demands on their time to realistically expect that on-the-job-training can go far to get them up to speed on the scientific basics.[1] The Carnegie Commission (Carnegie 1993) recommended that administrations should appoint senior regulatory agency officials with both policy and scientific expertise more frequently.

Many of the agency's political leaders have failed to understand or communicate what science can and cannot contribute to the regulatory decisionmaking process. Confronted with politically tough choices, agency decisionmakers understandably look to science to provide clarity or decision rules. But more often than not, policymakers are disappointed because they want definitive answers from uncertain science and they cannot avoid making value-laden policy judgments.

Whether unconsciously or wittingly, policymakers often seek legitimacy or political cover for their judgments from science or scientists. A former EPA official notes, "There is a tendency among Administrators and AAs [Assistant Administrators] to subterraneanly lump more risk management [policy decisions] into risk assessment [science] and to look for external sources of scientific credibility, through the SAB [Science Advisory Board] and external [academic] science." In some cases, policymakers are simply overly optimistic about what science or scientists can or will do. Faced with a tough policy choice, a former EPA Administrator asked the National Academy of Sciences (NAS) to advise him on the decision, only to be disappointed that the NAS report did not help. Predictably, it dealt in generalities and lent credence to both sides of the argument. "A lot of good scientists can't be found when decisions have to be made," protested the former Administrator. And, of course, commissioning an independent scientific review is often politically expedient. A former EPA Administrator explains, "Asking for an NAS report is a way to 'deep-six' a problem during your watch if you're a public official."

To create effective demand for rigorous, balanced, and transparent scientific analysis in regulatory decisionmaking, the demand for science must come from EPA's political leadership. EPA political appointees can create demand for science by being sophisticated consumers of scientific information. Largely, this means knowing what questions to ask and appreciating what science can and cannot contribute to decisions. For example, a senior careerist remarks, "What it takes is for the Administrator to walk into a meeting knowledgeable about the scientific issues and ask tough questions of the staff. Do that two or three times, and people will be responsive." The "sophisticated consumer" approach can elicit a rigorous staff assessment of the existing science and can generate more openness about and analysis of the scientific uncertainties. The EPA political leadership should acknowledge and publicize what science can and cannot contribute to environmental regulatory decisionmaking and guard against abdicating its responsibility to make tough policy choices either to scientists or to stakeholders.

EPA SCIENCE BUDGET

A drawback of consistently relying on the agency's political leadership for creating demand for science is revealed by the complaint of a former

EPA senior scientist: "One of the major frustrations with Administrators" is that they "focus too late on the major uncertainties that drive the decision. We could have done something about them if we had more warning." While political appointees can demand that staff "do their homework" conscientiously with existing research and "show their work" in transparent fashion, the policymaker's time horizon tends to be too short to yield original research. Most federal political appointees have a tenure of two to three years, a period generally too short to even initiate and complete a routine rodent carcinogen test for a particular substance. The short time frame of political appointees is reinforced by statutory deadlines and interest group expectations that give an aura of urgency to almost everything EPA does.

The short time horizon of policymakers (in both the executive and legislative branches) helps to explain the chronic failure of strategic planning and budgeting for environmental regulatory science. The root of this failure is a shortsighted tendency to "eat your seed corn," not allocating a sufficient portion of current resources to prepare for future decisionmaking. (Of course, the sometimes randomly introduced elements of the regulatory agenda also make it difficult to plan for a good fit between the science and regulatory agendas.)

Strategic investment in environmental research is one way of preparing for future regulatory decisionmaking. The EPA Science Advisory Board (U.S. EPA/SAB 1988) concluded that it would be reasonable to spend at least 1% of national environmental regulatory compliance costs on relevant research and development (R&D). The SAB's recommendation to double EPA's R&D budget (U.S. EPA/SAB 1988, 1990), however, does not follow directly from this conclusion. Assuming that the nation spends on the order of $120 billion per year on environmental regulatory compliance ($140 billion to $180 billion might be more accurate), the 1% target would mean an environmental regulatory science budget of at least $1.2 billion per year (about twice EPA's current Science and Technology appropriations account). However, the EPA Science and Technology account does not capture all EPA science activities in program and regional offices, nor does it include the approximately $100 million budget transfers from EPA to the Department of Health and Human Services (for the Superfund Basic Research Program and the Agency for Toxic Substances and Disease Registry). It also fails to capture other public and private environmental science expenditures that would be relevant if the target is based on national environmental compliance costs.

Therefore, rather than doubling EPA's science budget in gross terms, EPA's science budget as a percentage of the agency's total budget (less construction grants) should increase from approximately 8% to *at least* 15% over the next four years. This increased priority on science at EPA would be consistent with recent proposals (for example, by the Progressive Policy Institute [Knopman 1996]) for EPA to take the lead on setting national-level environmental standards, conducting research and environmental monitoring, and providing technical assistance while devolving increased responsibility for program implementation to the states. It would, however, most likely come at the expense of federal pollution control programs. A potential downside of increasing the scientific portion of the EPA pie is that environmental program costs could be shifted to state, tribal, and local governments. In the long run, however, strengthening the scientific basis for environmental policy decisions should improve the cost-effectiveness of the national pollution control system.

A number of blue-ribbon panels have recommended expanding the agency's long-term research at the expense of short-term regulatory support (NRC 1995; U.S. EPA/SAB 1988, 1990). EPA, however, has a regulatory mission, and its science agenda should broadly support that mission. The Carnegie Commission found the need to strengthen ties between regulatory agency research and regulatory policy (Carnegie 1993), while the National Commission on the Environment argued for making the environmental research agenda more relevant to future policy decisions (NCE 1993).

Even with a substantial increase in EPA's science budget, requests for science from the agency's regulatory programs will continue to outstrip the agency's science budget, necessitating careful science planning, coordination, and prioritization. EPA should develop a strategic science planning process that dedicates a substantial portion of its science budget to regulatory decisions that will predictably arise within two to five years. A reasonable division of EPA's science budget would be to allocate a specific percentage (for example, one-third each) for the near term (under two years), midterm (two to five years), and long term (over five years). (The time frames apply to the regulatory decisions for which the science is intended and not necessarily to the duration of the research activities. A long sequence of short-term scientific tasks, for example, may be required to thoroughly prepare for a regulatory decision that is likely to occur five years hence.) With respect to EPA's current science resource allocation, dividing the budget equally into three parts would probably result in diverting resources from both the near-term and long-term science efforts into the intermediate range, which appears now to be the most neglected. The institutional control over the science resource allocation process could vary, with largely unshared control for the near- and long-term science resources and mixed control for the midterm science resources. Each of the program offices could control its own near-term science resources. The Office of Research and Development (ORD) could *husband* the agency's midterm science resources to ensure that program offices do not divert them for near-term activities, but the program offices should *set* the midterm science agenda. ORD could broker an interoffice priority-setting process and should ensure that the agency remains committed to allocating sufficient funding to regulatory decisions that are anticipated during the next two to five years. ORD should be held accountable for ensuring that this process is particularly transparent and that the outcome produces a trackable midterm science strategy that directly supports the agency's regulatory decisionmaking needs. ORD could control the agency's long-term science resources and set priorities for the long-term science strategy in consultation with the program offices.[2] As a result, if the science budget were divided into three equal parts, a full two-thirds of the agency's science budget would be earmarked directly for regulatory support, but half of that would be conserved to address decisions that will arise in two to five years.

It would be difficult to say under which category (near-, mid-, or long-term) each EPA science activity fits in this scheme. For example, activities related to substance-specific reassessments might seem to fit under all three categories. Therefore, there needs to be room for flexibility. Nevertheless, some generalities seem possible. Major components of the short-term science agenda would likely include analyzing available information and conducting short-term scientific studies for regulatory development and maintaining existing environmental monitoring networks for assessing environmental exposures and regulatory compliance. Major components of the mid- and long-term science agenda should include developing and updating testing procedures and risk assessment guidelines; upgrading environmental monitoring systems; conducting long-term toxicological, epidemiological, and ecological studies; reassessing risks on the basis of new scientific information; and evaluating program impacts on environmental trends and conditions.

PEER REVIEW

The Office of Research and Development previously had a unit responsible for providing internal, independent review of program office use of science in regulatory development. The Browner administration eliminated the review function of this office and folded it into an ORD organization that operates on an interoffice consensus basis. The shift was apparently in response to mutual antagonisms between the operational and oversight functions and is consistent with the Browner administration's removal of other internal regulatory review functions. (The change may have been intended simply to smooth tensions between the reviewers and the program offices, but it is also consistent with some notions of organizational reengineering.)

For the change to be *effective* requires more than trust in the judgment of the regulatory program political leadership. It also suggests a high level of confidence in the science analytic capacity of the regulatory program offices. EPA may be able to better manage the inherent tensions under the current arrangement, but the need for internal peer review of the acquisition and use of science in regulatory decisionmaking cannot be avoided for long. If the function is ignored, in all likelihood an egregious episode at some point in the future will cause it to be reinvented. The internal regulatory science review function should be reestablished and revamped before this occurs. Reestablishing ORD's independent science review function would provide a quality control measure over the use of science by environmental program and regional offices and should focus the EPA Science Advisory Board agenda on the most important regulatory science issues. It would also strengthen ORD's standing within EPA and thereby reinforce the office's ability to force science issues to top decisionmakers, particularly late in the rulemaking process when decisions are being finalized. (See discussion in Chapter 2, "The Web of Communications.")

The internal review process should require the Assistant Administrator for the Office of Research and Development to issue a notice to the EPA Science Advisory Board prior to EPA's issuing any major regulatory proposal, final rule, or administrative guidance.[3] The notice could take one of three forms analogous to a red flag, a yellow flag, or a green flag. ORD's nonconcurrence—a "red flag"—would trigger an SAB review of the scientific criteria for the decision or guidance. ORD, however, could not simply object to the program office's analysis in its notice to the SAB. To trigger an SAB review, ORD would be required to provide an alternative analysis. This alternative analysis would provide an initial basis for SAB's review.

A "yellow flag" notice would be intended to draw the attention of SAB and the Administrator's Office to significant scientific issues that have arisen in the course of the regulatory or guidance development. A yellow flag would signal that the agency should consider some additional form of external peer input or review. Delaying the decision under active consideration may or may not be warranted. For example, ORD might anticipate that the scientific issue would recur in future regulatory decisions, but it may be sufficient to seek independent peer input prior to those decisions without delaying the rulemaking under consideration. Alternatively, the situation might warrant pausing before setting a precedent. In response to a yellow flag notice, and considering other board and agency priorities, SAB could initiate an advisory activity, or the Administrator could request that SAB advise her on how to best address the issue.

An ORD notice of concurrence—a "green flag"—would not preclude an SAB review but would indicate that ORD does not recommend a more thorough external science review.

Permitting ORD to trigger an SAB review does raise a thorny implementation issue. That is, which office should pay for the incremental costs of such board reviews? One can imagine, however, various pitfalls to any blanket rule covering this issue. For example, requiring ORD to fund any SAB review that the office triggers could create disincentives for providing the board with adequate resources through the normal budgeting process. Therefore, it might be best to apportion the incremental SAB review costs through case-by-case negotiations brokered by the Administrator or her designee.

ENDNOTES

[1] The National Academy of Public Administration recommended training EPA decisionmakers in risk analysis and risk communication (NAPA 1995). EPA recommended regular scientific briefings for the Administrator (U.S. EPA 1992).

[2] NRC (1995) recommended that ORD have sufficient budgetary autonomy to conduct an expanded long-term research program.

[3] This recommendation is similar in some respects to one made by the Expert Panel on the Role of Science at EPA (U.S. EPA 1992), which recommended that the Administrator appoint a science adviser whose role would be analogous to that of the general counsel and who would not release a document unless it is judged scientifically defensible. NRC (1995) also recommended that the Administrator designate the Assistant Administrator for ORD as the agency's chief science officer responsible for overseeing science policy, peer review, and quality assurance. My recommendation, however, specifically places an analytical burden on the Assistant Administrator for ORD to exercise a limited authority over the use of science by program and regional offices.

REFERENCES

Carnegie (Carnegie Commission on Science, Technology, and Government). 1993. *Risk and the Environment: Improving Regulatory Decision Making*. New York: Carnegie Commission, June.

Knopman, D. 1996. *Second Generation: A New Strategy for Environmental Protection.* Washington, D.C.: Progressive Policy Institute.

NAPA (National Academy of Public Administration). 1995. *Setting Priorities, Getting Results: A New Direction for EPA.* A NAPA Report to Congress. Washington, D.C.: NAPA, April.

NCE (National Commission on the Environment). 1993. *Choosing a Sustainable Future*. Washington, D.C.: Island Press.

NRC (National Research Council). 1995. *Interim Report of the Committee on Research and Peer Review in EPA*. Washington, D.C.: National Academy Press, March.

U.S. EPA (U.S. Environmental Protection Agency). 1992. Memorandum from F. Henry Habicht, Deputy Administrator. Washington, D.C.: U.S. EPA, February 26. Mimeographed.

U.S. EPA/SAB (U.S. EPA Science Advisory Board). 1988. *Future Risk: Research Strategies for the 1990s*. EPA-SAB-EC-88-040. Washington, D.C.: U.S. EPA.

———. 1990. *Reducing Risk: Setting Priorities and Strategies for Environmental Protection*. EPA-SAB-EC-90-021. Washington, D.C.: U.S. EPA.

Appendix A

The 1991 Lead/Copper Drinking Water Rule

BACKGROUND

In response to the 1974 Safe Drinking Water Act (SDWA), in 1975 EPA promulgated an interim drinking water regulation for lead that set a maximum contaminant level (MCL) of 50 ppb (parts per billion). Like rules for other drinking water contaminants, the lead rule was not specific as to where or how the water should be sampled to ensure compliance. For most contaminants it did not seem to matter where drinking water samples were taken, since roughly the same concentration was measured at the tap as at the drinking water treatment plant. However, the primary source of lead is not lead in the source water but lead leached from the delivery system. Therefore, removal of lead from water at the treatment plant would not prevent drinking water contamination by plumbing.

In a June 1984 case, which dramatized the exposure of children to lead through drinking water, a routine blood test run on a 24-month-old Massachusetts girl revealed a blood-lead (PbB) level (that is, concentration of lead in the bloodstream) of 42 µg/dl (micrograms per deciliter). This level was considerably higher than the Centers for Disease Control's (CDC's) screening level of concern. After eliminating paint, furniture, food, yard soil, and toys as possible sources of the lead exposure, public health officials discovered that the family's tap water contained up to 390 ppb lead and was being contaminated by lead solder in plumbing in the newly constructed home (Stapleton 1994, 89–90).[1] By November 1985, EPA had proposed to lower the Maximum Contaminant Level Goal (MCLG) for lead to 20 ppb (*Federal Register* 50: 46936).

However, the agency's proposed revision was overtaken by events. In January 1986, Massachusetts instituted the nation's first ban on the use of lead solder in lines carrying drinking water (Stapleton 1994, 89–90). In

May 1986, the U.S. Congress weighed in through the SDWA Amendments, which banned future use of materials containing lead (for example, solder, flux, and pipes) in public drinking water systems and residences connected to them and limited the lead content of brass used for plumbing (for example, in fittings). The 1986 amendments also listed eighty-three contaminants, including lead, for which EPA was to simultaneously develop MCLGs and National Primary Drinking Water Regulations (NPDWRs). MCLGs are nonenforceable health-based goals, while NPDWRs (which include either MCLs or treatment and/or monitoring requirements) set enforceable standards. SDWA directs that MCLGs be set at a level at which "no known or anticipated effects on the health of persons occur and which allows an adequate margin of safety." The House report on the 1974 bill stated that if there is no safe threshold for a contaminant, the recommended MCL (the health-based goal) should be set at zero. Prior to the 1996 SDWA Amendments, SDWA directed the agency to set the MCL for a contaminant as close to the MCLG as is feasible (*Federal Register* 56: 26460).[2]

In August 1988, the agency proposed to revise the lead in drinking water MCLG, MCL, and NPDWR (*Federal Register* 53: 31516), and the rule was finalized in June 1991 (*Federal Register* 56: 26460). The final rule established some precedents, including setting an MCLG of zero for a noncarcinogen and establishing an action level triggering specified treatment, as opposed to an MCL. In the proposed rulemaking, EPA had proposed an MCL of 5 ppb measured at the drinking water plant, but the agency adopted the action level of 15 ppb measured at the tap in order to capture all sources of lead in drinking water. Some members of Congress, notably Rep. Henry Waxman (D-Calif.), and environmental groups protested EPA's decision to forgo setting a numerical MCL because they believed an action level would be more difficult to enforce. The rule was also one of the most expensive drinking water rules ever adopted by EPA, with compliance costs estimated at $500–$790 million per year (*Federal Register* 56: 26539).

The 1991 lead/copper[3] drinking water rulemaking can be separated into three parts: (1) the decision to adopt a multimedia strategy for controlling lead, (2) the determination of the MCLG, and (3) the formulation of the NPDWR. Science played a substantial role in each of the three components of the rulemaking. First, the control of lead in drinking water is appropriately viewed as part of one of EPA's first concerted attempts to regulate a chemical as a multimedia pollutant. The agency's multimedia strategy for controlling lead was based on the recognition that lead levels measured in blood and bone were the sum result of exposures and uptake from various sources (industrial and automobile emissions, lead paint, lead plumbing) and media (air, water, food, soil, and dust). Second, although EPA would typically consider its classification of lead as a probable human carcinogen as sufficient to warrant an MCLG of zero, lead's carcinogenicity was rendered subordinate in the drinking water decision. It was overtaken by a coalescence of scientific opinion that there is no discernible threshold for lead's noncancer effects and that the noncancer effects—in particular, impaired learning ability and delays in normal mental and physical development in children—were more serious than the potential cancer effects. Third, the realization that the major source of lead in drinking water was lead corroded from materials in household plumbing and the distribution system, in combination with the observation that measurements of lead in drinking water display great variability at the same tap over time, persuaded EPA to adopt an NPDWR designed as an action level requiring water treatment to reduce the corrosiveness of water in the distribution system, replacement of water service lines containing lead, and/or source water treatment.

The science of water chemistry made important contributions in formulating the treatment requirements, but there remained considerable uncertainties about the efficacy of corrosion controls and the contribution of lead service lines to lead levels at the tap. Risk analysis also played an indirect—but crucial—role in the development of the NPDWR because to the extent that EPA could use scientific information to quantify and monetize greater regulatory health benefits, the agency could justify a more stringent regime of treatment and monitoring requirements.

Table A-1 provides background for the development of national lead regulation and policy. Table A-2 provides a time line of the lead in drinking water rulemaking. Nichols (1997) and Levin (1997) provide additional background on the evolution of lead regulation and focus on EPA's development and use of economic analysis in rulemakings for lead in gasoline and lead in drinking water, respectively. In terms of establishing health-based goals, the lead drinking water regulatory development essentially "piggybacked" on the scientific analysis performed by and for the EPA air quality program. In fact, although the venue shifted from air to drinking water, there was considerable continuity in the "cast of characters" that took part in the debate over the agency's lead strategy. Ultimately, a committed band of policy entrepreneurs within the agency effectively mobilized the scientific data, analysis, and legitimacy accumulated in one compartment of the agency (the air program) and transported it to another (the drinking water program).[4] Due to prior exposure to the science of lead's health effects in relation to the air quality program, senior agency officials were familiar with and sensitized to much of the relevant scientific information when the drinking water decision arose. This preexisting sensitivity enabled the science—and those responsible for transporting

Table A-1. Background on U.S. Lead Regulation and Policy

1946	Episode in England where workers cleaning gasoline tanks suffered neurologic effects from high-level exposure to tetra-ethyl-lead.
1960s	Public health officials consider blood-lead (PbB) levels up to 60 µg/dl in children and 80 µg/dl in adults acceptable, based on acute clinical effects from occupational exposure studies.
1971	Lead-Based Paint Poisoning Prevention Act directs the Consumer Product Safety Commission to establish a level of safety for lead in paint. The Department of Housing and Urban Development is made responsible for removing lead-based paint from public housing.
1972	EPA inherits the Lead Liaison Committee from the Public Health Service.
1973	Clean Air Act initiates phaseout of leaded gasoline. (EPA negotiations with industry over lead standards in gasoline are the first opened to public scrutiny under the 1972 Federal Advisory Committee Act.)
1975	Lead-intolerant catalytic converters required on new automobiles.
	EPA sets an "interim" MCL for lead in drinking water.
	Centers for Disease Control (CDC) guidelines set screening level for children's PbB at 30 µg/dl.
1977	Clean Air Act Amendments add lead to criteria air pollutants.
1978	EPA sets lead national ambient air quality standard (NAAQS) of 1.5 $\mu g/m^3$ (designed to consider children's multimedia exposures to lead).
1979	Needleman et al. report study of nonacutely toxic lead levels on children's IQ.
1980s	FDA encourages domestic food industry to decrease use of lead-soldered cans.
1981	Ernhart et al. dispute Needleman's results.
1982	EPA begins review of lead NAAQS, convenes a Clean Air Science Advisory Council (CASAC) subcommittee to review dispute between Needleman and Ernhart. Review panel concludes that Needleman's study neither supports nor refutes the hypothesis that low or moderate levels of lead exposure produce cognitive or behavioral impairments in children.
1983	National Health and Nutrition Examination Survey (NHANES II) shows decline in population PbB levels resulting largely from reductions in leaded gasoline.
1985	EPA accelerates phaseout of leaded gasoline, reducing the limit from 1.0 gram/gallon to 0.5 gram/gallon, based on assessment of noncancer health effects.
	EPA proposes revised lead in drinking water standard.
	CDC screening level for children's PbB reduced from 30 µg/dl to 25 µg/dl.

Table A-1. Background on U.S. Lead Regulation and Policy *(Continued)*

1986	Safe Drinking Water Act Amendments ban future use of materials containing lead in public drinking water distribution systems and residences connected to them.
	Superfund Amendments and Reauthorization Act requires the Agency for Toxic Substances and Disease Registry (ATSDR) to prepare a study of lead poisoning in children.
	Lead in gasoline limit further reduced to 0.1 gram/gallon.
	EPA revises Air Quality Criteria for lead. CASAC accepts inclusion of Needleman's reanalysis supporting his earlier conclusions.
1987	Reports of studies in Boston (Bellinger et al.) and Cincinnati (Dietrich et al.) confirm effects of low-level lead exposure on children's cognitive development.
1988	Lead Contamination Control Act requires monitoring and recall of lead-lined tanks from drinking water coolers in schools.
	Lead-Based Paint Poisoning Prevention Act authorizes federal funds for community PbB screening.
	ATSDR report estimates that about 17% of U.S. children in 1984 were exposed to lead at levels that pose the risk of adverse health effects; suggests a potential risk of developmental toxicity from lead exposure at PbB levels of 10–15 µg/dl or lower; and identifies paint and contaminated soil as the principal sources of lead for children most at risk.
	EPA proposes Lead/Copper Drinking Water Rule.
1990	Clean Air Act Amendments prohibit leaded gasoline by 1996.
	EPA updates Air Quality Criteria and staff paper for lead, but NAAQS for lead remains unchanged.
	CASAC recommends a maximum safe PbB level for children of 10 µg/dl and concludes that there is no discernible threshold for lead effects.
	Superfund suit pits Needleman and federal government versus Ernhart and industry.
1991	EPA formalizes its multimedia *Strategy for Reducing Lead Exposures.*
	EPA finalizes the Lead/Copper Drinking Water Rule.
	CDC lowers its PbB screening level to 10 µg/dl and its guideline for medical intervention to 20 µg/dl.
1992	Residential Lead-Based Paint Hazard Reduction Act requires EPA to develop standards defining hazardous levels of lead in lead-based paint, household dust, and soil.
1994	Based on a finding that abatement of lead in soil above 500 ppm can achieve "measurable" reductions in children's PbB, EPA issues Revised Interim Soil Lead Guidance for CERCLA sites and RCRA corrective action facilities.
1995	EPA proposes maximum achievable control technology for secondary lead smelters under the Clean Air Act Amendments of 1990.

Table A-2. Time Line of the 1991 Lead/Copper Drinking Water Rule

1974	Safe Drinking Water Act requires EPA to establish health-based goals and enforceable regulations to protect public health with an adequate margin of safety.
1975	EPA sets an "interim" lead in drinking water maximum contaminant level (MCL) of 50 ppb.
1977	National Research Council's Safe Drinking Water Committee suggests EPA's interim MCL for lead may not provide a sufficient margin of safety.
	Office of Air and Radiation (OAR) develops first Air Quality Criteria for lead.
1982	OAR begins review of lead national ambient air quality standard (NAAQS).
1984	ORD/ECAO/Cincinnati's Health Effects Assessment for Lead, prepared at request of OSWER, classifies lead as a probable human carcinogen (group B2). (Note that OSWER has interpreted Superfund ARARs (applicable or relevant and appropriate requirements) for groundwater contamination to include drinking water MCLs.)
1985	ODW draft Water Criteria Document developed (never finalized).
	ODW proposes revised lead in drinking water standard, a maximum contaminant level goal (MCLG) of 20 ppb.
1986	Safe Drinking Water Act Amendments list lead among 83 contaminants for which EPA required to develop MCLGs and National Primary Drinking Water Regulations.
	OAR's revised Air Quality Criteria for lead suggesting no discernible threshold for indicators of lead exposure endorsed by Clean Air Science Advisory Committee (CASAC).
	OPPE issues *Reducing Lead in Drinking Water: A Benefits Analysis* (revised in spring 1987).
	Science Advisory Board (SAB) Subcommittee on Metals suggests that EPA's Drinking Water Health Advisory for PbB of 15 µg/dl is too high.
1988	Draft provision in Lead Poisoning Prevention Act that would have contained a drinking water MCL for lead measured at the tap is deleted.
	Lead Contamination Control Act calls for monitoring and recall of water coolers with lead-lined tanks, especially from schools.

and supplementing it—to have considerable access and impact in the lead in drinking water rulemaking.

The agency's multimedia lead strategy, formally articulated in 1991, evolved over many years during which a band of EPA staff, acting as policy entrepreneurs, had a potent influence on agency policy regarding lead. Although one component of the agency's lead strategy—the phase-

Table A-2. Time Line of the 1991 Lead/Copper Drinking Water Rule *(Continued)*

1988 *(cont.)*	ORD's Cancer Assessment Group issues *Review of the Carcinogenic Potential of Lead Associated with Oral Exposure*, prepared at request of ODW and OSWER, characterizing lead as a probable human carcinogen (Group B2).
	EPA proposes an MCLG for lead of zero with an MCL of 5 ppb measured at the source and corrosion control triggered at 10 ppb based on tap samples.
	Debate within EPA over contribution of lead in water to blood-lead levels.
	SAB Drinking Water Subcommittee concludes that neither a zero MCLG or an MCL of 5 ppb had been justified on a public health basis.
	SAB Executive Committee forms ad hoc Joint Lead Study Group, including members of Environmental Health Committee and CASAC, to review agencywide lead issues.
1989	CASAC reviews lead NAAQS staff paper and criteria document addendum.
	SAB Joint Lead Study Group agrees with B2 carcinogen classification for lead; finds inconsistencies among research, air, and drinking water programs regarding threshold PbB of concern and sensitive population; and recommends that EPA develop an agencywide lead strategy based on preventing adverse neurological effects in children.
1990	With disagreement over health-based goal largely put to rest by CASAC's finding of no discernible threshold for effects, debate centers on costs and benefits of alternative regulations.
1991	Rep. Henry Waxman (D-Calif.) fails to pass legislation mandating an at-the-tap MCL for lead.
	EPA uses unpublished Centers for Disease Control (CDC) report to estimate a relationship that drives a substantial portion of the estimated regulatory benefits. A later version of the report reduces the estimated relationship by a factor of three. The CDC report is never published.
	EPA issues final rule. MCLG for lead set at zero and NPDWR requires corrosion treatment, lead service line replacement, source water treatment, monitoring, and education triggered at 15 ppb based on tap-sampling scheme.

out of lead in gasoline—has resulted in what many analysts regard as perhaps the most cost-effective environmental regulation on the books,[5] the scientific basis for, and implementation of, the strategy continue to be controversial. A primary focus of the current controversy regarding EPA's lead strategy is the cleanup of sites with lead contaminated soils (*Science*, October 15, 1993, 323).[6]

SCIENTIFIC ISSUES

In general, virtually all of the respondents (90%) believed that there was adequate scientific information available in 1991 to inform the decision to revise the lead drinking water standard. Most (80%) responded that the *quantity* of scientific information was abundant to very abundant, though the health effects information was consistently viewed as more abundant than the exposure information. A less robust majority (70%) responded that the *quality* of the scientific information was good to very good. A slim majority (60%) responded that the level of scientific uncertainty was small to very small. Information regarding lead exposure from drinking water (as opposed to health effects at given PbB levels) was perceived as contributing most to the uncertainty.

Early public health PbB screening levels were based (with safety factors added) on the levels associated with clinically observable lead intoxication, primarily of occupationally exposed adults. Over the past twenty years, health researchers have become aware of a suite of toxic effects occurring with lead exposures considerably below those associated with acute poisoning. The National Research Council (NRC 1993) reports that the health effects noted at PbB of approximately 10 µg/dl include:

- impaired cognitive function and behavior in young children; [7]
- increases in blood pressure in adults, including pregnant women;
- impaired fetal development; and
- impaired calcium function and homeostasis[8] in sensitive populations.

NRC (1993) also concludes that somewhat higher PbB concentrations are associated with impaired biosynthesis of heme (a substance required for blood formation, oxygen transport, and energy metabolism) and cautions that some cognitive and behavioral effects may be irreversible.[9]

Because EPA conventionally assumes that the dose-response relationship for suspected carcinogens contains no threshold, the health-based goals (MCLGs) for suspected carcinogens in drinking water have been set at zero (in accordance with the 1974 House report language). In a position endorsed in 1989 by the Science Advisory Board (SAB), EPA has classified lead as a probable human carcinogen since 1984. However, lead's carcinogenicity was rendered subordinate in the drinking water rulemaking by a consensus that lead's neurological effects in children and other noncancer effects were more detrimental to public welfare than its carcinogenic effects. But until the lead rulemaking, the goals for drinking water contaminants regulated on the basis of noncarcinogenic effects had been set using the reference dose (RfD) approach. (The RfD presumes a threshold below which adverse effects would not be anticipated.) In some cases, however, the appearance of a threshold for noncancer effects, if one exists, may be limited only by our technical detection capabilities. Acknowledging this, EPA's 1986 air quality criteria document for lead found that there was no apparent threshold for biomarkers of lead exposure (U.S. EPA 1986), and in 1990, the Clean Air Science Advisory Committee concluded that there is no discernible PbB threshold for some lead health effects (U.S. EPA/CASAC 1990).

The prevailing scientific consensus regarding both the PbB level at which particular adverse effects can be expected to occur and the severity of those effects at particular PbB levels was bitterly fought over in scientific journals and courtrooms, and some lingering dissent remains within the scientific community. However, many scientists and analysts who were previously skeptical or uncertain about the validity of reports of adverse health effects from low-level exposure to lead have since come to conclude that the weight of evidence supports the earlier conclusions, particularly regarding neurological effects in children. Nevertheless, the lead in drinking water case needs to be considered in the context of the scientific controversy that coincided with the regulatory development.[10]

The Controversy over Neurological Effects in Children

In a 1979 issue of the *New England Journal of Medicine*, lead researcher Herbert Needleman of the University of Pittsburgh and his colleagues reported a drop in IQ of three to four points associated with "high" but nonacutely toxic lead levels measured in children's teeth (Needleman et al. 1979).[11] In 1981, Claire Ernhart, a psychologist at Case Western Reserve University in Cleveland, and her colleagues argued in *Pediatrics* that Needleman had not done an adequate job of controlling for confounding variables that might explain the differences in cognitive performance (for example, poor schools, parental neglect) and had performed so many comparisons that he was bound to come up with a few that were statistically significant merely by chance.[12] Ernhart and colleagues were also concerned that a large number of subjects had been eliminated from the analysis without a well-described exclusion procedure. Ernhart and colleagues suggested the lead effects were too small to be detected by a crude measure like IQ, except at some of the highest levels of exposure, just below acutely toxic levels.

When EPA began a review of the national ambient air quality standard (NAAQS) for lead in 1982, Lester Grant, director of ORD's Environmental Criteria and Assessment Office (ECAO) convened an ad hoc subcommittee of CASAC to review Needleman's and Ernhart's work. Needleman claims that his refusal to share his data with industry precipitated the formation of the review panel. At the same time, according to Needleman, Grant was under pressure in a similar case because a carbon monoxide[13] researcher would not, or could not, produce his data. The review panel concluded that Needleman had used questionable measures to categorize lead exposure and had not provided sufficient justification for excluding particular subjects from the study. They also expressed concern about missing data, and some of the statistical analyses Needleman employed, all of which led them to conclude that the results neither supported nor refuted the hypothesis that low or moderate levels of lead exposure produce cognitive or behavioral impairments in children.[14] The panel reached the same conclusions about two of Ernhart's papers, which they also criticized for methodological flaws (*Science*, August 23, 1991, 842–844).[15]

According to Needleman, EPA then provided funding for him to reanalyze his original data based on more appropriate statistical methods (regression rather than analysis of covariance). By the time the review panel's report was presented to CASAC, both Grant and CASAC were convinced that Needleman's original conclusions were accurate, and they became part of the 1986 lead air quality criteria document (*Science*, August 23, 1991, 842–844). Studies in Boston (Bellinger et al. 1987) and Cincinnati (Dietrich et al. 1987),[16] as well as others, independently reported a relationship between low-level lead exposures and cognitive function in children. Needleman contends that EPA responded to the controversy "timidly" by stating, "Even if we disregard Needleman's study, you get the same take home message." Thus, publicly at least, Needleman's work only contributed to the "weight of evidence" underlying EPA's lead policy. Notwithstanding EPA's public position, according to EPA officials, Needleman remained active and vocal in the process and was invited to brief drinking water program management. Two former senior EPA political appointees specifically recalled Needleman's work as being influential in the lead in drinking water decisionmaking.

Ernhart, however, continued to criticize Needleman's work and argue that the link between low-level lead exposure and neurological problems was being overstated. She also testified in favor of industry positions on phasing out leaded gasoline. In 1990 (in the interim between proposal and finalization of the lead in drinking water rule), the Department of Justice brought a Superfund suit against Sharon Steel, UV Industries, and the Atlantic Richfield Company. Each company had had a financial interest in a closed lead smelter in Midvale, Utah. The government intended to show that mine tailings on the site posed a health risk to children living in the area and hired Needleman as a lead expert. The corporations' lawyers brought in Ernhart as an expert witness and University of Virginia psychologist Sandra Wood Scarr, who had served on EPA's special review panel that had examined Needleman's and Ernhart's research.[17] After briefly reviewing Needleman's original data analysis printouts, Scarr concluded that Needleman's first set of analyses failed to show any relationship between lead level and IQ, and that only by rerunning the analyses, eliminating important variables that might also cause changes in IQ scores, did the statistically significant relationship show up (*Science*, August 23, 1991, 842–844).

The Superfund case was settled out of court before Scarr and Ernhart had presented their conclusions. (For more on the story of lead in soil at Superfund mining sites, see Appendix H.) Before the settlement agreement was announced, however, Department of Justice lawyers asked the court to force Scarr and Ernhart to return their notes on the Needleman data and refrain from speaking about what they had found. Contending that there was no good cause to suppress data gathered with public funds and that the government's request was an abridgment of First Amendment rights, Scarr and Ernhart fought the gag order and won. In turn, Scarr and Ernhart submitted their report to the National Institutes of Health Office of Scientific Integrity (*Science*, August 23, 1991, 842–844), leading to a University of Pittsburgh inquiry. Ultimately, Needleman was found guilty of sloppy statistics but cleared of scientific misconduct charges.[18]

A former SAB member remarked, "There was controversy about the Needleman data because industry-sponsored scientists charged that Needleman falsified the analysis. But there were ample international data that supported Needleman's conclusions. His statistics may not have been ideal, but you can't fault the conclusions that he drew." Responding to accusations that she was merely an industry mouthpiece, Ernhart claims she objected to Needleman's work before she began accepting research support from the International Lead Zinc Research Organization.[19] Scarr reportedly had no ties to the lead industry (*Science*, August 23, 1991, 842–844). A meta-analysis of lead studies reported by Needleman in 1990 (*Journal of the American Medical Association* 263: 673–678) is sometimes regarded as confirming his earlier work. However, meta-analysis is controversial, in large part because assigning weights to different studies is notoriously subjective (see Mann 1990). An academic concludes that "Needleman's meta-analysis was subjective, with no explicit rationale for the weighting of various studies."[20] According to Sue Binder, former chief of CDC's Lead Poisoning Prevention Branch, it is extremely hard to find unaligned lead experts: "They will all go to their graves thinking the other side is made up of total idiots" (*Science*, August 23, 1991, 842–844).

Though perhaps not entirely responsible for their differences, disagreements over values contribute to the Needleman/Ernhart feud. While Needleman's career has centered almost exclusively on lead (he chairs the University of Pittsburgh Medical Center's Lead Research Department),

Ernhart's research has addressed other risk factors in early childhood development, such as fetal alcohol exposure.[21] Although many environmental advocates, researchers, and officials are quick to dismiss or vilify Ernhart and her associates for fraternizing with industry, perhaps they should not be so quick to dismiss the *sentiment*—based on judgment, not science—that concern over low-level lead exposures may not warrant the same attention and resources as more manifest risk factors, such as poor nutrition, abuse, and random violence. As one academic commented, "There are plenty of risks that we know about that are certain and are certainly large. These are the things we should go after, rather than small, uncertain risks." Viewed from this perspective, the fight is over trade-offs and about priorities.

Effects Related to Blood Pressure

In the 1980s, EPA researcher Joel Schwartz (see discussion below) determined a relationship between blood lead and blood pressure among adult males in the United States on the basis of data from NHANES II, the second National Health and Nutrition Examination Survey (for example, Schwartz 1988).[22] An independent researcher reported, "There's a general consensus now among the nonoccupational hygiene folks that there's a blood pressure effect" due to lead.[23] However, the specific relationship between increased blood pressure and more serious cardiovascular outcomes (such as hypertension, stroke, heart attack, and mortality) remains highly uncertain. One reason that there is less confidence in the outcomes for effects related to blood pressure than there is for the neurological effects in children, according to the researcher, is that whereas the epidemiological studies of cognitive impairment in children measured the same children over time (longitudinal), the studies relating blood pressure to other effects compare different groups in a given period (cross-sectional). The effects of lead on blood pressure may also be hard to detect given the comparatively large effects of other variables, such as diet and smoking, on blood pressure.

The 1991 rulemaking did not attempt to quantify the effects of lead on fetal development and other health endpoints due to the uncertain state of the science. (See further discussion below concerning unquantified health effects.)

Exposure Analysis

In assessing the magnitude of risks associated with lead in drinking water, EPA had to evaluate the various sources of lead exposure, the occurrence of lead in drinking water, drinking water consumption patterns, and the water-lead-to-blood-lead relationship. Overlapping the debate over lead's health effects and the PbB threshold of concern was another regarding the relationship between PbB levels and lead exposures from various media, including air, food, paint, dust and soil, and water. As an EPA official formerly in the Office of Drinking Water (ODW) recalls, "Background levels of lead were lowered due to the phaseout of lead in gas, and population blood-lead levels were declining measurably. As a result, drinking water was a larger portion of a smaller total." While the lead in drinking water rule was being formulated, a multimedia uptake model developed by EPA as a tool for the lead NAAQS was available, though unvalidated.[24] Using this model, the agency estimated in the proposal that, on average, the typical drinking water contribution to total lead exposure for a two-year-old child is about 20% (U.S. EPA 1988). At the same time, ATSDR (1988) concluded that lead paint and contaminated urban soils were the main sources of lead exposure for children with PbB above the screening level (10 µg/dl).

Estimating exposure was complicated because survey data on the presence of lead in drinking water at the tap were crude and scarce, and data on drinking water consumption patterns were even more so. According to an EPA official, researchers that were key in the area of lead occurrence in drinking water included Peter Karalekas, formerly an EPA Region I (Boston) engineer, and Michael Schock, a chemist and metal corrosion expert then with the Illinois State Water Survey and now with EPA/ORD in Cincinnati (see, for example, Karalekas et al. [1976] and Schock and Wagner [1985]).[25] Given the complexity of water chemistry and its importance in assessing lead occurrence in drinking water, Schock's rare corrosion expertise was particularly valuable. According to an EPA official formerly in ODW, "Water chemistry is among the hardest to do. Water is the universal solvent, so everything is a contaminant, many of which then affect the physical properties of the water. How aggressive the water is in attacking lead in pipes depends on the water composition."

Given these complexities, modeling estimates of lead occurrence were of limited value, and quality monitoring data were most needed to accurately and precisely assess lead exposures via drinking water. In 1981, ODW reported a national study of lead levels in drinking water, but it was based on partially flushed (30 seconds) convenience (that is, non-probabilistic) samples (Patterson 1981). Early in the analytical process, there were limited metropolitan survey data regarded as statistically more reliable, including studies in Boston and Chicago. After the proposal, individual drinking water utilities submitted survey data as public comments, and in 1988, the American Water Works Service Company conducted a national survey. Although some respondents viewed the survey data as important, it did not yield a rock-solid foothold for the exposure assessment. A former ODW official commented, "The utilities conducted their own peer review, the agency hired external statisticians to evaluate the information, and there was further analysis by in-house staff. The agency spent a lot of time and money on it, and, ultimately, different people had different takes on it."

Drinking water consumption patterns were an important consideration in the exposure assessment because lead measurements in tap water display considerable variability. For example, lead from plumbing tends to accumulate in the standing water over time, so that the first flush from a faucet in the morning contains overnight accumulations. There is also substantial variability even with repeated samples (for example, same house, same tap, first flush). As a result, "any particular measurement should be taken very cautiously," concludes EPA biostatistician Allan Marcus, who contributed to the lead in drinking water analysis as an EPA contractor with Battelle.[26] The drinking water office, according to a former ODW official, contracted out for a "quick survey" on children's drinking water consumption patterns, but it was not used because it suggested that "children rarely drink the first flush from the tap in the morning." (On the other hand, the survey may have been disregarded because it was viewed as statistically unreliable.) In any event, the final rule did not specify the presumed underlying drinking water consumption pattern.

In addition to considering the drinking water consumption profile, another aspect of the exposure analysis is the relationship between the concentration of lead in drinking water and resultant blood-lead levels in various subpopulations. In the proposed rule, EPA relied on a 1983 study supported by the Food and Drug Administration that correlated lead in milk (canned formula and breast fed) with infants' PbB (Ryu et al. 1983) and the "Glasgow Duplicate Diet Study" (Lacey et al. 1985) to develop a "correlation" factor (that is, a nonthreshold linear relationship) for predicting PbB from drinking water lead concentrations. EPA's water-lead-to-blood-lead relationship developed for the proposal was criticized by public commenters. The agency contracted Allan Marcus to reanalyze the Ryu and Lacey studies, along with a study of Edinburgh school-age children (Laxen et al. 1987). Marcus developed a nonlinear relationship between water-lead and blood-lead levels in children that is more consistent with what is known about lead pharmacokinetics. In promulgating the final rule, EPA also concluded that it was better to rely on the Glasgow study for indicating responses among infants because it relied on exposure through drinking water (*Federal Register* 56: 26469). EPA relied on a British study (Pocock et al. 1983) for responses in adults. For responses in children older than six months, the final rule relied not on the reviewed Edinburgh study but on another study (Maes et al. 1991), which had been submitted for publication—and therefore not published in a peer-reviewed journal—by staff of CDC's Center for Environmental Health and Injury Control. The CDC study evaluated Hawaiians exposed to lead in drinking water. The Edinburgh and Hawaiian studies were noteworthy in that they both attempted to control for other sources of lead exposure (for example, dust and soil) in addition to lead from drinking water, but they provided differing estimates of the relationship between water lead and blood lead in older children. (See discussion below regarding these studies.)

THE PROCESS WITHIN EPA

Setting the Agenda

Several factors contributed to getting the issue of lead in drinking water on the agency's agenda. Certainly, it was part of an overall strategy and an extension of the agency's program addressing lead in air. An academic expressed the view that EPA's entire lead strategy was developed in response to the blow to the agency's reputation in the wake of the scandals of the early 1980s. In this view, "lead was a convenient target," and lead in drinking water was simply a continuation of the agency's attempt to burnish its image with a formula that had proven effective in the lead in gasoline phaseout. On the other hand, an agency insider who was skeptical of the health benefits from reducing lead in drinking water found the agency's lead strategy to be "an honest attempt by the administration to find the intersection between high-risk problems and things people care about."

The factors most frequently mentioned by respondents as keys to getting lead in drinking water on the agency's agenda were congressional pressure and a statutory deadline for review of the drinking water regulations imposed by SDWA. According to a former senior agency official, "My first meeting in [the agency] was with a congressional staffer who was concerned that we were very far behind our statutory obligation on drinking water standards. Of the eighty-three contaminants that Congress had set in the '86 Safe Drinking Water Act Amendments, only thirty-five had been regulated. Too many things were in the pipeline."

However, congressional forces had allies among a core group of agency staffers. According to a former ODW official and others, a band of EPA policy entrepreneurs was primarily responsible for pushing the lead in drinking water issue higher on the agency's agenda. Informally leading this group in the case of lead in drinking water was Ronnie Levin of the Office of Policy, Planning and Evaluation (OPPE). Levin's 1986 benefits analysis estimated substantial net economic benefits from reducing lead in drinking water (Levin 1986).[27] Joel Schwartz, who is Levin's husband and was also an OPPE official at the time, had been conducting statistical and economic analysis of lead epidemiological data for many years and, according to a former drinking water official, "did work that was so complex that it was unintelligible to most within EPA."[28]

According to press reports, an executive summary of Levin's analysis was made public in November 1986 to preempt another study being prepared by a group led by Ralph Nader (*Inside EPA*, November 14, 1986, 13). The public release of the benefits analysis sparked citizen group criticism and media attention of then Assistant Administrator for Water Larry Jensen's decision to conduct further analysis prior to promulgating revised lead drinking water standards. After an internal review of her first draft conducted by ORD, Levin produced a "draft final" benefits analysis that was released in December 1986 (and revised in spring 1987).[29] Internal allies allowed that Levin's regulatory analysis relied on limited exposure data[30] and her water chemistry analysis was "not as precise as some would have liked" (see further discussion below). Furthermore, approximately half of the monetized health benefits in Levin's analysis were associated with highly uncertain reductions in hypertension, heart attacks, strokes, and mortality in males aged forty to fifty-nine.[31] One EPA official speculates that the studies linking exposure to blood pressure in middle-aged men were salient to policymakers. "It was no longer them, those poor, highly exposed urban children; it became 'our' problem." Whether or not linking lead to cardiovascular effects in fact made the issue more salient to policymakers, and despite the warts and uncertainties in Levin's estimates, her analysis secured a spot high on the agency's agenda for lead in drinking water.

Assessing the Science

Apart from a 1985 ODW water criteria document that was never finalized, the risk analysis specifically for lead in drinking water was conducted primarily as input to the economic analysis. "The scientific work on lead was very well known to Levin and Schwartz from the lead in gasoline debate," according to a senior EPA official, "but the drinking water office was largely ignorant of the scientific findings on lead due to the agency being compartmentalized." The "stovepipes" delineating people along programmatic lines also "extended beyond the agency to the environmental community whose drinking water experts were unaware of lead issues." As a result, comments this official, "the drinking water office and the environmental community jumped on the bandwagon late in the process." Richard Cothern, a former ODW scientist, recalled that the office spent a couple of years reassessing lead before it became aware of the air criteria documents. A former ODW official laments that Bill Marcus, the office scientist responsible for the lead review, "was collecting rat data," mechanically following the protocol for developing an RfD from bioassay data. According to Cothern, once aware of the epidemiological data on neurological effects, and confident because "so much bright light had been shone on the studies by then," some of the ODW technical staff, including Marcus and himself, promoted lead on the drinking water agenda because they "felt that cancer got too much attention. We looked at lead . . . as a serious noncarcinogen."[32]

However, others in ODW were in no apparent hurry to tighten the standards for lead and other contaminants. A senior careerist stated that within EPA "the Drinking Water Office is regarded as captured" by the regulated community. As evidenced by the deadlines imposed by the 1986 SDWA Amendments, Congress grew impatient with the drinking water program's pace of issuing and revising regulatory standards. An academic remarks, "EPA's drinking water office is a peculiar institution. I don't understand how they get through conflict of interest guidelines. Members [of ODW] held positions in the AWWARF [American Water Works Association Research Foundation]. If someone in the air office were a member of EPRI [the Electric Power Research Institute], that would raise some eyebrows." If these perceptions are accurate, a possible explanation for the close relationship between the EPA drinking water program and its regulated community is that many drinking water utilities are government institutions with historic links to local public health agencies, and, traditionally, there has been considerable federal deference to local public health officials. (In fact, EPA owes its existence in no small measure to policy reformers who believed that state and local officials were allowed to exercise too much discretion in environmental health protection.)

Lee Thomas, who took over the agency during President Reagan's second term, replaced then Director of ODW Vic Kimm with Mike Cook, who had worked for Thomas previously in the Office of Solid Waste and Emergency Response. An EPA official formerly in ODW stated, "Cook got sick of his in-line staff not finishing any of the regulations they were working on." He assigned Jeanne Briskin, who came to ODW from OPPE as a special assistant, to lead the Lead Task Force charged with developing a proposed revision to the lead in drinking water regulations. While these personnel changes enabled the reconsideration of the scientific issues, they also resulted in an interpretation and use of science by people more inclined to regulatory action. "Personnel," as the adage says, "is policy."

An ad hoc organizational arrangement that coincided with the personnel shakeup also led to conflicting interpretations of the scientific data within EPA. Prior to initiating her lead in drinking water benefits analysis, Levin gained approval from Joe Cotruvo, then Director of the Criteria and Standards Division for ODW, and a member of Schnare's staff was assigned to be a liaison with OPPE, according to EPA officials. The arrangement took advantage of the expertise on lead issues OPPE had accumulated in the context of the air program. Unlike other program offices that often relied on OPPE for economic analysis, however, ODW had its own economic analysis unit, and David Schnare was the section chief. According to an EPA official formerly in ODW, Schnare's unit was viewed as being "an internal OMB" (Office of Management and Budget). During development of the proposal, a bitter struggle between Schnare and Levin within EPA paralleled that between Ernhart and Needleman outside the agency.

Regarding the agency's risk analysis, typical comments by respondents suggested that agency guidelines and standard operating procedures were not applicable: "This wasn't a cookbook exercise." "EPA didn't have guidance for noncancer risk assessment." In fact, however, there were standard operating procedures from which the agency departed in this case. As indicated above, the lead in drinking water rule eschewed the convention of a reference dose for noncancer effects based on bioassay data because there appeared to be no discernible threshold for lead's noncancer effects. In addition, the exposure analysis was unconventional. For example, the exposure duration described in the agency's *Exposure Factors Handbook* (EPA 1989) for the standard exposure scenario for residential drinking water is in units of days, and given the infrequency of drinking water sampling, a drinking water contaminant concentration would more typically be treated as a constant over a period of months or years. However, the concentration of lead in drinking water was known to vary markedly within days. As well, lead concentrations from pipes newly joined with lead-containing solder (that is, in new construction or as a result of repairs) remained high for a few years after installation. Thus, highly aggregated average concentrations would not capture these "pulses" in the lead exposure profile, and the standard operating procedure was inaccurate. The departure from standard procedure resulted in protracted haggling inside and outside the agency over the appropriate sampling measures (for example, first flush, second flush, fully flushed) and concentrations to employ for the analysis and monitoring of lead in drinking water.

According to an EPA official, behind this outwardly technical debate over the monitoring scheme were value judgments about the relative seriousness of false positives (Type I errors) versus false negatives (Type II errors). "There was a policy fight about how much to spend on monitoring." It boiled down to "how much uncertainty you are willing to live with," considering that costs started escalating at a sampling intensity that provided a fifty-fifty chance of a false negative. Conventionally, this rate of false negatives might seem unacceptable. On the other hand, if one's assessment of the science is that the standard provides a large margin of safety, then one might be willing to live with a relatively high rate of false negatives.

This debate over the monitoring scheme was an outgrowth of the agency's early decision to pursue a lead in drinking water control strategy based on required treatment actions instead of a numerical maximum contaminant level measured at the tap. According to an EPA research official, Michael Schock and his water chemistry colleagues in ORD laboratories in Cincinnati convinced members of the band of EPA policy entrepreneurs that they had made overly optimistic assumptions about the efficacy of water treatment technologies and that, consequently, their desired MCL was infeasible. The achievable MCL (at the tap) that the "water engineering faction" favored, however, was considered too high by the lead policy entrepreneurs, according to this source. Thus, to accommodate the limitations and uncertainty of the efficacy of the water treatment processes, the action level/treatment strategy was crafted as a staff-level compromise prior to the 1988 regulatory proposal.[33]

While virtually all of the respondents (90%) characterized EPA's treatment of the available science in the final rulemaking as good to very good, and a like number responded that the communication of the risk analysis to agency decisionmakers was very good, a minority felt that internal risk communications were poor as a result of the distorting influence of interoffice and interpersonal struggles. Others felt that the internal controversy, while unpleasant, forced many analytical assumptions to be disclosed and discussed at senior levels in the agency.

Something that went undisclosed to EPA decisionmakers, however, was a reanalysis of an unpublished CDC study (Maes et al. 1991) for evaluating the relationship between lead in drinking water and blood lead in children over the age of six months. In the final rule, EPA relied on an intermediate iteration of the CDC study that estimated that at water lead levels above 15 ppb (the water-lead level triggering the treatment requirements), a 1-ppb increase in water lead is associated with an increase of 0.06 µg/dl blood lead (*Federal Register* 56, 26470). However, a later version of the CDC study available in the EPA drinking water docket estimates the same relationship as 0.02 µg/dl blood lead—a figure lower by a factor of three.

According to a Public Health Service official, the CDC report underwent a number of iterations based on comments from statistical reviewers. An EPA official alleges that the final analysis of the CDC report that contained the 0.02 figure was available prior to the finalization of the drinking water rule, but that it was not published or shared with the EPA Lead Task Force because the people who had access to it and could block its publication as peer reviewers were members of the band of EPA policy entrepreneurs. The Public Health Service official confirmed that the report has never been published.

The version of the CDC report filed in the docket suggests that the reanalysis was concluded *prior* to the final rulemaking. The abstract concludes, "These findings could have implications for the regulatory standards of water lead levels, currently being revised by the Environmental Protection Agency" (Maes et al. 1991, 3). However, the date when the paper was logged in to the docket was not registered. There is also no addendum in the docket explaining the discrepancy with the figure used in the final rulemaking.

A member of the band of EPA policy entrepreneurs recalled reviewing the CDC report in draft and providing comments suggesting that the estimated coefficient should have been higher. (According to this source, the blood-lead level of the CDC study population was not equilibrated, and the authors had included too many separate variables in the statistical model in an attempt to control for the drinking water consumption rate.) This source stated that the authors of the CDC report could not accommodate their comments for reasons related to their statistical model. Also, this source claimed to be unaware of any discrepancy between the version of the report used in the final rulemaking and that in the drinking water docket.

Other sources have also criticized the CDC study design and analysis. The CDC analysis, for example, included both children and adults exposed to lead in drinking water. But this raises the question of why EPA used the earlier version of the study if it, too, was so flawed. In his review and reanalysis of the scientific literature as EPA's contractor, Marcus (1990) had concluded that the study of schoolchildren (ages six to nine) in Edinburgh, Scotland, provided "the most useful data set" for estimating the relationship of water lead to blood lead in older children. Like the CDC Hawaiian study, the Edinburgh study (Laxen et al. 1987) attempted to control for other sources of lead exposure (that is, dust) in addition to lead in drinking water. Marcus's reanalysis of the Edinburgh data estimated that at water-lead levels above the action level, a 1-ppb increase in water lead is associated with an increase of 0.03 µg/dl blood lead (Marcus 1990). This figure is closer to that in the later version of the CDC study (0.02) than to the estimate that EPA used in the final rulemaking (0.06). A reasonable interpretation of the various estimates is that they simply indicate that there is a range of uncertainty in the relationship between water lead and blood lead in older children.

Determining whether the earlier or later version of the CDC paper or the Edinburgh study is technically superior is beyond the scope of this report. What is germane to this study is that there was a reanalysis of an important study that, for whatever reason, was not available to the regulatory development working group or to decisionmakers. If the information had been available, it might have prompted some assessment of the effect of the uncertainty in the relationship between water lead and blood lead on the estimated health impacts of the regulation prior to finalization. This episode underscores the *potential* for key intermediaries within EPA to withhold scientific information from decisionmakers. This potential exists due to the multiple, overlapping roles of intermediaries as producers of scientific information, peer reviewers, regulatory development working group members, and policy entrepreneurs. In light of the availability of the reviewed and reanalyzed Edinburgh study, this episode also emphasizes the need for EPA to consistently apply independent peer review to studies that substantially affect major regulatory decisions.

Role of Agency Science Advisers

Most respondents concluded that nonagency scientists played a significant role in the decisionmaking process and in legitimizing the decision; however, some were clearly more influential than others. A few respondents regarded the SAB Drinking Water Subcommittee as unimportant in the process. The subcommittee, which was chaired by Gary Nelson of Purdue, did not endorse the 1988 proposal. Instead, the group that carried weight was the ad hoc Joint Lead Study Group. The group, chaired by Arthur Upton, was formed at the request of the SAB Executive Committee and included members of the Executive Committee, the Environmental Health Committee, and CASAC.[34] The group addressed the carcinogen classification of lead; pointed out differences among the research, air, and drinking water offices in target blood-lead levels and the definition of populations at risk; and recommended that EPA regulate lead exposures on the basis of neurological effects in children. The broad purview of the Joint Lead Study Group may have limited the extent to which it could conduct a detailed, comprehensive review of the proposed drinking water regulations.[35]

Officially, the drinking water panel, then a subcommittee of the SAB Environmental Health Committee, was requested by EPA to review the procedure for determining what an adequate tap sample is for measuring the drinking water lead concentration. However, the agency "didn't get much feedback from [the committee] on that," according to an EPA official. Instead, members of the drinking water panel commented that they found the health benefits analysis of the proposed MCL for lead "unconvincing" and characterized Levin's 1986 analysis as "sort of an advocacy type of document with a lot of stretching of notions."[36] According to one EPA official, "People [in EPA] blew off the Drinking Water [Subc]ommittee review. They [the subcommittee] made ridiculous noises. It [the benefits analysis] was presented to them as a *fait accompli*." Another EPA official responded, "Nobody on their [Drinking Water] [Sub]committee knew anything about the relevant subject matter. There were microbiologists, but no statisticians, no health effects experts, no corrosion control experts." Another succinctly summed the staff's approach to the drinking water panel, "We tried to bulldoze it." Eventually, it was agreed to let the SAB Joint Lead Study Group review the health effects. The group's "view of the health effects was going to take precedence," according to an EPA official. "We ended up supplementing the drinking water panel with Allan Marcus, our contractor," who provided the panel with statistical expertise.

An EPA staffer noted that the lack of endorsement by the Drinking Water Subcommittee "didn't make any difference. [EPA Administrator] Reilly didn't ask what the SAB thought of it." The response of one former senior EPA official was particularly telling, "Was the SAB involved on this? Did they approve or disapprove?"

However, another EPA official concluded that the lack of detailed external review at the proposal stage created problems later in the regulatory development process. "The Drinking Water Office tried to ignore the SAB Drinking Water [Subc]ommittee when it didn't tell us what we wanted to hear. They gamed it . . . to come up with a predetermined answer." However, it resulted in "wasting a lot of time later in the drinking water office. Because they didn't get proper review, they had to go back and review" the occurrence, exposure, and benefits analyses in response to public comments prior to promulgating the final rule. In the end, therefore, it appears that the agency paid a price for shortchanging the quality control function during the proposal stage. For its part, the drinking water panel suffered an incursion on its jurisdiction by commenting on an area (that is, benefits analysis) beyond its charge where it lacked acknowledged expertise.

Science through the Lens of Policy

In addition to ignoring the SAB drinking water panel, the band of EPA policy entrepreneurs overcame misgivings from colleagues in other agencies in pursuing their drinking water agenda. An EPA official remarked, "Some of the people in the lead health community who cared about the effects looked at lead in drinking water as a small exposure, a distraction. Vernon Houk from CDC [then Director of the Center for Environmental Health and Injury Control] railed against doing much in drinking water because he didn't want to disarm lead in paint." An NIEHS official who served as a member of the Joint Lead Study Group was also mentioned in this regard.

After the 1988 drinking water proposal, Jeff Cohen, who transferred from the Office of Air Quality Planning and Standards where he had managed the review of the lead NAAQS, took over the reins of the Drinking Water Office's Lead Task Force from Jeanne Briskin. Many of the exposure and benefits analyses were performed or revised under the direction of the task force after the proposal. Many of the assumptions used during the proposed rulemaking were revisited and replaced. According to an EPA official, this was especially true regarding the analysis of the effectiveness of corrosion control treatment and lead service line replacement in reducing lead levels at the tap. Also, the benefits of reduced cardiovascular effects from lead in adult males were characterized as less certain. Ronnie Levin continued to be involved from her new post in the Office of Research and Development's regulatory evaluation group, but the aggressive tactics she successfully employed in setting the regulatory agenda were less welcome in the negotiations leading up to the final promulgation. From the Office of Drinking Water, David Schnare continued to attack the assumptions underlying the estimated health benefits of the rule and highlight the expense of compliance with the complex, prescriptive regulations.

A former ODW official commented that the internal advocates of the lead regulation acted in an "extremely partisan" manner on the basis of nothing definitive but rather a "gut feeling" about the science. "History," the official believes, "has shown them to be right." However, a minority remains steadfast in opposition to the EPA policy entrepreneurs' foray into drinking water. The lingering disagreement might be construed as questioning whether the policy entrepreneurs within the agency "got the science right" or attempted to inflate the quantifiable health benefits of the rule in order to compensate for an inability to quantify other benefits to the satisfaction of reviewers.[37] Like the disagreement between Ernhart and Needleman, the dispute boils down to policy differences. Regarding the lead in drinking water benefits analysis, one EPA official notes, "Of the 23 million kids with benefits, 18 million of them got benefits of 0.1 IQ points. They were contending that by shifting the entire distribution [of children's PbB] slightly, that these small changes in the mean [of the distribution] are important. This is not an at-risk strategy. The way to manage risks is to attack the upper tail of the distribution!"

The lead in drinking water rule, however, was one component of the agency's multimedia lead strategy. In pursuing this policy, EPA took the *policy* position that "drinking water should contribute minimal additional lead to existing body burdens of lead" (*Federal Register* 56: 2469). According to one EPA official, "We decided that every source should contribute as little as possible" due to what the agency considered a narrow margin of safety between typical levels of lead exposure in the general population and those associated with adverse effects. Recognizing that drinking water was not the primary source of exposure for the children at greatest risk, the official acknowledges, "We knew that we weren't going to bring lead-poisoned kids relief. The drinking water rule was a general population strategy rather than an at-risk strategy." In this view, pursuing small increments in risk reduction aggregated over a large population is a legitimate risk management strategy and can be just as worthy as targeting large risk reductions for a small subpopulation under some circumstances.[38]

The success of EPA's band of policy entrepreneurs in setting the regulatory agenda in this case can be attributed in large part to an alliance of health effects scientists from ORD coupled with economists, statisticians, and policy analysts from OPPE. These offices and the technical disciplines from which they drew were able to trump drinking water experts inside and outside the agency. According to one EPA official, the drinking water staff has some "talented analytical chemists; the health effects area is where they are weaker." This primacy of health-related disciplines was also reflected in the pecking order and makeup of the agency's official science advisory panels. The complexity and statistical sophistication of the EPA policy entrepreneurs' regulatory analysis also appears to have been rewarded by considerable deference from decisionmakers and others without the time or ability to penetrate it.[39]

As a group of entrepreneurial staff that mobilized scientific arguments to advance a preferred policy, EPA's band of policy entrepreneurs is not unique. The principal reasons that such networks form are shared views and values and an agreement on the strategy of using science as a means to achieve the desired ends. If the direction of policy change that policy entrepreneurs seek to effect is guided by substantive concerns, they can represent a positive force in a political milieu that is responsive to constituent demands and ideological arguments. Such groups gain influence not only when their arguments are compelling on the merits but also when they resonate with policymakers. In this respect, the EPA policy entrepreneurs had the wind at its back: the population of primary concern was children, agency policymakers were familiar with lead issues, and the initial regulatory analysis had suggested substantial net economic benefits.

SCIENCE IN THE FINAL DECISION

Of those responding (seven), all interviewees agreed that the level of consideration by agency decisionmakers to the scientific issues was thorough to very thorough. The means of risk communication to agency decision-makers were diverse. According to an EPA official, "Every possible path—memos, options meetings, review meetings, policy meetings, briefings, et cetera—I've ever seen was used in that case." LaJuana Wilcher, who became Assistant Administrator for the Office of Water in 1989, was given staff briefings and traveled to ORD labs in Cincinnati to be briefed on corrosion research being conducted there. Deputy Administrator Henry Habicht was directly and deeply involved through a series of briefings and meetings. According to an EPA official, Administrator William Reilly got involved just before signing the rule because he was "hauled up before [Rep. Henry] Waxman's Subcommittee."

The factor most frequently cited by respondents as facilitating the consideration of science in this case was the agency's accumulated experience with lead issues in the air program. One EPA official recalled, "Agency decisionmakers had had a lot of time to think about the issues. We were able to benefit from the agency history on the issue with all of the experience on the air side. We didn't have to start at ground zero." Another agency official felt that the high stakes in terms of compliance costs associated with the rule "caused the decisionmakers to focus on the substance . . . and devote a lot of time to the substance of lead." But others pointed to the receptivity of decisionmakers, particularly Deputy Administrator Habicht, to being involved in substantive matters. Other factors mentioned as facilitating the consideration of science were the estimated net economic benefits of control, the internal advocacy and analysis roles of the band of EPA policy entrepreneurs, and the broad consensus among health scientists inside and outside the agency about the effects of lead.

A former senior EPA official found that "there was more science on lead than on most things that we regulated, direct epidemiological data on the relationship between blood lead and children's IQs." The fact that "the science was based on epi [epidemiological] data rather than animal studies . . . made it more reliable" in this official's judgment. Another former senior official commented that the decision was "driven by real-world epi evidence on lead" and stated that science played a particularly large role in the lead/copper rule. "There was adequate science to inform the decision to revise the standard," according to this official. "The big question was how far to go. Should all lead pipes be replaced, including the service lines? Was the science good enough to justify large expenditures?" The answer, of course, depended on the estimated benefits of the rule, and to the extent that the agency's use of science increased the regulatory benefits estimate, it could rationalize more stringent and costly monitoring and treatment specifications. During the final stages of the rulemaking, the lead/copper drinking water rule was detained by OMB reviewers, but, ultimately, in a departure from normal procedure, Deputy

Administrator Habicht signed off on the final rule as acting Administrator without OMB clearance.

Although the high stakes in the lead rule may have been responsible for causing some agency decisionmakers to focus on scientific matters, the factor most frequently cited as impeding a thorough consideration of the science was the high cost of compliance ($500–$790 million per year). A former senior agency official concluded that "the high level of emotion impeded the use of detached scientific information." Others pointed to the personalized controversies, that is, Levin versus Schnare inside the agency and Needleman versus Ernhart outside, and turf battles between ORD and OPPE, on the one hand, and ODW, on the other. Another factor was that much of the exposure and benefits analysis remained to be done after the proposal but prior to finalization, and, given the level of congressional scrutiny, the 1991 deadline was not one that could be missed without consequences.

CONCLUDING OBSERVATIONS

Deputy Administrator Habicht's willingness to engage in substantive debates effectively created high-level demand for analysis of the existing science. However, given the rate of turnover of agency political appointees (whose tenure is typically limited to two or three years), there are limits to the role of agency leadership in creating the demand necessary to generate original research. Basic research has the potential of filling data gaps in regulatory science analysis but often requires several years to complete. The short time horizon of politicos helps to explain the chronic failure of strategic planning for environmental regulatory science. In terms of a fate and transport analogy, there is frequently no feedback loop between the ultimate sources of science (that is, basic researchers) and the endpoints (that is, regulatory decisionmakers).

Continuing with the fate and transport analogy, an EPA official formerly in ODW observed that the epidemiological studies regarding neurological impairment in children exposed to lead that had been relied on so heavily in the air program formed "an existing body burden" of science in a separate "body compartment" within the agency that was unusual. This body burden was formed by a series of longitudinal epidemiological studies of children involving some measure of cognitive function that included the work of Needleman and colleagues, as well as studies by Bellinger et al. (1987) and Dietrich et al. (1987).

The major source of support for the domestic studies was NIEHS. According to an independent scientist, the institute's involvement in lead dates back to the early 1970s when NIEHS was under the directorship of

David Rall. This researcher suggested that EPA had little or no role in creating the demand for this science: "EPA wasn't in a position to be making demands on its sister agencies until the early '80s." EPA's initial "exposure" to the accumulating epidemiological database was through ORD's Health Effects Research Laboratory and Environmental Criteria and Assessment Office in Research Triangle Park, N.C. The information accumulated first in the air program "compartment" of the agency, and the further analysis, review, and legitimacy it received there made it both "bioavailable" and "potent." It was also in this context that agency decisionmakers first became exposed and "sensitized" to the information.

The basic science regarding the health effects of lead was supported by other agencies. However, EPA's official science advisers had to confer legitimacy on the science before it was available to the agency for use in regulatory decisions. EPA also had to expend considerable resources "metabolizing" (reanalyzing) existing or newly generated exposure information to render it useful for regulatory purposes. It also required the data gathering, analysis, and advocacy of a group of internal policy entrepreneurs to "transport" the information into the drinking water program and overcome "barriers" (in terms of standard operating procedures and culture) to "assimilation" in this compartment. At the same time, these barriers were becoming more permeable, promoting interchange of information among the agency's body compartments, as a result of high-level personnel changes stemming ultimately from external political pressures to increase the drinking water office's rate of developing and revising drinking water standards.

The controversies inside and outside the agency over the science appear to have had both distorting *and* illuminating effects. In addition, certain data either were never available or were "excreted" in the process. For example, ODW derailed plans for a national lead in drinking water survey under NHANES III. The "quick" (and presumably dirty) ODW drinking water consumption pattern survey was rejected. In some cases, data that were used were "selectively available." For example, both Needleman and Schwartz refused to allow their opponents access to raw data. Also, a CDC manuscript that could have lessened the estimated benefits of the final rule was unavailable to the regulatory development working group and to decisionmakers.

Figure A-1 illustrates the fate and transport dynamics for some of the key sources of scientific information in the lead in drinking water rulemaking.

Outside the agency, the 1986 SDWA Amendments ban on lead in plumbing reduced the incentives for the lead industry to mount a serious counterargument to applying the epidemiological data in the drinking water program area. Although the drinking water utilities generally

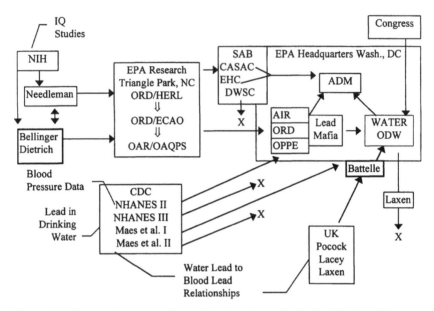

Figure A-1. Fate and transport dynamics for science in the lead in drinking water decision

protested the lead regulations, their ability to critique, minimize, or distort the epidemiological findings were extremely limited by the sector's lack of expertise on the health effects of lead and by the preexisting legitimacy conferred on the science by the environmental health scientists operating in the air program.

EPILOGUE

Conflict over the lead in drinking water rule continued after its promulgation. For example, shortly after finalization, the California Department of Health Services informed EPA that the state would not initiate the process for implementing the lead/copper rule due to a lack of resources (*Inside EPA*, November 22, 1991, 1). In response to a 1994 D.C. circuit court remand resulting from Natural Resources Defense Council and American Water Works Association challenges to the 1991 lead and copper rule, in April 1996, EPA proposed revisions to the rule to eliminate a number of requirements and to clarify conditions under which lead service line replacement would be required (*Federal Register* 60: 16347–16371).

ENDNOTES

[1] A reviewer notes that documented cases of lead poisoning from tap water date back to the 1920s and that this particular case was not reported in the scientific literature until 1989. This episode is not cited to imply that it motivated EPA to propose to revise the maximum contaminant level goal in 1985; however, Stapleton (1994) suggests that this particular episode dramatized the danger of lead solder in Massachusetts, which instituted the nation's first ban on lead solder in lines carrying drinking water shortly thereafter.

[2] Prior to the 1996 amendments, SDWA did not explicitly direct EPA to consider costs; however, on the basis of 1974 House report language and 1986 statements in the *Congressional Record*, EPA interpreted feasibility of MCLGs to mean that "the technology is reasonably affordable by regional and large metropolitan public water systems" (*Federal Register* 54: 22093–22094). While EPA followed this interpretation of feasibility in the case of lead in drinking water despite concerns voiced by some about the costs of regulatory compliance for small systems, the agency's decisionmaking regarding arsenic in drinking water has been greatly influenced by the projected costs that would be borne by small suppliers if the MCL for arsenic were substantially lowered. (See the arsenic in drinking water case study in Appendix B.) The 1996 SDWA Amendments direct EPA to consider a feasible MCL and define feasible as affordable to large systems but permit the agency to adopt another standard if the benefits of the feasible standard do not warrant the costs.

[3] EPA promulgated the rules for lead and copper concurrently because both occur in drinking water as corrosion by-products; however, the rulemaking for copper was relatively noncontroversial, and this case study focuses exclusively on lead issues.

[4] As used in the public policy literature (for example, Kingdon 1984), the concept of policy entrepreneurs refers to members of the policy community who blend the qualities of the expert analyst and the advocate to effect policy change, particularly through advancing ideas onto the policy agenda.

[5] See, for example, Portney (1990), 62–63.

[6] See Appendix H for a discussion of lead in soil at Superfund mining sites. Another issue of contention is the trade-offs between promoting recycling and limiting emissions from lead battery recycling.

[7] More recent studies by Needleman and others suggest an association between cumulative low-level lead exposures and delinquent behavior in youths (for example, Needleman et al. 1996).

[8] This refers to maintaining calcium levels at the appropriate dynamic equilibrium.

[9] In an unprecedented decision, acting National Research Council Chairman Robert White threatened to withhold the academy's endorsement and distribution of the report of the Committee on Measuring Lead in Critical Populations (chaired by University of Maryland toxicologist Bruce Fowler) because the draft

report addressed lead abatement, economic, regulatory, and policy issues outside the agreed-on scope of the study sponsored by the Agency for Toxic Substances and Disease Registry (ATSDR) (*Science*, July 30, 1993, 539). According to a member of the committee, the members took a "broad view" of how to address the ATSDR charge to the committee "and had the expertise to do so." This presumes, of course, that scientific expertise qualifies one to make judgments regarding regulatory policy issues. Ultimately, the NRC study was issued with slight modifications after ATSDR intervened (*Science*, July 30, 1993, 539).

[10] Some may feel that the discussion of scientific issues gives too much weight to the arguments of a few skeptical scientists. As indicated above, many scientists who were previously skeptical about the health effects from low-level exposure to lead have since come to conclude that the weight of evidence supports the earlier conclusions of other scientists. My intent here is, to the extent possible, to avoid making substantive scientific judgments that I am unqualified to render, but as a disinterested party, to present the scientific issues and debates that are germane to the case. Even if one takes the view that the charges were trumped up by parties with financial or personal interests (an argument that I am not making), it would be wishful thinking to believe that the controversy over the science was irrelevant to the 1991 lead in drinking water decision while one of the principal scientist/advocates in the field was under investigation for scientific misconduct by the Department of Health and Human Services Office of Scientific Integrity.

[11] According to Needleman, the bulk of his epidemiological studies were supported by the National Institute of Environmental Health Sciences (NIEHS), but his relationship with EPA began early. EPA first became aware of his work in the early 1970s via a "group of MDs" working in the Office of Research and Development (ORD) Health Effects Research Lab in Research Triangle Park, N.C. Needleman's first paper on tooth lead levels was published in 1972, and EPA sent Needleman to Amsterdam to present it. EPA later provided additional research support.

[12] To maintain the global Type I error rate (probability of a false positive), only a limited number of significance tests can be run. Consequently, if a researcher were to perform a large number of statistical tests, some positive results are expected to occur by chance. Ernhart et al. (1981) were supported by a grant from the National Institute of Environmental Health Sciences.

[13] Carbon monoxide is a criteria air pollutant for which ECAO would prepare criteria documents.

[14] Ernhart et al. (1993) cite the panel's conclusions from p. 38 of EPA report 600/8-83-028A. A member of the expert panel states that the panel's report remained interim and was not intended for citation. It is not uncommon for some agency reports never to be finalized, either because they are controversial, superseded by more current reports, or simply not worth the time and effort to finalize.

[15] According to Ernhart et al. (1993), Ernhart submitted all of her data to the panel. Ernhart's Cleveland study, which was funded primarily by the National Institute of Alcohol and Drug Abuse, hypothesized that prenatal alcohol and lead

exposures had an interacting effect on children's intellectual development, but the analysis did not confirm the effect. Although the inclusion of pregnant women with histories of alcoholism might have been regarded as a strength of the Cleveland study if it had detected a synergistic effect between lead and alcohol, according to an independent researcher, any *independent* effects of lead on children's neurological development may have been masked or swamped by effects of alcohol on fetal development, parental neglect, and so forth.

[16] According to an independent researcher, primary support for the Boston and Cincinnati studies was provided by NIEHS.

[17] Scarr has retired from UVA and is currently chief executive officer of Kindercare Learning Centers.

[18] The Office of Research (formerly Scientific) Integrity (ORI) of the Department of Health and Human Services (DHHS) accepted a University of Pittsburgh panel of inquiry's finding of no scientific misconduct on the part of Dr. Needleman. The ORI concluded that Needleman et al. (1979, 1990) inaccurately reported the methods and criteria for selection and exclusion of subjects, but found no resultant bias in the analytical results (ORI 1994).

[19] Ernhart's lead research also had support from the March of Dimes Birth Defects Foundation (see, for example, Ernhart et al. [1986]).

[20] In a more recent review of twenty-six epidemiological studies since 1979, Pocock et al. (1994) determined that there is a statistically significant association between children's blood lead and IQ, with a doubling of PbB from 10 µg/dl to 20 µg/dl associated with a mean reduction in IQ of around one to two points. The finding is perhaps most noteworthy because Pocock has been a noted skeptic of a causal relation between lead and IQ. Pocock et al. (1994) conclude that while low-level lead exposure may cause a small IQ deficit, "the degree of public health priority that should be devoted to detecting and reducing moderate increases in children's blood lead, compared with other important social detriments that impede children's development, needs careful consideration."

[21] In general, the community of lead researchers has been shaped to a great extent by a self-selection process in the academic community. An environmental lawyer suggests that as a result of sustained federal support for research into lead's health effects, it is one of the few areas of pollution control for which there is a sizable number of scientists not affiliated with industry.

[22] NHANES is conducted by CDC's National Center for Health Statistics. NHANES II covered 1976–1980. Schwartz first reported the analysis in a 1986 memorandum to ECAO's Lester Grant in relation to the lead air quality criteria.

[23] The field of occupational hygiene is viewed by some as being industry oriented.

[24] According to Needleman, the model has since been tested and does "a pretty good job in relation to empirical data." However, the validity of the model remains an issue under debate, says an EPA official. In large part, the question of model validation hinges on the extent to which the generic model has to be modified for site-specific applications. See Appendix H for further discussion of EPA's

Integrated Exposure Uptake Biokinetic Model for Lead in Children (or IEUBK model).

[25] Schock had been with ORD-Cincinnati prior to joining the Illinois State Water Survey. He later rejoined the EPA laboratory and was a member of the regulatory development working group on lead in drinking water after the 1988 proposal.

[26] Marcus currently works in ORD/ECAO.

[27] Levin (1986) estimated materials benefits from reduced corrosion damage ($525.3 million per year) alone exceeded estimated compliance costs ($239 million per year).

[28] Schwartz is currently an associate professor of biostatistics at the Harvard School of Public Health. Schwartz was also involved as an analyst in the particulate matter case study. See Appendix C.

[29] The internal review resulted in increased exposure and cost estimates, reducing Levin's benefits-to-cost ratio from seven-to-one to four-to-one but maintaining the net economic benefits at about $800 million (*Inside EPA*, December 12, 1986, 3).

[30] According to a former EPA official, one reason that Levin had to "scrounge around" for lead in drinking water exposure data was that EPA/ODW vetoed a survey. Although ODW had previously reported a crude national survey (Patterson 1981), during the mid-1980s while CDC's National Center for Health Statistics was planning the first phase of NHANES III (covering 1988–1991), the EPA Administrator's Office was convinced to commit additional funds for a more rigorous lead drinking water survey. However, says the former EPA official, the administration acquiesced when an ODW representative insisted that the funds be allocated to other research. An EPA official commented that the drinking water program management felt that sufficient information was available on the occurrence of lead in drinking water and that waiting for new survey data from the NHANES III study could have led to unnecessary delay in the rulemaking.

[31] The ODW final regulatory impact analysis (Wade Miller Associates and Abt Associates 1991) characterized the health benefits associated with cardiovascular effects as "conceivable" but not within the "estimated range" of benefits. According to an EPA official formerly in ODW, part of the reason that the office regarded the cardiovascular effects as "speculative" was that Joel Schwartz, who had conducted the analysis, would not permit access to the NHANES II data because Schwartz had accessed them in draft under an agreement that they would not be distributed. Instead of sharing the data with other analysts, according to a colleague, Schwartz responded to methodological criticism of his analysis by rerunning the analysis until his critics "ran out of objections." EPA's Information Law Division is unaware of any current agencywide policy regarding the internal sharing of scientific data that are not designated confidential business information.

[32] Marcus and Cothern presented a paper in May 1985 discussing the various noncarcinogenic effects of lead and the blood-lead levels at which they may occur (Marcus and Cothern 1985). They concluded that the lowest blood-lead levels at

which adverse effects occur is "about 10 μg/dL," the same level to which CASAC referred five years later.

[33] The 1988 proposal contained both an MCL measured at the source and a triggering level measured at the tap, but in the final rulemaking, EPA dropped the MCL measured at the source as an unnecessary complication.

[34] For the group's report, see U.S. EPA/SAB (1989).

[35] A member of the Joint Lead Study Group claims that the group did not sacrifice depth for scope, saying, "We had several intense days of thrashing out the issues." As discussed below, however, an EPA official concluded that the lack of detailed external review of the proposed lead in drinking water rule created problems later in the regulatory development process.

[36] See the Lead Industries Association's summary of the transcripts of the June 2–3, 1988, meeting of the Drinking Water Subcommittee in the EPA Drinking Water Docket.

[37] It is interesting to note that some spillover health benefits of corrosion control were not captured by EPA regulatory analyses. According to an agency research official, EPA realized that there could be some ancillary health benefits because implementing the corrosion controls would permit drinking water suppliers to use less chlorine to achieve the same level of disinfection and thereby reduce the formation of hazardous disinfection by-products. What EPA failed to recognize at the time, however, was that corrosion control would also lead to reduced microbial formation in the drinking water delivery system, for example, by reducing pitting of the pipes that provides microbes with tiny refuges where they are safe from contact with disinfectants.

[38] An EPA research official points out that the decision of which control strategy is adopted—attacking the upper tail of the risk distribution or shifting the entire distribution—has consequences regarding research design and risk assessment tools. EPA's Integrated Exposure Uptake Biokenetic Model for Lead in Children, for example, predicts changes in mean PbB that would be difficult to convert into changes in the numbers in the tail of the distribution of children at risk from lead exposure. This official observes that the agency is sometimes not explicit regarding which strategy it is pursuing, making it more difficult to develop data and tools that will be useful for decisionmaking.

[39] An EPA research official comments that those with the time and ability to penetrate the band of EPA policy entrepreneurs' statistical analysis also found it impressive. "It's not always clear when complexity is ornamentation and when it is necessary," says this official, "but there were good reasons for complexity in this case." A member of the band of EPA policy entrepreneurs observes, however, that the increased sophistication of EPA's use of scientific information has "created a black box" that makes the risk assessment process less transparent to regulatory program managers and policymakers. Thus, the increased complexity of risk assessment makes it easier for undisclosed assumptions to be buried, consciously or unconsciously, in the analysis and, in some cases, may shift decisionmaking power within EPA from program managers and policymakers to risk analysts.

REFERENCES

Bellinger, D., A. Leviton, C. Waternaux, H. Needleman, and M. Rabinowitz. 1987. Longitudinal Analyses of Prenatal and Postnatal Lead Exposure and Early Cognitive Development. *New England Journal of Medicine* 316: 1037–1043.

Dietrich, K., K. Krafft, R. Bornschein, P. Hammond, O. Berger, P. Succop, and M. Bier. 1987. Low-Level Fetal Lead Exposure Effect on Neurobehavioral Development in Early Infancy. *Pediatrics* 80: 721–730.

Ernhart, C., B. Landa, and N. Schell. 1981. Subclinical Levels of Lead and Developmental Deficit—A Multivariate Follow-Up Reassessment. *Pediatrics* 67(6): 911–919.

Ernhart, C., et al. 1986. Low Level Lead Exposure in the Prenatal and Early Preschool Periods: Early Preschool Development. *Neurotoxicology and Teratology* 9: 259–270.

Ernhart, C., S. Scarr, and D. Geneson. 1993. On Being a Whistleblower: The Needleman Case. *Ethics and Behavior* 3(1): 73–93.

Karalekas, P., et al. 1976. Lead and Other Trace Metals in Drinking Water in the Boston Metropolitan Area. *Journal of the New England Water Works Association* 90 (2, February).

Kingdon, J. 1984. *Agendas, Alternatives, and Public Policies*, Boston: Little, Brown, and Co.

Lacey, R., et al. 1985. Lead in Water, Infant Diet, and Blood: The Glasgow Duplicate Diet Study. *Science of the Total Environment* 41: 235–257.

Laxen, D., et al. 1987. Children's Blood Lead and Exposure to Lead in Household Dust and Water. *Science of the Total Environment* 66: 235–244.

Levin, R. 1986. *Reducing Lead in Drinking Water: A Benefit Analysis*. Draft Final Report. Washington, D.C.: Office of Policy, Planning and Evaluation, U.S. EPA, December (revised Spring 1987).

Levin, R. 1997. Lead in Drinking Water. In *Economic Analyses at EPA: Assessing Regulatory Impact*, edited by R. Morgenstern. Washington, D.C.: Resources for the Future.

Maes, E., et al. 1991. *The Contribution of Lead in Drinking Water to Levels of Blood Lead*. Prepublication draft available from the EPA Drinking Water Docket.

Mann, C. 1990. Meta-Analysis in the Breech. *Science* 249: 476–479.

Marcus, A. 1990. *Contributions to a Risk Assessment for Lead in Drinking Water*. Draft Final Report. Prepared by Battelle under contract to the EPA Office of Toxic Substances, June 15.

Marcus, W. L., and C. R. Cothern. 1985. The Characteristics of an Adverse Effect: Using the Example of Developing a Standard for Lead. *Drug Metabolism Reviews* 16(4): 423–440.

Needleman, H., et al. 1979. Deficits in Psychological and Classroom Performance in Children with Elevated Dentine Lead Levels. *New England Journal of Medicine* 300: 689–695.

————. 1990. The Long-Term Effects of Exposure to Low Doses of Lead in Childhood. *New England Journal of Medicine* 322: 83–88.

————. 1996. Bone Lead Levels and Delinquent Behavior. *Journal of the American Medical Association* 275: 363–369.

Nichols, A. 1997. Lead in Gasoline. In *Economic Analyses at EPA: Assessing Regulatory Impact* edited by R. Morgenstern. Washington, D.C.: Resources for the Future.

NRC (National Research Council). 1993. *Measuring Lead Exposure in Infants, Children, and Other Sensitive Populations.* Washington, D.C.: National Academy Press.

Patterson (J. Patterson Associates). 1981. Corrosion in Water Distribution Systems. Prepared for Office of Drinking Water, U.S. EPA.

Pocock, S., et al. 1983. Effects of Tap Water Lead, Water Hardness, Alcohol, and Cigarettes on Blood Lead Concentrations. *J. Epidemiol. Comm. Health* 37: 1–7.

————. 1994. Environmental Lead and Children's Intelligence: A Systematic Review of the Epidemiological Evidence. *British Medical Journal* 309: 1189–1196.

Portney, P., ed. 1990. *Public Policies for Environmental Protection.* Washington, D.C.: Resources for the Future.

Ryu, J., et al. 1983. Dietary Intake of Lead and Blood Lead Concentration in Early Infancy. *Am. J. Dis. Child.* 137 (September): 886–891.

Schock, M., and I. Wagner. 1985. Internal Corrosion of Water Distribution Systems. In *The Corrosion and Solubility of Lead in Drinking Water.* AWWA-RF/DVGW Report. Denver, Colo.: American Water Works Association.

Schwartz, J. 1988. The Relationship between Blood Lead and Blood Pressure in the NHANES Survey. *Environmental Health Perspectives* 78: 15–22.

Stapleton, R. 1994. *Lead Is a Silent Hazard.* New York: Walker and Company.

U.S. ATSDR (U.S. Agency for Toxic Substances and Disease Registry). 1988. *Nature and Extent of Childhood Lead Poisoning in Children in the U.S.: A Report to Congress.* Washington, D.C.: U.S. Department of Health and Human Services, July.

U.S. EPA (U.S. Environmental Protection Agency). 1986. *Air Quality Criteria for Lead*, Volumes I–IV and Addendum.. Research Triangle Park, N.C.: Environmental Criteria and Assessment Office, June.

————. 1988. Memorandum to Greg Helms, ODW, from Rob Elias, ECAO/RTP, regarding the Total Human Exposure Model for Lead, March 29.

————. 1989. *Exposure Factors Handbook.* EPA/600/8-89/043. Washington, D.C.: U.S. EPA.

U.S. EPA/CASAC (U.S. EPA Clean Air Science Advisory Committee). 1990. *Review of the OAQPS Lead Staff Paper and the ECAO Air Quality Criteria Document Supplement.* Washington, D.C.: U.S. EPA, January.

U.S. EPA/SAB (U.S. EPA Science Advisory Board). 1989. *Report of the Joint Study Group on Lead. Review of Lead Carcinogenicity and EPA Scientific Policy on Lead.* EPA-SAB-EHC-90-001. Washington, D.C.: U.S. EPA, December.

U.S. ORI (U.S. Office of Research Integrity, DHHS). 1994. Letter from Lyle Bivens, Director of ORI, to Dr. Claire Ernhart, Case Western Reserve University, March 3.

Wade Miller Associates, Inc., and Abt Associates, Inc. 1991. *Final Regulatory Impact Analysis of National Primary Drinking Water Regulations for Lead and Copper*. Prepared for Office of Drinking Water, U.S. EPA, April.

Appendix B

The 1995 Decision Not To Revise the Arsenic in Drinking Water Rule

BACKGROUND

Inorganic arsenic is a naturally occurring toxic substance found in drinking water supplies of the United States and other countries. Arsenic is also found in food, but most of the arsenic in food occurs in organic forms that are much less toxic than inorganic forms of arsenic (Abernathy and Ohanian 1992).[1] Prior to the development of synthetic pesticides in the 1940s (and for a considerable time thereafter), inorganic arsenical compounds (for example, lead arsenate and copper acetate arsenate, Paris green) were widely used in agriculture. Arsenic is still used commercially in wood preservatives and is also released into the environment as a result of smelting nonferrous metal ores, particularly copper. Because arsenic does not degrade in the environment, contamination from historical releases is cumulative.

In 1942, the Public Health Service (PHS) set a 50-ppb standard for arsenic in drinking water based on information about its acute poisonous effects.[2] In 1962, PHS recommended a limit of 10 ppb. In response to the 1974 Safe Drinking Water Act (SDWA), EPA adopted an "interim" standard of 50 ppb in 1975. The 1986 SDWA Amendments required EPA to finalize the maximum contaminant level goal (MCLG) and the enforceable maximum contaminant level (MCL) for arsenic by 1989. After the deadline passed, a group of citizens in Oregon, commonly referred to as the Bull Run Coalition, sued EPA for failure to comply. The agency negotiated, and missed, a subsequent series of court-decreed milestones for revisiting the standard. At each step, the agency cited the need for further research. In January 1995, Robert Perciasepe, EPA Assistant Administrator for the Office of Water, filed a declaration with the U.S. District Court of Oregon stating that the agency would be unable to propose a new standard for arsenic by the court-ordered deadline of November 30, 1995. Perciasepe cited remaining uncertainties in the health risk assessment for ingested arsenic and the arsenic control technologies and the "extremely high costs" that some regulatory options would impose on public water systems, "especially on a large number of smaller systems" (Declaration of Robert Perciasepe, Amended Consent Decree, *Donison, et al. v. EPA*, No. 92-6280-HO [and consolidated cases] U.S.D.C. Oregon, January 9, 1995).

Over the past twenty years, EPA has addressed other sources of arsenic in the environment. In 1978, EPA began a review of remaining inorganic arsenical pesticide uses.[3] By 1988, EPA had banned all pesticidal uses of inorganic arsenicals except wood-preservative products. Since September 1993, all agricultural uses of the inorganic arsenicals have been prohibited (www.epa.gov/docs/fifra17b/Arsenic_Acid.txt.html). Arsenic was also one of the original seven hazardous air pollutants (HAPs) listed by EPA between 1970 and 1984 under Section 112 of the Clean Air Act. In 1983, EPA proposed a national standard for arsenic emissions from copper smelters.[4]

Arsenic is also an important issue in EPA contaminated site and waste management programs. Arsenic is a key contaminant at many abandoned mining, milling, and smelting sites. Under current regulatory practices, current *or proposed* drinking water MCLs or MCLGs are frequently the operative Superfund remedial objectives for groundwater contamination.[5] The reference dose (RfD) for ingested arsenic provided in EPA's Integrated Risk Information System (IRIS), may be used in establishing remedial objectives for soil contamination. Arsenic also represents approximately half of the estimated carcinogenic potential in coal fly ash (SEGH 1994), a waste generated by coal-fired power plants. Since 1976, electric utility wastes have been exempted from the hazardous waste provisions (Subtitle C) of the Resource Conservation and Recovery Act (RCRA) (TNCC 1995). Under current hazardous waste program practices, however, there is a linkage between the MCL for a substance in drinking water and the RCRA regulations covering its treatment, storage, and disposal.[6] As a result of the considerable spillover effects that could result from changes in the MCL for arsenic (or the RfD for arsenic), the attentive regulated community is not limited to public drinking water suppliers.

Recent debate over the arsenic in drinking water standard also occurs in the broader context of the debate over "unfunded mandates."[7] The regulatory compliance burden on local governments from the Safe Drinking Water Act is a prime focus of this debate. Most of the public drinking water supplies that violate the current arsenic in drinking water standard of 50 ppb are well water systems serving fewer than 500 people.[8] Many of the public drinking water systems with high arsenic levels occur in western states with a tradition of resisting federal authority.[9] Also, the 1991 lead/copper drinking water rule was one of the most expensive drinking water rules ever adopted by EPA and one that did not consider the affordability for small systems (see Appendix A). While the agency was considering whether to tighten the standard for arsenic, a former EPA drinking water official points out, "We were taking a lot of crap over the lead in drinking water standard."[10]

Legislative negotiations on amending SDWA also began in earnest during the 103rd Congress, and, ultimately, the 1996 SDWA Amendments established a state revolving fund for drinking water investments (analogous to the existing revolving fund for wastewater treatment and surface water) and overhauled the process for selecting drinking water contaminants for regulation, as well as the criteria for standard setting.[11] However, during the legislative negotiations, controversies regarding four standards EPA had been working on—arsenic, radon, sulfate, and disinfection by-products—were helping to frustrate efforts to form the necessary legislative coalition. As a former EPA official recalls, "Industry and the local groups wanted arsenic to go away, but the environmental groups and Waxman's people [Rep. Henry Waxman (D-Calif.), former chair of the House Health and the Environment Subcommittee] were pushing arsenic." For EPA, the case of arsenic in drinking water had particularly ominous parallels to the proposed drinking water standard for radon, which Congress had severely criticized, because drinking water is not considered to be the major source of exposure for most people in either case.[12] If EPA were to propose to dramatically lower the level of (inorganic) arsenic permitted in drinking water, it would appear to the nonspecialist as being inconsistent with the higher levels of (mostly organic) arsenic that the Food and Drug Administration permits in seafood that it is responsible for regulating. A direct comparison, however, is inaccurate due to the differential toxicities of organic and inorganic arsenic forms. Recalling the radon debacle and mindful that such nuances tend to get overlooked in political debates, EPA drinking water officials might wonder, "Do we want to go through the same thing on arsenic?"

Finally, recent debates over arsenic in drinking water occur in the broader context of efforts to instill greater biological sophistication and realism into EPA risk assessment methods. Because there is some evidence that suggests that human metabolism may have some capacity to detoxify the naturally occurring metalloid, arsenic represents to some a potentially high-impact test case for departure from the agency's linear, no-threshold cancer risk model that has been the target of much external criticism. To others, arsenic is an equally important test case in establishing the hurdle that human epidemiological studies must clear in the absence of supporting evidence from experimental toxicology to form the scientific basis for environmental regulation.

Table B-1 provides a summarized background of the 1995 decision to pursue additional research rather than revise the drinking water standard for arsenic. Since Assistant Administrator Perciasepe petitioned the court in 1995, the 1996 SDWA Amendments required EPA to develop a plan for additional research on cancer risks from arsenic, propose a standard for arsenic by January 2000, and promulgate a final standard by January 2001.

SCIENTIFIC ISSUES

Responses among interviewees were evenly split (six yeas–six nays) as to whether there was adequate scientific information available in 1995 to inform a decision to revise the MCL for arsenic in drinking water.[13] According to an independent risk analyst, "Virtually everyone agrees that ingested [inorganic] arsenic causes human skin cancer, and there is persuasive evidence that it causes cancer in the bladder and other internal organs as well. The debate is on the shape of the dose-response curve" in the low-dose region. The split opinion arises from a disagreement about what EPA can or should do in the face of this uncertainty.

Interviewees who believe that the available evidence is at least adequate maintain that the information used to quantitatively estimate skin cancer risks from arsenic in drinking water is "as strong or stronger" than the scientific basis for most EPA regulatory decisions. An EPA official summarizes this argument, "We have human data and exposures in the range of general population exposure. We're not extrapolating from high doses based on limited animal studies."[14] Some of these respondents also point to the evidence of noncancer health effects of chronic arsenic exposure (such as vascular or neurological damage in the extremities, noncancerous skin lesions, and potential reproductive effects) that were not quantitatively estimated.

Interviewees who believe that the available evidence provides an inadequate basis for modifying the arsenic in drinking water standard claim that the strength of the human data from epidemiological studies has been mischaracterized and point to the evidence of the human body's capacity to methylate, and subsequently eliminate through excretion, some ingested inorganic arsenic. Methylation has been suggested as a detoxification pathway for arsenic that becomes increasingly inefficient or perhaps saturated with increasing exposure. While these respondents would agree that EPA often makes regulatory decisions on the basis of weaker scientific information, they believe that the available information is inadequate in the context of the arsenic in drinking water standard.

The principal scientific issues include (1) interpretation of epidemiological evidence of skin cancer, (2) the epidemiological evidence and assessment of internal cancers, (3) the possibility of an arsenic detoxification pathway in humans, (4) arsenic in drinking water exposure assumptions, (5) the lack of understanding of arsenic's mechanism of toxicity, and (6) interpretation of the evidence of arsenic's nutritional essentiality. There was general consensus among respondents that the greatest source of scientific uncertainty was the correct form of the dose-response curve (that is, linear or nonlinear, no threshold or threshold) for estimating the risk of cancer from low levels of arsenic in drinking water. However, only one respondent, an EPA water program scientist, believes that there is currently sufficient evidence to justify departing from the agency's default procedure of employing the no-threshold linear model. Another EPA scientist stated that arsenic is "the best case I know of where you might challenge the default. Whether it's good enough, I don't know."[15]

Epidemiological Studies of Skin Cancer

EPA relies on the results of animal studies to assess the toxicity of most regulated carcinogens. For assessing skin cancer risks of ingested arsenic, however, scientists have "lousy animal models," according to an independent toxicologist. "You wouldn't expect skin cancer in a rat because it's histologically different from humans" (that is, the anatomy of its tissues differ). Therefore, scientists have focused on human epidemiological studies to assess arsenic skin cancer risks.

One such study (Tseng et al. 1968) conducted in rural Southwest Taiwan where blackfoot disease[16] is endemic formed the cornerstone of the 1984 Office of Research and Development health assessment document (HAD) for arsenic (U.S. EPA/ORD 1984) and the 1988 Risk Assessment Forum (U.S. EPA/RAF 1988) arsenic report. The Taiwan database has considerable strengths. First, it is extraordinarily large.[17] This is important because epidemiological studies must be very large to detect even substantial cancer rates.[18] Second, given equal sample sizes, an epidemiological skin cancer study of an Asian population is more powerful (that is, it can detect smaller effects) than one of a Caucasian population because sun-induced skin cancer is relatively uncommon in Asian populations.[19] Third, the types of skin cancers (mainly on the extremities, rather than on sun-exposed surfaces) observed in the Taiwan study population are believed to be diagnostic of arsenic exposure. Fourth, the well water tested in the study area displayed a broad range of arsenic concentrations (<10–1,820 ppb), permitting coverage of the relevant concentrations of concern in the United States (<50 ppb), as well as higher levels that would be expected to result in higher than background levels of skin cancer. Fifth, the entire study group was physically examined and pathological studies confirmed over 70% of the observed skin cancer cases (U.S. EPA/RAF 1988). Therefore, there is high confidence that

Table B-1. Background on the 1995 Arsenic in Drinking Water Decision

1942	Public Health Service (PHS) sets a 50-ppb standard for arsenic in drinking water based on its acute poisonous effects.
1962	PHS recommends a 10-ppb standard; 50 ppb is grounds for rejection of the supply.
1968	Tseng et al. report association between arsenic in drinking water and skin cancer in Taiwan.
1975	EPA sets "interim" standard for arsenic in drinking water of 50 ppb.
1977	National Research Council (NRC) *Drinking Water and Health*, Volume 1, suggests the 50-ppb standard may not provide an adequate margin of safety.
1978	EPA initiates review of arsenical pesticides.
1980	NRC *Drinking Water and Health*, Volume 3, recommends further investigation of possible beneficial nutritional effects of arsenic at low doses.
	International Agency for Research on Cancer (IARC) concludes that there is sufficient evidence that inorganic arsenic in drinking water is a skin carcinogen in humans based on the Taiwan study.
	EPA Office of Water prepares draft arsenic water quality criteria document.
1981	Southwick et al. report Utah epidemiological study finding no cancer in a group of 145 people consuming drinking water containing arsenic levels of approximately 200 ppb.
1983	NRC *Drinking Water and Health*, Volume 5, concludes that U.S. epidemiological studies fail to confirm the Taiwanese results and states, "It is therefore the opinion of this committee that 0.05 mg/liter [50 ppb] provides a sufficient margin of safety...." In the absence of new data, arsenic should be presumed an "essential" nutrient for humans based on mammalian animal studies.
1984	EPA/ORD reports assessment of inorganic arsenic. A 50-ppb drinking water standard results in an upper-bound skin cancer risk estimate of 2%, based on Taiwanese database.
1985	EPA/ODW relies on evidence of essentiality of arsenic to human nutrition in a proposed rulemaking (*Federal Register* 50: 46959).
	Chen et al. report association between arsenic and internal cancers in Taiwan.
1986	SDWA Amendments require EPA to review arsenic in drinking water standard by 1989.
	SARA includes ARARs provisions, broadening implications of arsenic drinking water standard.
	NRC *Drinking Water and Health*, Volume 6, states EPA should consider metabolism and pharmacokinetics in assessing the risks of drinking water carcinogens.

Table B-1. Background on the 1995 Arsenic in Drinking Water Decision
(Continued)

1986 (cont.)	Peer review workshop of draft EPA/RAF *Special Report on Ingested Inorganic Arsenic.*
1987	NRC *Drinking Water and Health,* Volume 8: Pharmacokinetics in Risk Assessment.
1988	June 21: EPA Administrator Lee Thomas's memo permits managers to down-weight ingested inorganic arsenic risks by an "uncertainty" factor of ten because skin cancer is generally nonlethal.
	July: EPA/RAF *Special Report.* A 50-ppb drinking water standard results in a skin cancer risk estimate of 0.25%, based on Taiwanese database. Down-weighting yields skin cancer "risk" on the order of 10^{-4}.
1989	September: Bull Run Coalition of Oregon sues EPA for failing to meet 1986 SDWA deadlines.
	September 28: EPA/SAB/DWC recommends the agency revise its arsenic risk assessment to consider "the possible detoxification mechanism that may substantially reduce cancer risk from the levels EPA has calculated using a linear quadratic model fit to the Tseng [Taiwan study] data." Recommends that arsenic's nutritional essentiality should not be an influential factor.
1990	EPA agrees to propose arsenic rule by November 1995 and finalize the rule by November 1997.
	Various ad hoc groups inside and outside EPA begin to formulate arsenic research agendas.
1992	Smith et al. estimate 50-ppb standard represents U.S. internal cancer risks on the order of 10^{-2}.
	EPA sets a reference dose for arsenic noncancer risks using a range of values.
1993	EPA begins SDWA negotiations with Democratic 103rd Congress.
	Brown raises problems with Taiwanese epidemiological study dose reconstruction.
	Hopenhayn et al. find no consistent evidence for arsenic threshold hypothesis in humans.
	EPA/SAB/DWC finds an association between internal cancer and exposure to high levels of arsenic in drinking water but suggests evidence of nonlinear arsenic pharmacokinetics.
1994	November: Republicans gain majority in 104th Congress.
	December: Decision briefing for AA/OW Perciasepe at EPA headquarters.
1995	January: Perciasepe petitions USDC Oregon for more time to conduct research on arsenic.
1996	SDWA Amendments require EPA to develop a plan for additional research on cancer risks from arsenic and promulgate a final standard by January 2001.

the health effects have been accurately measured. (Most epidemiological studies rely on available or easily collected information [such as medical records, surveys, or public health registries] to estimate the incidence of health effects. Such data are generally inaccurate and often suffer from the bias of underreporting.)

Like all environmental studies, however, the Taiwan study has weaknesses. Dietary and food preparation sources and activity patterns that may have contributed to total arsenic exposure were not assessed, and the reported health effects of arsenic may also be confounded with those of a number of additional substances (for example, humic acids) found in the well water that were not taken into account.[20] Some have argued that other aspects of the Taiwanese study, while not deficiencies of the study per se, limit the relevance of the study to U.S. populations or its utility for estimating risks from low-level exposures. For example, the diet of the Taiwan study population was lower in protein and some amino acids than the typical U.S. diet, and these nutrients are required for biomethylation. The implications of this observation for the U.S. arsenic in drinking water standard, however, remain unclear.[21]

The main limitation of the Taiwanese data set for regulatory purposes, however, is the "dose reconstruction," according to an independent toxicologist. In the course of research conducted *after* the 1988 EPA arsenic reassessment was reported, independent statistical consultant Ken Brown discovered previously undetected problems with the Taiwan arsenic in drinking water exposure database.[22] While investigating the association between arsenic in drinking water and *internal* cancers in the Taiwan study population with Taiwanese investigator C. J. Chen using a different exposure database than had been used by Tseng and colleagues (see Brown and Chen 1993), Brown discovered that within Taiwan villages the wells often varied in arsenic concentration by a wide range.[23] Consequently, it was recognized that it was impossible to precisely estimate the levels of arsenic in drinking water to which individuals in Tseng's study population were chronically exposed prior to medical examination.[24] Measurement errors in the "dose reconstruction" would undeniably produce inaccurate skin cancer risk estimates; the rub is whether any resultant bias is severe enough to warrant discounting the best available human data.[25] The lack of precision in dose or exposure estimation is generally more serious for epidemiological studies than for toxicological studies.[26] There is a trade-off, however, between experimental control and relevance to real-world human health.

Other epidemiological studies in Germany, Mexico, Argentina, and Chile (reviewed in NRC 1983 and U.S. EPA/RAF 1988) also suggest an association between ingested arsenic and a variety of skin diseases, including skin cancer, but these studies were weaker than the Taiwanese study. U.S. EPA/RAF 1988 compared the predictions from the quantitative analysis of the Taiwan study to the observations in Mexico and Germany and concluded that the results were consistent. Three epidemiological studies conducted in the United States (in Alaska, Oregon, and Utah, reviewed in NRC 1983) failed to detect any positive relationship between arsenic in drinking water and disease. Southwick et al. (1981), arguably the best domestic study, did not find any statistical differences in cancer incidence or death rates in Millard County, Utah. However, the study's small sample size and the frequent occurrence of skin cancer in the United States limit the study's power to detect differences that could be considered substantial.[27] The principal strength of the epidemiological studies in developed countries is that medical records are better than those in developing countries.

Internal Cancers

U.S. EPA/RAF 1988 quantitatively analyzed the risks from arsenic in drinking water in terms of skin cancer, which rarely metastasizes (that is, spreads to other organs) and is rarely fatal in the United States. However, Chen et al. (1985, 1986, 1988) provided evidence from Taiwan of a relationship between arsenic in drinking water and internal cancers (bladder, kidney, colon, liver, and lung). U.S. EPA/RAF 1988 concluded that the summary data in Chen's published reports were insufficient to quantitatively assess dose-response for internal cancers. Applying a linear model to the raw Taiwanese data, Smith et al. (1992) estimated the risk of internal cancers in the United States due to consuming drinking water containing 50 ppb of arsenic to be on the order of 10^{-2} (one in one hundred), an order of magnitude *higher* than the Risk Assessment Forum's 1988 estimated skin cancer risk.[28]

In reviewing EPA's 1993 Draft Drinking Water Criteria Document on Inorganic Arsenic, the EPA Science Advisory Board's Drinking Water Committee[29] agreed that the Taiwanese data demonstrate an association between internal cancer and exposure to high levels of arsenic in drinking water. However, the panel pointed to evidence of nonlinearities in the pharmacokinetics of arsenic and to dietary sources of arsenic exposure in Taiwan (discussed below) that represent uncertainties in directly extrapolating the results to low-level drinking water exposures in the United States (U.S. EPA/SAB/DWC 1993).[30]

Detoxification

There is both human and animal evidence suggesting that methylation constitutes a detoxification pathway for ingested inorganic arsenic.[31]

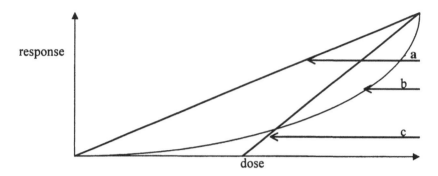

Figure B-1. Comparison of (a) no-threshold, linear, (b) sublinear, and
(c) threshold dose-response models

There is also some evidence suggesting that methylation becomes
increasingly inefficient or, perhaps, even saturated with increasing doses
of ingested arsenic. If there is a discrete point at which methylation
becomes saturated, then a threshold cancer model may be appropriate. If,
as seems more likely, methylation becomes increasingly inefficient with
increased exposure, then a nonlinear (sublinear) cancer dose-response
model may be appropriate. Both the nonlinear and the threshold models
(or a hybrid of the two resulting in a curve with an inflection point)
would indicate lower risks in the low dose region than would be estimated by a no-threshold, linear model (see Figure B-1).

"In the absence of adequate information to the contrary," EPA's 1986
"Guidelines for Carcinogen Risk Assessment" called for using a linear statistical model to estimate risks in the low-dose region by extrapolation
from the available data on higher doses (U.S. EPA 1986). The National
Research Council (NRC 1986, 1987) recommended that EPA generally
consider the absorption, metabolism, distribution, and elimination of
toxic compounds in the body (that is, pharmacokinetics or toxicokinetics)
when assessing drinking water carcinogens.[32]

In 1989, the EPA Science Advisory Board concluded "that at dose levels below 200 to 250 µg As^{3+}[trivalent arsenic]/person/day, there is a possible detoxification mechanism [methylation] that may substantially reduce
cancer risk from the levels EPA has calculated using [a] linear quadratic
model fit to the Tseng [Taiwan skin cancer study] data" (U.S. EPA/SAB
1989).[33] Some participants in the arsenic debate (for example, Beck et al.
1995; Carlson-Lynch 1994) have argued that the hypothesized saturable
detoxification pathway suggests that EPA should employ a nonlinear
shaped dose-response model. However, as indicated above, a majority of
respondents believe that the current scientific findings are insufficient to

override the default. A brief discussion of the most frequently cited evidence follows.[34]

Valentine et al. (1979) observed that blood arsenic was elevated only when drinking water arsenic was above 100 µg/l. Mushak and Crocetti (1995) argue, however, that the "threshold" observed by Valentine et al. may simply represent the detection limit of the analytical tests available at the time.[35] Buchet et al. (1981) monitored the urinary excretion of arsenic in four volunteers exposed to arsenite at dose levels of 125, 250, 500, and 1000 µg/day for five days and found that the arsenic methylating capacity was hampered in the two high-dose subjects. If the results are generalizable, this would suggest that total urinary excretion of arsenic may be compromised at high doses leading to increased deposition in body tissues. However, even at the daily dose of 1000 µg/day, the arsenic methylation capacity was not completely saturated. Consequently, Smith et al. (1995) conclude that "the evidence of *any* metabolic saturation from this study is not conclusive." Furthermore, there may be important differences among individuals in methylation efficiency that prevent generalization from four individuals to the general population. Failure to consider parameter uncertainties and interindividual variability in pharmacokinetic modeling can lead to misleading results, or at least, overconfidence in the results obtained.[36]

Work by How-Ran Guo and colleagues (for example, Guo 1993) has also been cited as evidence of a possible nonlinear dose-response relationship between arsenic exposure and urinary cancer.[37] The study is based on data collected from 243 townships in Taiwan (including a population of 11 million). As described by Beck et al. (1995), a strength of the Guo study is that it considers the distribution of arsenic concentrations in wells within a township, rather than simply applying the median arsenic concentrations for all wells in a village to represent exposure to each village inhabitant, as was done by Tseng et al. (1968). Guo's work reportedly has two principal shortcomings. First, the study used records of individual family ownership of wells to determine which families drank from which wells (*Risk Policy Report*, September 16, 1995, 11). According to an ORD scientist, well ownership is a crude surrogate for arsenic exposure.[38] Second, the study assessed health outcomes using death certificates and a Taiwan cancer registry begun in 1979 (*Risk Policy Report*, September 16, 1995, 11). The ORD scientist points out that death certificates do not permit an assessment of nonfatal cancers and suggests that the cancer registry is incomplete.[39]

While there is some evidence of methylation efficiency decreasing with increasing doses of ingested inorganic arsenic, Hopenhayn-Rich et al. (1993) conclude that current human studies do not support the methylation *threshold* hypothesis. Smith et al. (1995) elaborate that "if a methylation threshold for arsenic does exist, the epidemiological and experimental evidence suggest that it must be at exposure levels well above 2000 µg/day, making it completely irrelevant to usual human exposures." A former SAB member finds the current evidence of arsenic detoxification to be "highly suggestive of the value of further information, but not conclusive with respect to the regulatory decision." An industry scientist concurs, "There's not sufficient evidence available to depart from the low-dose linearity default. The disagreement is over whether it's worthwhile to try to answer that question." While there is considerable agreement among respondents that the existing evidence of arsenic detoxification does not meet the criterion of "adequate information to the contrary" required by 1986 guidelines to depart from standard operating procedures, it remains unclear what constitutes the necessary level of information.

Considering the evidence of a gradual decline in arsenic methylation efficiency with increasing dose levels, the scientific disagreement regarding arsenic detoxification may not be so much about the plausibility of a nonlinear dose-response curve for ingested arsenic as it is about what course of action to take in the absence of knowledge of the precise form of a hypothesized nonlinearity.[40] Although the current discussion in the larger scientific community tends to focus on the degree, if any, of nonlinearity in the dose-response relationship for arsenic, some of the combatants on both sides of the issue have caricatured the scientific debate as "threshold versus no threshold," frequently using the terms *nonlinear* and *threshold* interchangeably. As illustrated in Figure B-1, however, a sublinear dose-response curve need not indicate a threshold dose prior to toxicity.

For scientists who appreciate the distinction, the use of the term *threshold* may be a form of shorthand for "so nonlinear as to present a 'virtual' threshold, for all practical purposes." As opposed to the notion of a "virtual" threshold, however, the unqualified threshold hypothesis permits one to defend the current drinking water standard (or even higher levels) as presenting zero risk and needing no revision whatsoever. The uncompromising position of the "default camp" is summed up in the challenge presented by an ORD scientist, "If it is nonlinear, where is the threshold, where is the nonlinearity?" The poor signal-to-noise ratio in the low-dose region, however, may preclude defining a nonlinear curve form with absolute precision at *de minimis* risk levels.[41]

Exposure Assumptions

Risk estimates are often sensitive to changes in exposure scenarios. EPA generally assumes a drinking water consumption rate of two liters per day (l/day) for U.S. adults (U.S. EPA 1992), and ORD did so in its 1984 arsenic health assessment document (U.S. EPA/ORD 1984). Reasoning that people performing heavy labor outdoors in a subtropical climate would consume more water, the EPA interoffice Risk Assessment Forum assumed Taiwanese water consumption rates of 3.5 l/day for men and 2 l/day for women (U.S. EPA/RAF 1988). Later, in deriving the agency's reference dose for arsenic, an interoffice IRIS working group assumed Taiwanese drinking water consumption rates were 4.5 l/day (for men and women) (Abernathy et al. 1989).[42] Adopting a higher drinking water consumption rate for the Taiwanese than that which is applied to the U.S. population lowers the estimated risk from arsenic ingestion in the United States. However, there have been no specific measurements of water intake by persons in the Taiwan study area. According to an EPA official, "There was some limited anecdotal evidence that these individuals would consume more than the average for the U.S. Staff members with experience thought this was reasonable."[43] Indicating the sensitivity of the risk estimate to the assumed drinking water intake, Beck et al. (1995) calculate that by using a Taiwanese water consumption rate of 4.5 l/day, as opposed to 3.5 and 2 l/day for males and females, respectively, EPA's current U.S. skin cancer risk estimate would be decreased by approximately 50%.[44]

Similarly, if food contributes much more to inorganic arsenic exposure in Taiwan than it does in the United States, the estimated risk from arsenic in U.S. drinking water would be significantly reduced (Brown and Abernathy 1995). Using estimates of Taiwanese inorganic arsenic dietary intake, Yost et al. (1994) surmised that EPA's arsenic cancer risk estimate would be reduced by up to an order of magnitude. An ORD scientist, however, dismisses this evidence as flimsy. "We still don't have any data on food. ARCO paid to have" a very small number of rice and yam samples taken from Taiwan and analyzed.[45] As a result of the small sample size, the uncertainty about the inorganic arsenic contribution from food in Taiwan remains large.[46]

Mechanism of Toxicity

Many observers argue for a better understanding of arsenic's biological mechanism of toxicity in humans as a means of definitively reducing uncertainty in risk assessment. The lack of good animal or *in vitro* models for mechanistic research, however, presents a considerable obstacle.[47] According to an industry scientist, "Mechanistic information is the missing information, but arsenic is a mystery compound. It's not something that rips up DNA or is an [extremely powerful] clastogen."[48] Mass (1992) suggests that methylation of arsenic may itself present a biological dilemma, protecting against acute, short-term effects but possibly contributing to cancer over the long term.[49] As a result, according to an independent toxicologist, "research establishing the biomethylation pathway, while interesting, would not nail down whether low-dose nonlinearity occurs." An EPA official estimates that pursuing a mechanistic research agenda would take fifteen to twenty years. If the costs of delay are substantial (that is, if the high risk estimates for arsenic in drinking water are accurate), this would represent a disadvantage to waiting for definitive mechanistic results.

The time required to perform the necessary research depends, however, on the weight of evidence deemed by the *decisionmaker* as sufficient. Although mechanistic research tends to be fundamental and long term in nature, it is conceivable that a policymaker might accept as sufficient mechanistic evidence that would require less than fifteen to twenty years to produce. An independent risk analyst, for example, believes it may be possible that mechanistic research resulting in a test of the linear dose-response hypothesis for arsenic, something short of a complete mechanistic understanding, could be conducted within a few years. Of course, as indicated above, questions would remain about what precisely to do about the arsenic in drinking water standard if the linearity hypothesis were rejected.

Nutritional Essentiality

Numerous studies have indicated that arsenic is an essential nutrient for rats, hamsters, goats, and other animals. Extrapolating from these animal studies, a possible human arsenic nutritional requirement would be 12 µg/day (Uthus 1992). According to an independent toxicologist, however, "As often happens with some new elements that putatively show this [nutritional essentiality] in lab experiments, it's often hard to show this in humans." If arsenic is, in fact, both essential for human nutrition and a human carcinogen, there may not be much separation, if any, between essential and cancer-causing doses. This source points out that this overlap is not unusual for micronutrients. What is apparently unusual in the case of arsenic, is that the same chemical form that is carcinogenic may also be essential.[50] U.S. EPA/SAB 1989 concluded that attributing a prominent role to the essentiality of arsenic in human nutrition in evaluating health risks is unfounded as a consequence of the lack of convincing evidence in humans.

THE PROCESS WITHIN EPA

Setting the Agenda

The 1989 suit filed by the Bull Run Coalition against EPA for failing to comply with deadlines under the 1986 SDWA amendments was the event most frequently mentioned (five) by respondents as being responsible for forcing arsenic in drinking water on the agency's regulatory agenda. However, activity within the agency dates back to the review of arsenical pesticides begun in 1978. Over the twenty years since EPA set its "interim" drinking water standard, a number of factors elevated arsenic's position on the drinking water regulatory agenda. The 1984 and 1988 skin cancer risk estimates focused attention on arsenic but caused the drinking water office, according to a former staff scientist, to question, "Is it really that bad?" An independent toxicologist noted that the reported findings of internal cancers (Chen et al. 1985) "notched up the issue." The subsequent U.S. risk estimate for internal cancers (Smith et al. 1992) caused an even greater sense of urgency that penetrated the drinking water program. According to a former EPA official, the conclusion by the EPA Risk Assessment Forum in 1988 that Chen's 1985 report on internal cancers was insufficient for quantitative risk assessment "became the basis of the agency twiddling its thumbs until the Smith report came out in 1992."

The Health Assessment Process

The process of assessing the health risks of arsenic in drinking water pits risk assessors in the EPA Office of Water (OW), who appear to support the current standard or something close to it as reasonable, against risk assessors in the Office of Research and Development headquarters, who estimate that the arsenic standard does not provide the same level of health protection provided by many other EPA regulations. The ORD risk assessors, among others within EPA, appear concerned by the prospects of case-by-case departures from the agency's default risk assessment procedures, which represent the implementation of EPA science policy statements, and from the standards of scientific proof that the agency customarily requires to take regulatory action.[51]

With limited scientific resources at their command, the EPA water program analysts have enlisted the aid of the agency's science advisers and leveraged the resources of the regulated community to challenge ORD's arsenic health assessment. The OW staff also found an ally within ORD in the Health Effects Research Lab in Research Triangle Park, North Carolina. According to a former EPA official, there were some staff within the EPA Office of Ground Water and Drinking Water who advocated making the arsenic standard more stringent. However, ORD's principal internal ally in the process appears to be the Office of Prevention, Pesticides, and Toxic Substances (OPPTS). ORD scientists have staked the legitimacy of their analysis on the interoffice Risk Assessment Forum (RAF) "consensus-building" process that approved the 1988 assessment. Three-quarters of the responses (nine) rated EPA's treatment of the available scientific information as good to very good (independent of the quality of the underlying data). However, some respondents distinguished between the water program's analysis (U.S. EPA 1992) and the ORD analysis endorsed by the RAF. (One respondent rated the water analysis as good and the ORD analysis as poor.) Most respondents viewed the ORD analysis as representing the agency's "official" scientific assessment, but some respondents were unaware of the existence of the competing OW analysis.

The 1974 SDWA (Section 1412 [e]) required EPA to enlist the advice of the National Academy of Sciences to identify health effects associated with specified drinking water contaminants and research needs. The apparent intent was to have the academy conduct an independent scientific assessment for EPA to use in developing drinking water regulations. In 1977, the first report issued by the National Research Council's Drinking Water and Health Committee suggested that the 50-ppb drinking water standard for arsenic may not provide an adequate margin of safety (NRC 1977).[52] However, in 1980, the NRC committee changed its posture somewhat, recommending further investigation of possible beneficial nutritional effects of arsenic at low doses (NRC 1980).

In the same year, the EPA Office of Water developed ambient water quality criteria documents covering sixty-five contaminants, including arsenic (U.S. EPA/OW 1980).[53] This early arsenic criteria document, according to an independent toxicologist, never got out of draft.[54] This source suggests that it was during this exercise that senior careerists in Office of Drinking Water (ODW) were exposed to and may have become skeptical of the Taiwan skin cancer study (Tseng et al. 1968). In the same year, however, the International Agency for Research on Cancer concluded on the basis of the Taiwan study that there was sufficient evidence that ingested inorganic arsenic causes human skin cancer (IARC 1980). This international scientific consensus presented a challenge to those who viewed the interim drinking water standard as sufficiently protective. When Southwick et al. (1981) reported no association between arsenic in drinking water and cancer in a Utah population, their EPA-supported study provided an opportunity for another review of the epidemiological evidence of the adverse health effects of arsenic in drinking water.

As part of its fifth report (NRC 1983), the NRC Drinking Water and Health Committee conducted a review of the epidemiological evidence and concluded that the U.S. studies failed to confirm the Taiwan study. Shifting from its earlier questioning of the protection provided by the interim arsenic standard, the committee stated, "It is therefore the opinion of this committee that 0.05 mg/liter [50 ppb] provides a sufficient margin of safety. . . ." Building on its 1980 recommendation, the committee also recommended that, in the absence of new data, arsenic should be presumed an "essential" nutrient for humans based on the results of multiple mammalian animal studies. The charge to the NRC committees is negotiated between the academy staff and the sponsoring institution. Joe Cotruvo, then Director of the Office of Drinking Water Criteria and Standards Division,[55] and William Marcus, a drinking water program staff scientist, served as EPA project officer and liaison, respectively, to the NRC Drinking Water and Health Committee. The committee's membership changed over time, but members who advocated replacing risk assessment default models and assumptions with more biologically sophisticated methods were prominently represented.[56]

Taking advantage of scientific material accumulated by the pesticides and air toxics program offices, ORD issued its health assessment document (HAD) for inorganic arsenic in 1984 (U.S. EPA/ORD 1984).[57] Based on the Taiwan study, the HAD estimated that the 50-ppb arsenic in drinking water standard resulted in an upper-bound skin cancer risk estimate of approximately 2×10^{-2} (or 2%), much higher than the 10^{-4}–10^{-6} (1-in-10,000 to 1-in-1,000,000) range to which EPA customarily regulates health risks.[58] Rejecting ORD's assessment, ODW relied on the evidence of arsenic's nutritional essentiality in a 1985 proposed rulemaking (*Federal Register* 50: 46959). Congress interceded the next year with the SDWA Amendments, which set a 1989 deadline for EPA to review and finalize drinking water standards for eighty-three specific contaminants, including arsenic.

However, the 1984 RCRA Amendments and SARA of 1986 expanded the arsenic "scope of conflict" to include the hazardous waste community. Until this point, the capability of the EPA drinking water program to counter ORD's arsenic in drinking water assessment was limited by its own modest scientific resources and those of the drinking water suppliers. A by-product of the hazardous waste legislation was that it enabled ODW and the drinking water suppliers to subject ORD's arsenic assessment to increased scientific scrutiny by capitalizing on the resources of new stakeholders—companies with hazardous waste liabilities (for example, ARCO) and the research arms of affected industrial sectors (for example, ILZRO and EPRI, the Electric Power Research Institute).

ODW also extended the scope of the scientific debate by focusing on the role of pharmacokinetics in risk assessment. In 1986, the NRC *Drinking Water and Health* report recommended that EPA consider pharmacokinetics in assessing the risks of drinking water carcinogens. A year later, the committee published its final report, the proceedings of a conference on pharmacokinetics in risk assessment (NRC 1986, 1987).[59] Thus, ODW scientists had an external source of legitimacy for challenging the agency's no-threshold, linear cancer risk model. According to a former EPA official, the water program's health risk assessors "needed a smoking gun before they saw anything as a contaminant" and are "pushing for arsenic as a test case" to establish a precedent for a "threshold carcinogen." The threshold issue was particularly salient to the drinking water program, because consistent with its no-threshold default cancer model, the agency has implemented SDWA's goal of preventing any known or anticipated health effects with an adequate margin of safety by setting the MCLG at zero for all drinking water carcinogens.[60] (The threshold issue may be less central to standard setting now because the 1996 SDWA Amendments authorize EPA not to promulgate the feasible MCL if the benefits do not warrant the costs. However, the estimated benefits of regulatory controls at low-dose levels will continue to be sensitive to the presumed form of the dose-response curve.)

Meanwhile, in response to interoffice disagreements, EPA had referred revision of the 1984 ORD arsenic HAD to the Risk Assessment Forum for reassessment. RAF is designated as the agency's interoffice forum for resolving scientific disputes. Its membership in 1986–1987 consisted of four scientists from ORD, four from the Office of Pesticides and Toxic Substances (OPTS), and one each from OW and Region 1. The 1986–1987 RAF was chaired by Peter Preuss, former director of the ORD Office of Health and Environmental Assessment (OHEA). The principal authors of the 1988 RAF report were Herman Gibb and Chao Chen (ORD/OHEA); Tina Levine, Amy Rispin, and Cheryl Siegel-Scott (OPTS); and William Marcus (ODW).[61] Gibb and Chen were primarily responsible for the quantitative risk estimation and contracted with Ken Brown, then a statistician at Research Triangle Institute.[62] Gibb is the EPA scientist most commonly associated by respondents with the RAF-endorsed reassessment.

In December 1986, a draft report was peer-reviewed at a public workshop at which the statistical reviewers endorsed the quantitative analysis of the Taiwan study. A revised draft was presented to RAF in March 1987, but the forum was unable to bring the reassessment to closure by consensus. In July, the final report was elevated to the EPA Risk Assessment Council (RAC), a management level interoffice forum for resolving science policy disputes within the agency. ODW requested that RAC address the issue of whether the risk of skin cancer from ingested arsenic should be down-weighted (that is, downgraded or depreciated) relative to internal cancer risks because skin cancer is generally treatable and nonfatal. There appears to have been little scientific basis, however, for deriving the proposed down-weighting factor (that is, dividing the skin cancer risk by ten). An ORD scientist charges, "The drinking water office would never admit it, but in the original [1988] risk assessment, they said, this is much too high, we've got to do something about this. Let's say that skin cancer is fatal 10% of the time." A former EPA drinking water official allows, "The Office of Drinking Water's assumption that 10% of skin cancers were health-threatening was arbitrary."

The ORD scientist points to an inconsistency in down-weighting skin cancer relative to internal cancers. "Other cancers, including some internal cancers, are not 100% fatal. Should we be applying [down-weighting] factors to all nonfatal cancers?" Noting the inconsistency and precedent-setting potential, RAC responded to ODW that the agency would not explicitly identify the factor of ten as down-weighting due to nonlethality. Instead, it was negotiated that the factor of ten would be called an "uncertainty factor" that would not be part of the risk assessment but would be presented in a memo from Administrator Lee Thomas to EPA offices.

In its charge to the SAB Environmental Health Committee to review the RAF arsenic reassessment, EPA asked the committee to review the propriety of the "uncertainty factor." SAB's Executive Committee perceived this as a policy judgment thinly veiled as science, and the committee declined to review the report. According to a respondent who was an SAB member at the time, Board Chairman Norton Nelson was angered that EPA would even ask SAB to review the matter.[63] In June 1988, Administrator Lee Thomas endorsed RAC's recommendation in a memorandum permitting managers to down-weight ingested inorganic arsenic risks by an "uncertainty" factor of ten (Thomas 1988). EPA released the RAF *Special Report* a month later, but without the benefit of SAB's imprimatur.

Based on the Taiwanese database, the RAF reassessment concluded that the 50-ppb arsenic in drinking water standard resulted in a skin cancer risk estimate of 2.5×10^{-3} (or 0.25%). To attain a 10^{-4} cancer risk using the 1988 estimate, the MCL for arsenic would have to be lowered to 2 ppb (U.S. EPA/OW 1994). However, applying the "uncertainty" factor of ten yielded a skin cancer "weighted risk" on the order of 10^{-4}, or just within the bounds of what EPA ordinarily considers an "acceptable" risk (10^{-4}–10^{-6}). After the RAF report was issued, press reports suggested that ODW was considering proposing an MCL of between 30 and 35 ppb, but other offices were seeking a lower standard (*Inside EPA*, October 7, 1988, 6).

On a separate track, the Drinking Water Subcommittee of the SAB Environmental Health Committee met days prior to Thomas's June memo to review arsenic in drinking water. Richard Cothern, then executive secretary for the committee, was a former colleague of Bill Marcus's in ODW.[64] During a June 1988 meeting in Cincinnati, Cothern introduced to SAB evidence synthesized by Marcus and Amy Rispin (OPTS) during the course of the RAF arsenic reassessment. Marcus and Rispin had brought to light information concerning methylation as a possible detoxification pathway for arsenic.[65] The RAF report concluded that the issue merited further investigation but found that the available studies were inconclusive regarding a methylation saturation point or the relevance of the pathway to carcinogenesis. A former drinking water program scientist expected the SAB Drinking Water Subcommittee to be similarly equivocal.

Foreshadowing things to come, however, in 1988, SAB Director Terry Yosie observed that the board's "continuing efforts to persuade EPA to use pharmacokinetic data in risk assessment has begun to see results" (Yosie 1988). In September 1989, SAB recommended that EPA revise its 1988 arsenic assessment considering the potential reduction in cancer risk due to detoxification (U.S. EPA/SAB 1989). "Unlike its own scientists, EPA couldn't ignore the 1989 SAB report," said the former drinking water program scientist. In the same month, EPA was sued by the Bull Run Coalition.[66]

Setting the RfDs

After SAB advised EPA to revise its skin cancer risk estimate for ingested arsenic, the agency set up an interoffice working group to develop an oral reference dose for arsenic for EPA's Integrated Risk Information System. The RfD is designed as an allowable exposure (mg/kg–body weight/day) for noncancer risks.[67] The ensuing internal fight over the RfD value was a proxy battle over the drinking water standard, according to a former ORD official. (See the discussion below regarding the drinking water equivalent level.) Charles Abernathy, who had assumed the arsenic portfolio from William Marcus in 1989, was the primary Office of Water representative on the working group.

During the course of the RfD negotiations, the assumed Taiwanese drinking water consumption rate was increased from 3.5 l/day for men and 2 l/day for women to 4.5 l/day for both sexes, effectively lowering the risk estimate for arsenic (see exposure assumptions discussion above). Other negotiations concerned the appropriate uncertainty factor to employ. EPA's standard procedure in deriving an RfD is to ascertain the no observed adverse effect level (NOAEL) and then divide that by an uncertainty factor to provide a margin of safety. With the proposed values of one (corresponding to no safety factor), three, and ten on the table, the staff could not come to agreement. (Recall that the agency had adopted an "uncertainty factor" of ten to down-weight the risk of skin cancer.) The issue was elevated to RAC and, according to an ORD scientist, ultimately set at three by then Deputy Administrator Henry Habicht.

The reason there was such a fight over the RfD value for arsenic is that following standard agency procedures would have suggested an allowable arsenic intake considerably lower than the existing MCL. The Taiwan study (Tseng et al. 1968) identifies a NOAEL of 0.8 µg/kg/day and thus suggests an arsenic RfD value of 0.3 µg/kg/day (using the uncertainty factor of three). Assuming water consumption of 2 l/day by a 70-kg adult yields a drinking water equivalent level (DWEL) of 10.5 ppb (U.S. EPA/OW 1992). Because this value for the RfD would suggest an "allowable" oral arsenic ingestion level nearly fivefold *lower* than the current drinking water MCL,[68] a member of the IRIS workgroup says, "the arsenic RfD is not a number, like all others, but it's a range. It is a compromise. . . .The point estimate became nonpalatable due to the economic and political implications." According to a former EPA official, the endpoints of the "range" for the arsenic RfD listed in IRIS are, in fact, two separate point estimates derived by irreconcilable factions on the working group.[69] Because the RfD for a substance may also be used in establishing remedial objectives for soil contamination, the lack of consensus on the value for arsenic presents a decisionmaking challenge to Superfund site managers.

Developing Research Agendas, Buying Time

In response to the 1989 SAB recommendation, the EPA Office of Water developed the *Draft Drinking Water Criteria Document on Arsenic* in September 1992 (U.S. EPA/OW 1992) and submitted it to the SAB Drinking Water Subcommittee in March 1993.[70] The committee found that Taiwan data (Chen et al. 1985) provided evidence of an association between internal cancer and exposure to high levels of arsenic in drinking water but again suggested that evidence of nonlinear arsenic pharmacokinetics should be considered. Members also raised questions about the precision of analytical procedures to detect arsenic in drinking water in the 2-ppb range (U.S. EPA/SAB/DWC 1993). With Democrats controlling both the Congress and the executive branch for the first time in more than a decade, environmentalists anticipated a steady stream of regulations to be released from the queue. Some in the EPA water program and the regulated community, however, worked to buy time on arsenic, perhaps in expectation of SDWA legislative reforms or in an effort to build scientific consensus for a precedent-setting departure from standard EPA risk assessment procedures. The vehicle for their holding pattern was research agenda setting. An ORD scientist alleges that the Office of Water is merely "buying time, waiting for a new administration, so somebody else will have to make the decision. Their ulterior motive for developing a research agenda is buying time, to put something before the court."

In the 1988 reassessment, the RAF identified a broad set of future research directions covering epidemiology, mechanism of skin carcinogenesis, pharmacokinetics, and nutritional essentiality.[71] Since SAB advised EPA to revise its assessment, at least three ad hoc groups inside and outside the agency have formulated arsenic research agendas. However, little original research was funded by EPA or the regulated community. Much of the analysis supported or promoted by OW and the regulated community did not resolve scientific issues but instead served to expand the recognized scientific uncertainty by casting doubt on the Taiwan database or the analysis used in previous assessments. An ORD scientist observes that "Brown's work [for example, Brown and Chen (1993)] threw a cloud over the exposure assessment part of the risk assessment, but it's not the type of information that would have led to a decision" (that is, a decision to revise the standard).

In the absence of new, original research, OW in its court negotiations has pointed to the development of research agendas as evidence of progress toward the goal of finalizing an MCL for arsenic. However, there has been no consensus among the groups that have formulated arsenic research agendas on these two issues: (1) what the most critical research needs are, and (2) whether the identified research is likely to substantially alter the agency's risk estimate or do so within an appropriate time frame. A drinking water official says, "People are still debating on how they would address the uncertainties, on what the research needs are. Some say $1 million wouldn't improve the database incrementally because it's so good. Others say it wouldn't make a dent because it's so bad." This lack of scientific agreement suggests that the problem "may be too intractable to take a formal analytic approach to developing a research agenda," says an EPA scientist.[72]

After the 1989 SAB report, EPA formed an ad hoc arsenic research recommendation working group, headed by ORD/HERL's Jack Fowle. In early 1991, after conferring with the SAB Drinking Water Subcommittee, Fowle's working group issued a research agenda addressing the two areas that it felt contributed most to the uncertainty regarding arsenic risk assessment: (1) mechanism of cancer and (2) metabolism and detoxification. Because the agency was being sued for not meeting its regulatory deadline, the working group focused on research that could be conducted in three to five years but concluded that such research was unlikely to reduce uncertainty in the level below 50 ppb by more than a factor of a fewfold. If time were not a constraint, the working group considered the development of a suitable animal model for evaluating arsenic's cancer mechanism to be the top research priority (Fowle 1992).

An ORD scientist claims that ODW "wanted someone to counter" the Fowle working group's conclusion that no near-term research was likely to sufficiently reduce the uncertainty in the arsenic risk assessment. This source suggests that "ODW wants to believe that somebody, somewhere can come up with the answer in three years." Whatever its intentions, ODW asked a group of ORD scientists who specialized in drinking water health effects research at HERL in Research Triangle Park, North Carolina, to develop a new arsenic research agenda. The HERL group's research agenda, coordinated by Fred Hauchman, addressed a variety of issues, including epidemiological study, but the initial focus was laboratory-based research on the role of pharmacokinetics and interindividual genetic variation affecting arsenic metabolism.

"Hauchman's group wanted to do the research, but it couldn't get clearance to devote ORD money to the issue," remarked a drinking water program scientist. However, one ORD scientist claims that "some of the research won't amount to a hill of beans as far as reducing the uncertainty" in the near term and views HERL's interests in lab-based research as "parochial." The EPA research and water offices have also disagreed about which office bears responsibility for funding the research. Another ORD scientist suggests that "there are some tremendous opportunities for fostering collaborations with [industry]. They have a lot at stake, and we have an interest in getting the best research done."

In May 1991, the agency made plans to reassess arsenic based on current data (Fowle 1992). Shortly before this announcement, as part of an administrative reorganization instituted by then Assistant Administrator for Water LaJuana Wilcher, Jim Elder was shifted from the Office of Water's surface water regulatory program to replace Mike Cook as director of the renamed Office of Ground Water and Drinking Water. Elder would be responsible for getting the proposed rule for arsenic out under the court-ordered deadline of November 1995.

In December 1991, the Society of Environmental Geochemistry and Health (SEGH) launched its Arsenic Task Force. SEGH is a scientific society including researchers from the public and private sectors, as well as academics. Fairly or unfairly, some EPA officials perceive that it is dominated by representatives of the regulated community. (An independent toxicologist who is a long-time member of SEGH thinks that the lack of balance may be with the society's issue-specific task forces.) SEGH created the Arsenic Task Force in response to EPA's court agreement with the Bull Run Coalition to propose an arsenic in drinking water rule and in recognition of the impacts that an adjustment of the risk estimate would have not only on drinking water but also on contaminated site remediation and coal fly ash disposal. SEGH appointed Willard Chappell (University of Colorado at Denver) and Charles Abernathy (EPA Office of Water) as Task Force cochairs. Task force members include EPA officials, academics, and environmental consultants.[73]

The SEGH Arsenic Task Force secured sponsorship for a 1993 conference on arsenic exposure and health effects from EPA, the Agency for Toxic Substances and Disease Registry (ATSDR),[74] the American Water Works Association, the American Mining Congress (AMC), the International Council on Metal in the Environment (ICME), the Electric Power Research Institute, and U.S. Borax. At the conference, two industry-supported researchers cast doubts on EPA's previous risk assessment (U.S. EPA/RAF 1988). Ken Brown presented the problems he had uncovered regarding the Taiwan exposure data and discussed their implications for interpretation of the original skin cancer study. How-Ran Guo also claimed to have evidence of a nonlinear relationship between arsenic in drinking water concentrations and urinary cancers.[75] However, academic researchers Allan Smith and Claudia Hopenhayn-Rich also discussed their skepticism of the methylation threshold hypothesis and presented their internal cancer risk estimates. The running scientific battle was continued at SEGH and Society of Toxicology meetings in 1994 and 1995.

As EPA considered tightening the arsenic in drinking water standard to as low as 2 ppb in light of the internal cancer evidence (Smith et al. 1992), the regulated community sought to persuade the agency to pursue a research agenda. During this period, SEGH Arsenic Task Force members Willard Chappell and Warner North met with ORD Assistant Administrator Robert Huggett to urge additional arsenic research. Afterwards, representatives of the American Water Works Association and its research foundation (AWWARF) met with Huggett to seek ORD's participation in the planning of a workshop to develop a research agenda.[76] AWWARF convened the workshop in May 1995 and identified two research projects that it would pursue immediately, an analytical method for detecting arsenic at low levels in blood and urine and a feasibility study for an epidemiological study. The workshop was held five months after Assistant Administrator for the Office of Water Robert Perciasepe declared that EPA needed to pursue additional research and would be unable to meet its court-ordered deadline of November 1995. According to press reports, the Office of Water planned to use the AWWARF list of research needs in its negotiations with the plaintiffs to demonstrate that significant areas of research must be completed before a new standard can be proposed (*Risk Policy Report*, May 16, 1995, 16–17). (See further discussion of ongoing research activities in the concluding section.)

Things Left Undone

The major frustration of a former drinking water official concerning arsenic was the lack of new research available when the time for decision-making arrived: "The political appointees should have never been put in that type of position." Interviewees offered a variety of reasons why substantial new research had not been done over the last ten years. Some simply feel that the existing database was more than adequate for the rulemaking and that further research is unwarranted. Others point to the multimillion dollar costs of sufficiently large, sophisticated epidemiological studies or the cost and potentially long delay associated with basic research into arsenic's toxic mechanisms. Another possibility is that the agency simply regards arsenic as a lower priority than other drinking water contaminants, such as pathogenic microorganisms and disinfection by-products, to which it is devoting its scarce research resources. In the view of some, arsenic suffers from being "an orphan material," affecting a large number of small communities and having no single, deep-pocketed firm or sector being affected greatly enough to adopt it and fund the research. AWWARF, for example, offered to provide some funding for additional research, but some EPA officials believed that the level of resources available from the drinking water suppliers (on the order of $1 million) would be insufficient to reduce the scientific uncertainties by the necessary order of magnitude.

A former EPA drinking water official, however, rejects the "orphan argument" and suggests, instead, that "a lot of people were comfortable not knowing." This source believes, "It was a conscious act on the part of several people over time. I never could figure out how much of it was from within EPA or from OMB. They resisted properly funding the 1986 Safe Drinking Water Act program. It seemed like they wanted to starve the program.... There was tension between ORD and ODW on who should fund the research ... but no matter who the Administrator was, ODW didn't compete well within the agency.... The Office of Water was a program that was heavily earmarked by Congress with the Chesapeake Bay, the Great Lakes, and construction grants. That took away much of our discretion, and there was no chance of a net increase in the budget.... They were relying on how the academic literature was going to turn out for them, and the agency was left holding the bag." Thus, the agency could not rely on others to spontaneously provide it with the scientific information required for regulatory decisionmaking. The question remains, however, whether the research would have been done even if the drinking water program had greater discretionary resources, given the comfort of "not knowing."

For several years leading up to EPA's 1995 decision, various groups had developed and debated arsenic research agendas. Two schools of thought disagreed about whether research that could be conducted in a reasonable amount of time would reduce the scientific uncertainties to the extent that the new findings would alter the decisionmaking calculus. In effect, both sides were presuming the sort of evidence that was needed to sway regulatory decisionmakers. The clashing camps were operating in the absence of any formal articulation by EPA policymakers as to what they would consider sufficiently compelling scientific evidence to warrant any particular level of the drinking water standard. It should not be surprising, then, that EPA and the regulated community have been, as an environmental lawyer notes, "strategizing on research rather than doing it."

Effects of the 1991 Reorganization

As indicated above, then Assistant Administrator LaJuana Wilcher re-organized the Office of Water in 1991. An important change was the separation of the office's analytical staff from the surface water and groundwater/drinking water programs. The health and ecological risk assessors, along with economists and engineers, were consolidated under the new Office of Science and Technology (OST). According to a former EPA drinking water official, the reorganization was intended, at least in part, to institutionally separate the risk assessment and risk management functions. One result of the reorganization, says this source, was that the OST risk assessment staff used scientific argu-ments to pursue an arsenic *policy* independent from the Office of Ground Water and Drinking Water where some staff, principally on the basis of emerging evidence for internal cancers, advocated tightening the drinking water standard for arsenic. Prior to the 1991 reorganiza-tion, the situation within EPA appeared basically as a case of dueling risk assessors, with ODW pitted against ORD. After the reorganization, the internal dynamic became more complex and fragmented, and the agency was speaking to itself and to others with multiple voices. This, in turn, had important effects on the communication of scientific infor-mation up the chain of command to policymakers.

Communicating the Science to Agency Leadership

Half of the responses rated the communication of the science to agency decisionmakers as good to very good (five); the other half rated it as poor (five). In December 1994, a decisional briefing was held for Assistant Administrator Perciasepe, who had been briefed prior to the meeting by his own staff. According to a former drinking water official, "It was a packed room. Meeting attendance is a good gauge of the importance of a decision. When the GC [then General Counsel Jean Nelson] and an AA [Lynn Goldman, Assistant Administrator for OPPTS] show up, you know you have a biggy." The water office staff critiqued the available scientific studies and recommended pursuing additional research. The former drinking water official commented that "Abernathy was very good at identifying the holes in the Taiwan database" and suggested that in the minds of drinking water program management, the problems with the dose reconstruction identified by Brown were "incredibly important."

Assistant Administrator Goldman and Peter Preuss, representing ORD, both reportedly suggested that there was no research that could be done in the near term that would substantially alter the 1988 risk estimate. Many of those who gave the communications poor marks faulted the water office staff for undermining the credibility of the epidemiological evidence. Whether that constitutes a distortion, of course, depends on one's assessment of the evidence.

Apparently, however, there was a serious miscommunication regarding the issue of internal cancers. The internal cancer evidence—and the risk assessment by Smith et al. (1992) in particular—had caused some within the Office of Ground Water and Drinking Water to believe that action was warranted. Although it remains unclear who is responsible for starting it, a rumor got started that Smith later reversed his position regarding the evidence of internal cancers. According to a former EPA official, OGWDW management unwittingly conveyed this misinformation to Assistant Administrator Perciasepe, as well as to members of Rep. Waxman's (D-Calif.) staff to support the recommendation that more research was needed prior to revising the arsenic standard. The rumor was not limited to inside EPA. An environmental lobbyist reported hearing that Smith had changed his mind from sources in the drinking water industry.

The details surrounding the miscommunication within EPA remain somewhat sketchy. According to an EPA official in the Office of Water, at a 1994 scientific workshop on arsenic held in Annapolis, Maryland, Smith presented preliminary findings from data collected in South America that had not produced statistical evidence of a correlation between arsenic levels in drinking water and internal (that is, bladder) cancers. OST staff briefed OGWDW management on Smith's presentation. According to the former EPA official, the OST staff told the OGWDW management that "Smith had changed his mind." The OW official denies this and suggests that Smith has never reversed his position. Smith also became aware of the rumor, and although he "has questioned the rationale for setting the arsenic in drinking water MCL below 10 ppb," a level at which food appears to become the dominant pathway of ingested inorganic arsenic, he denies backing away from his 1992 analysis. In fact, while the abstract of Smith's 1994 Annapolis presentation suggests what many scientists would consider, by itself, marginal evidence for an association between arsenic in drinking water and internal cancer,[77] it concludes by saying, "These findings provide new support for the evidence from Taiwan that arsenic ingestion increases the risk of bladder cancer" (Smith et al. 1994). "If anything," Smith now says, "the evidence [of internal cancers] has gotten stronger" as a result of evidence from the completed epidemiological

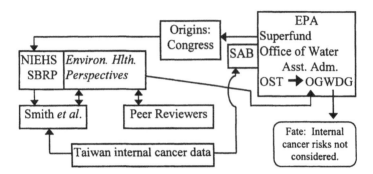

Figure B-2. Fate and transport of the assessment of internal cancer risks

studies in Argentina and Chile (Allan Smith, Professor of Public Health, University of California, Berkeley, personal communication).

Senior EPA drinking water management read the internal cancer risk analysis by Smith et al. (1992) as reported in the NIEHS journal *Environmental Health Perspectives*. In terms of a fate and transport analogy, management was directly exposed to scientific information that had not first passed through a filter (that is, the internal scientific gatekeepers in OST). Initially, there was a strong reaction to the information at the management level in proximity to the political decisionmaker. If OST staff did indeed misrepresent Smith's Annapolis presentation to OGWDW managers, they may have been seeking to "repair the damage" caused by the unfiltered exposure to the scientific information at this intermediate level. It is worth noting that the distinction between (1) an observed association between internal cancer and exposure to high levels of arsenic in drinking water that SAB endorsed in 1993 and (2) the ability to confidently estimate the risk of internal cancer at low levels of exposure, also seems to have been blurred somewhere in the communications process. According to the former EPA official, the lack of evidence for internal cancers was the pivotal basis for the OGWDW staff's recommendation to Assistant Administrator Perciasepe that the agency pursue additional research instead of immediately proposing a revised arsenic in drinking water standard.

Figure B-2 illustrates the fate and transport of the internal cancer risk assessment. Note the circuitous route that the flow of information takes. It begins with high-level demand for independent basic research relevant to contaminated site remediation, the congressionally designated interagency transfer of resources from EPA to the NIEHS Superfund Basic Research Program (SBRP). It ends with the decision not to consider the potential risks of internal cancers.

The Role of External Scientists in the Process

Two-thirds (eight) of responses rated the role of external scientists in the decisionmaking process as significant to very significant. According to an EPA drinking water official, the lack of unity among the agency scientists provided an opportunity for external scientists to play prominently. The most significant aspect of the participation of external scientists in the process appears to have been legitimizing OW's call for more research prior to revising the arsenic in drinking water standard. As discussed above, the NRC Drinking Water and Health Committee and the SAB drinking water panel both promoted consideration of arsenic pharmaco-kinetics and departure from the linear cancer model. As was the case in the 1991 lead in drinking water rulemaking (see accompanying case study), some EPA scientists asserted that the SAB drinking water panel had exceeded its legitimate scope of review by commenting on the health assessment. (The SAB Health and Executive Committee chairs, however, signed off on the 1989 review since the drinking water panel was, at the time, a subcommittee of the SAB Health Committee.)

A number of external scientists who have played particularly large roles in the process were, to a greater or lesser extent, "insiders." For example, many observers associate Warner North closely with the 1989 SAB report. At the time, North was vice chair of the SAB Environmental Health Committee through which the Drinking Water Subcommittee reported. When the subcommittee's EPA staffer Richard Cothern approached North with the panel's report from its 1988 meeting in Cincinnati, the controversial issue of a possible detoxification pathway for arsenic was framed as a false dichotomy between the linear, no-threshold dose-response model and the threshold or "hockey stick" model. By pointing out that there are any number of scientifically plausible non-linear dose-response curves lying between these two extremes, North helped negotiate the report through the internal SAB review process. (Note, however, that despite the carefully crafted language of the 1989 SAB report, the false dichotomy between the linear, no-threshold, and threshold models has proved to be a hardy perennial.)

Beck, a former EPA scientist and the Region I representative to the 1986–1987 Risk Assessment Forum, has been a prominent critic of the 1988 RAF arsenic reassessment from her new position with Gradient Cor-poration. Ken Brown, an EPA contractor on the RAF reassessment, later, serving as a consultant to AWWA, unearthed evidence that damaged con-fidence in the 1988 skin cancer risk estimate. Paul Mushak, a principal external author of the 1984 ORD arsenic health assessment document, has prominently defended the agency's risk assessment against its critics. In addition to EPA's official science advisers, "there has been an opportunity by those outside the agency to present information to the leadership of the Office of Water," according to an industry scientist. An EPA drinking water official says Perciasepe met to discuss scientific issues with "outside bodies that lobbied him, drinking water utilities (primarily from California), the AWWA, and also some people from ARCO concerned about Superfund."[78]

SCIENCE IN THE FINAL DECISION

Mirroring the polarized opinion about the adequacy of the available science for decisionmaking, there were two factors most frequently mentioned by respondents as impeding the use of science in the decision of whether to revise the arsenic in drinking water standard or pursue additional research. These factors were economic and political considerations (five responses) and reluctance to consider new scientific findings (that is, the evidence for a nonlinear dose-response model and/or the problems with the Taiwan study dose reconstruction) (five responses). A narrow majority of responses (seven of twelve) rated the impact of the scientific information on the decision to call for further research as low.

By most accounts, the decisionmaking authority was concentrated in the hands of Assistant Administrator Perciasepe. Assistant Administrator Lynn Goldman, a credentialed scientist,[79] however, "was satisfied that the information justified lowering the standard," according to a former EPA official. But Perciasepe's staff did not agree. "Perciasepe had a conversation with Goldman, and they decided that they wouldn't elevate it. It was brokered at the AA [Assistant Administrator] level rather than bringing in the DA [Deputy Administrator Robert Sussman]." This source observes that decisionmaking authority vested in the Assistant Administrator was typical of EPA regulatory decisionmaking in the Browner administration.

Based on Assistant Administrator Perciasepe's January 1995 court declaration, not only the scientific uncertainties but also the costs and distribution of those costs were important in the decision to pursue additional research (Declaration of Robert Perciasepe, Amended Consent Decree, *Donison et al. v. EPA*, No. 92-6280-HO (and consolidated cases) U.S.D.C. Oregon, January 9, 1995). A 1994 draft water office staff analysis provides estimates of the regulatory impacts under five MCL options that were available at the time of the 1995 decision (Table B-2).[80]

A variety of political forces were also aligned against making the arsenic standard more stringent. The administration wanted to avoid disrupting ongoing SDWA negotiations with Congress. "We were afraid that it [revising the arsenic standard] would hurt legislative negotiations" on amending the SDWA, according to a former drinking water official. An

Table B-2. Summary of Regulatory Impacts for Arsenic under Five MCL Options

MCL Options	2 ppb	5 ppb	10 ppb	20 ppb	50 ppb
Total annual cost ($ millions)	2,086	617	266	74	24
Annual household cost for small systems ($)[a]	261–1,454	252–1,423	266–1,412	275–1,301	269–1,266
Number of systems needing treatment	12,386	4,924	1,949	595	160
Number of people with reduced exposures (millions)	31.7	11.0	4.0	1.0	0.2
Annual skin cancer cases avoided	127.3	73.9	34.3	17.7	7.6

[a] Small systems serve between 25 and 3,300 people.

Source: U.S. EPA/OW. 1994. *Summary of Regulatory Impacts.* Draft. Washington, D.C.: U.S. EPA, November 30. Mimeographed.

EPA official observes, "It was yet another imposition on the small [drinking water] providers with whom EPA already has its biggest problems." Although it may be secondary to the broader issue of unfunded mandates, the EPA official suggests that the arsenic standard "is also tangled up in western water supply issues. The costs are pretty big, not astronomical, but the distribution of the costs is disproportionately on vocal, organized, western small systems. . . . They are a more homogeneous constituency than the eastern municipal providers, and they're already stressed—those are deserts that they're watering." It should also be recalled that the court declaration came on the heels of the November 1994 elections in which Republicans gained a congressional majority. Although in September 1994 the House of Representatives passed a bill (H.R. 3392) to amend the SDWA that called for a National Academy of Sciences study on arsenic and postponed the deadline for proposing a new standard until the end of 1996, after the fall elections, EPA was firmly in the crosshairs of a hostile Congress, and the tide for dramatic regulatory reform seemed unstoppable. If it had not been a foregone conclusion, the tenor of the early days of the 104th Congress closed the door on EPA's proposing a revised arsenic in drinking water standard by November 1995.

According to a former EPA drinking water official, "We didn't think the evidence was strong enough on the risk assessment to feel comfortable that we could withstand outside review. In the environment we had, we felt we couldn't get away with anything but an ironclad proposal. There was some chance that it [a proposed rulemaking] would be remanded, but there was a larger context. . . . There had been a crush of criticism of the scientific validity of our decisions. We didn't feel that we could say it's affordable to go below 50 ppb. The decision had a bigger context than arsenic in drinking water." An industry scientist says, "Many nonagency scientists feel that additional research is needed, but that's not the primary cause for calling for more research. It is driven by the large economic stakes" that extend beyond the drinking water sector to include hazardous waste and contaminated site remediation issues.

An ORD official asserts, "The situation is really very simple. The science indicates a high risk from current arsenic levels. But the problem is expensive to deal with and is mostly a western problem, so the agency doesn't want to act. However, it doesn't want to say that it doesn't want to act. So, the managers turn back to science for a way out. On arsenic, [Assistant Administrator] Perciasepe wants to do more research to punt—[this source] and others have told him that more research won't change the situation. The science was not used because the risk managers didn't want to make the decision called for by the risk information. This is often an issue."

While not discounting the political expediency of "punting" on arsenic, the situation may not be quite so clear-cut. A factor that appears to have been decisive in the thinking of some EPA water program managers was that the perceived health costs of delay to pursue additional arsenic research were not unduly large because of the limited health threat posed by skin cancer. "Skin cancer is curable. So, delay would not be that harmful for that many people," says a drinking water official. In fact, a number of EPA sources noted that Margaret Prothro, who was acting Assistant Administrator for Water prior to Perciasepe's confirmation, was successfully treated for skin cancer and minimized its gravity.

Initially, the evidence of a link between internal cancers and arsenic in drinking water had caused considerable concern. A former EPA drinking water official recalled his initial response to the "Smith paper" (Smith et al. 1992), "My god, we've got to do something!" According to this source, the senior program management were prepared to recommend "a zero MCLG" if they had been given "convincing science that there were internal cancers." Agency scientists, however, disagreed about the interpretation of the evidence. In the face of conflicting opinion, the message that "Smith had changed his mind" was apparently sufficient for decisionmakers to discount the consideration of internal cancers (at least at levels below the current standard).

The precedent that EPA established in interpreting the SDWA appears to have been another factor impeding a more candid discussion of the role of science in the decision to pursue additional research. According to an ORD official, arsenic is "an example of where the statute's wrong because it doesn't allow the decisionmaker to be honest." An EPA drinking water official says, "If the statute had been more flexible, we could have gone to a midrange. But the statute forced you to a low level."[81] Feeling "hamstrung by the act," water program decisionmakers had an incentive to focus attention on the scientific uncertainties to rationalize their decision. However, prior to the 1996 SDWA Amendments, implementation of the statute had been conditioned by the precedent EPA set in interpreting what is economically "feasible." In implementing the SDWA consistently with its default, no-threshold cancer model, EPA has set the unenforceable health goal (MCLG) for carcinogens at zero. The enforceable standard (MCL) is to be set as close to the MCLG as feasible, considering available technology and cost. Treatment technology for arsenic exists (that is, reverse osmosis) but is capital intensive, making it less affordable for the small suppliers who would be most affected by a more stringent standard. However, EPA has interpreted economic "feasibility" to mean reasonably affordable for large public water systems.[82] (Citing this interpretation, EPA downplayed questions of small system affordability in the case of the 1991 lead/copper drinking water rule. See Appendix A.)

An ORD official reports that the some agency scientists recommended that EPA grant waivers for small communities for whom treatment might be unaffordable instead of characterizing the science as inadequate for regulatory decisionmaking. However, "attorneys from the OGC [Office of General Counsel]," says this source, "were troubled that this [waivers] was not a viable option."[83] The agency's lawyers may have been troubled by this option because it was inconsistent with precedented statutory interpretation.

CONCLUDING OBSERVATIONS

Representatives from other EPA offices are concerned with the scientific precedent that OW is setting, fearing that the hurdle of scientific proof will become so high that virtually all rulemaking could be challenged on the basis of insufficient science. Many also view arsenic in drinking water as a test case for what EPA considers sufficient epidemiological evidence for regulatory decisionmaking. EPA's 1986 cancer risk assessment guidelines state that "if available, estimates based on adequate human epidemiological data are preferred over estimates based on animal data" (U.S. EPA 1986). To determine what constitutes adequate data, some would like to see a consistent minimum standard applied. However, the "value of information" depends on the decisionmaking context. The adequacy of data will depend upon the compliance costs and the consequences of inaction. EPA drinking water officials considered the consequences of inaction—as they understood them, that is, skin cancer—not to be too severe. A drinking water official acknowledges the conditionality on compliance costs, saying, "The costs were too high given the uncertainties in the science."

Many observers also see arsenic as an important test case for determining the weight of evidence that is necessary to justify departing from EPA's default linear cancer model. Unresolved, however, "is the question of what's adequate data for departing from the default," according to a drinking water official. "We have resolved to pursue additional research but are holding off the decision as to what constitutes enough." Essentially, this indecision places researchers in the position of not knowing what hypothesis to test, makes it less likely that further research will facilitate future decisionmaking, and also provides little incentive for research supporters to allocate resources to the problem. There has been some diffuse feedback from the EPA decisionmakers to scientists that more research is desired, but the investigators have no idea what evidence the policymakers will consider compelling.[84] Therefore, while decisionmakers may find it difficult to specify what evidence they (or their successors) would consider compelling, and while it may be irrational to establish across-the-board guidance on what constitutes "adequate" science for all regulatory purposes, not answering the question in a specific case also presents problems.

Some respondents claimed that EPA careerists had overstepped their boundaries into policymaking. It is clear from their comments that the opposing scientists and analysts are heavily invested professionally and intellectually in their points of view. Whereas the lead in drinking water case study (Appendix A) provides an example of a group of EPA staff outside the water program using science as policy entrepreneurs to push through a regulatory rulemaking, the arsenic in drinking water case gives the impression of a group of agency water program staff strategically employing science to promote departure from the standard risk assessment procedures (in this case, the linear, no-threshold cancer model) that originate outside the program. The pharmacokinetic information the regulatory program staff mustered, however, was not "actionable." Using a fate and transport analogy, an EPA scientist suggests that "somewhere [in the agency] there are barriers to exposure" that prevent consideration of the information. "The information is not being excreted or actively attacked, but it's being stored in the fat." It remains to be seen what is necessary to strengthen and mobilize the information for use in regulatory decisionmaking.

EPILOGUE

The 1996 SDWA Amendments required EPA to issue a research plan for arsenic in drinking water by February 1997. The legislation also authorized

$2.5 million for the research. EPA allocated $2 million and AWWARF and the Association of California Water Agencies contributed a combined $1 million to a joint request for applications (RFA) for research grants. In a December 6, 1996, *Federal Register* notice, EPA solicited public comments on the proposed research topics.[85] The proposed joint RFA and ORD's in-house arsenic research plans were both submitted for review to the Board of Scientific Counselors (BOSC), a newly formed EPA official science advisory panel that reports directly to the Assistant Administrator for ORD. BOSC completed its review in May 1997. In addition, AWWARF is supporting University of California's Allan Smith and colleagues to evaluate possible epidemiological studies that might be done on arsenic.

The National Research Council formed a committee in 1997 to consult with EPA on arsenic in drinking water. In March 1999, the committee issued its report (NRC 1999). The committee found that recent human epidemiological studies in Chile (Smith et al. 1998) and Argentina (Hopenhayn-Rich et al. 1998) confirmed earlier observations from Taiwan of increased internal (bladder and lung) cancers, as well as skin cancer, after chronic ingestion of inorganic arsenic. Furthermore, the committee concluded that EPA's current drinking water standard for arsenic does not achieve the agency's goal for public health protection (limiting cancer risks to 1-in-10,000), and therefore requires downward revision as promptly as possible. The committee estimated that the current MCL for arsenic represents a cancer risk in the range of 1-in-100 to 1-in-1000, although two of the sixteen panel members disagreed with the 1-in-100 estimate. Responding to the release of the NRC report, EPA and the American Water Works Association acknowledged that the standard for arsenic in drinking water would be tightened (Warrick 1999). After 57 years, the question that remains is by how much. Under the 1996 Safe Drinking Water Act amendments, EPA is required to propose a standard by January 2000 and finalize it by January 2001.

ENDNOTES

[1] Seafood, in particular, contains high levels of organic arsenic. For the majority of the U.S. population living in areas with low arsenic levels in their drinking water, food is most likely to be the major source of inorganic arsenic ingestion.

[2] Parts per billion (ppb) is considered equivalent to micrograms per liter (μg/l). In this context, acute poisoning refers to high doses administered over a short duration, typically to calculate a lethal dosage for a given percentage of test animals.

[3] Many uses of inorganic arsenicals as agricultural pesticides and defoliants were banned in 1967 prior to EPA's establishment (CAST 1976, 26).

[4] The only plant in the nation that would have been affected by the standard was the ASARCO copper smelting plant outside Tacoma, Washington. After con-

siderable public debate and input regarding the balancing of economic and health considerations, however, declining world copper prices forced the ASARCO plant to close. As a result, EPA never issued final regulations (Landy et al. 1994, 253–254).

[5] This practice arises from the ARARs (applicable or relevant and appropriate requirements) provision of the 1986 Superfund Amendments and Reauthorization Act (SARA) (Walker et al. 1995). See *Superfund* (U.S. GAO 1996) for a discussion of the variety of means used by states in setting groundwater standards for contaminated site cleanups.

[6] Approximately 70% of coal combustion waste products are currently managed in surface impoundments, landfills, mines, and waste piles. States are authorized to regulate coal ash under the RCRA nonhazardous waste provisions (Subtitle D), resulting in considerable interstate variation in waste management standards (TNCC 1995). The 1984 overhaul of RCRA strongly discouraged land disposal of hazardous wastes in response to concerns about groundwater contamination (Dower 1990). EPA was under a court-ordered deadline of February 1997 for issuing a final hazardous waste identification rule (HWIR, to identify wastes exempt from the management standards of RCRA Subtitle C). However, the EPA Science Advisory Board judged the multipathway risk assessment model proposed by the agency for the HWIR to be inadequate (*Environment Reporter*, March 15, 1996, 2131–2132). In early 1997, EPA secured a one-month extension in order to renegotiate a new deadline for finalizing the HWIR (EPA's RCRA Hotline, March 1997). EPA and industry litigants agreed to delay final promulgation of the HWIR until 2001 (*Inside EPA*, March 14, 1997, 1).

[7] Unfunded mandates refers to federally required actions by state and local governments that are unaccompanied by transfer of resources required for implementation.

[8] Between 1989 and 1991, twenty-seven public drinking water supplies reported arsenic levels above the current standard of 50 ppb. All twenty-seven supplies had groundwater sources and most of the supplies served fewer than 500 people (U.S. EPA/OW 1993).

[9] Arsenic in drinking water is not strictly a western issue. For example, parts of New England also experience higher than normal levels. According to a former EPA drinking water program official, the pressure exerted on EPA from the "Sagebrush Rebellion" movement was much higher in the case of radon than for arsenic.

[10] Shortly after its passage, the California Department of Health Services informed EPA that the state would not initiate the process for implementing the lead/copper rule due to a lack of resources (*Inside EPA*, November 22, 1991, 1). During the 103rd Congress, Senators John Glenn (D-Ohio) and Dirk Kempthorne (R-Idaho) introduced S. 993, the "Community Regulatory Relief Act." According to a National Governors Association report (*Backgrounder*, July 17, 1994, 2), then Office of Management and Budget (OMB) Director Leon Panetta and President Clinton endorsed the bill.

[11] The 1996 SDWA Amendments dropped the requirement that EPA issue drinking water standards for twenty-five new contaminants every three years. Instead, EPA is now required to develop a list of unregulated contaminants and make determinations of whether or not to regulate at least five of these contaminants every five years.

[12] It should be noted that SDWA required EPA to regulate radionuclides and arsenic under its standard setting provisions (and currently under Section 109 of the 1996 SDWA Amendments), but the agency has no authority to regulate radon in indoor air, and while EPA establishes permissible pesticide tolerances under the Federal Food, Drug and Cosmetic Act, the Food and Drug Administration regulates the safety of seafood, which can contain high levels of organic arsenic.

[13] The twelve respondents in no way represent a statistically valid sample.

[14] Although the arsenic exposures observed in some epidemiological studies include levels considerably higher than the current U.S. drinking water standard of 50 ppb, standard procedures for chronic animal studies include the maximum tolerated dose, which would be much higher, relatively speaking, than the doses observed in human populations.

[15] In January 1998, EPA's Science Policy Council approved a new policy on thyroid tumors that supports a threshold response for a cancer effect, indicating that certain pesticide-caused thyroid tumors in rodents operate through a different mechanism of action that is not relevant to humans (*Risk Policy Report*, February 20, 1998, 21–22). According to an Office of Research and Development (ORD) scientist, however, arsenic is generally viewed with greater interest and has greater likelihood of affecting agency procedures because the stakes are higher.

[16] Blackfoot disease is gangrene of the extremities caused by damage to the peripheral vasculature.

[17] The study group contained 40,421 individuals and the control group was 7,500.

[18] There were 428 cases of skin cancer in the study group and no cases in the control group. Based on the skin cancer rate for Singapore Chinese from 1968 to 1977, the expected skin cancer rate in the control population of 7,500 was 3.

[19] According to an EPA official, in Caucasian populations, any skin cancer effect of arsenic is going to be dwarfed by sun exposure, whereas in an Asian population, arsenic-induced skin cancer would "stick out like a sore thumb."

[20] Mushak and Crocetti (1995), however, dismiss the role of humic acids and suggest that arsenic present in crops may have been primarily in less toxic organic forms. Paul Mushak of PB Associates, which specializes in toxicology and health risk assessment of metals, is a former faculty member of the University of North Carolina, was a principal external author for the 1984 ORD health assessment document for arsenic, contributed an issues document on carcinogenicity and essentiality to the 1988 Risk Assessment Forum assessment, and has served as a member of and consultant to the EPA Science Advisory Board and Clean Air Act Science Advisory Committee. Slayton et al. (1996) respond to Mushak and Cro-

cetti (1995) by citing some evidence that inorganic arsenic accounts for a considerable portion of total arsenic in some foods. See Mushak and Crocetti (1996) for their rebuttal.

[21] Mushak and Crocetti (1995) estimate that the study group's nutritional status relevant to biomethylation is more than adequate. See Slayton et al. (1996) and Mushak and Crocetti (1996) for more on this debate.

[22] Brown's work was supported by American Water Works Association (*Risk Policy Report*, September 16, 1994, 11). AWWA is the trade association of public drinking water suppliers. Brown also coauthored a paper with EPA water program scientist Charles Abernathy suggesting that food may be a greater source of arsenic ingestion in Taiwan than in the United States, indicating that previous assessments may have overestimated risks in the United States (see discussion of exposure assumptions below).

[23] C. J. Chen is a Taiwanese researcher trained at Johns Hopkins University.

[24] Tseng et al. roughly estimated the level of arsenic in drinking water at the village level (that is, all individuals in a village were assumed to drink water with the same arsenic concentration), and villages were assigned to one of three broad dose levels. The problem lies in then unrecognized variable arsenic concentrations between the wells within a village, largely due to a mix of shallow wells (with low arsenic levels) and deep artesian wells (with high arsenic levels), and in having only one well test for twenty-four (40%) of the sixty villages. An EPA scientist suggests that "the epidemiological exposure measures are so poor at lower levels that you can't distinguish between 5 and 100 µg/l [ppb]."

[25] It is not self-evident that the dose-reconstruction problems discovered by Brown revealed an overestimation of the cancer risk. Measurement error in dose reconstruction does serve to bias the risk estimates, and Brown et al. (1997b) suggest a plausible scenario under which prior risk estimates for the low-exposure group would be overstated. However, neither the magnitude nor even the direction of the *true* bias may be estimable because reliability measures of the proxy measure for true exposure are absent. (If good reliability measures were present, reanalysis of the Taiwan data could have laid the issue to rest. Instead, new epidemiological studies are being planned.) Therefore, the dose-reconstruction problems cast doubt on the reliability of the Taiwan study for risk estimation in the low-dose region but do not necessarily indicate that risks derived from the study are overestimated. If there is a statistical bias in either direction, its magnitude could be either negligible or nonnegligible.

[26] It should be noted that even animal experiments present problems controlling administered dose levels. For example, because rodents eat their feces, they may be "redosed" with excreted substances.

[27] The Utah study compared 145 people in a community consuming drinking water with 200 ppb arsenic and 105 in another community whose drinking water levels of arsenic averaged 20 ppb. The study was done under a research grant from ORD's Health Effects Research Laboratory (HERL).

[28] Prof. Allan Smith is an epidemiologist with the School of Public Health, University of California, Berkeley, whose work was supported by the National

Institute of Environmental Health Sciences Superfund Basic Research Program (www.niehs.nih.gov/sbrp/newweb/resprog). This underscores the linkage of the drinking water standard to contaminated site programs. Carlson-Lynch et al. (1994) criticized Smith and colleagues for not using a curvilinear statistical model to reflect the hypothesized detoxification of arsenic in well-nourished humans below ingestion exposures of 200–250 µg/day and for assuming Taiwanese water consumption rates below those used by EPA in deriving the agency's reference dose for ingested arsenic. Both the nonlinear model and the higher drinking water consumption figures would yield lower risk estimates. Smith et al. (1995) responded that they employed the Taiwanese drinking water intakes used by the Risk Assessment Forum (U.S. EPA/RAF 1988). Brown et al. (1997a) find that the Taiwan internal cancer data are statistically consistent with either a linear or a nonlinear dose-response model (a hockey stick–shaped curve formed by two linear segments with different slopes). (See discussion below regarding detoxification, exposure assumptions, and setting the RfD.) Lynch and colleagues are with the environmental consulting firms ChemRisk and Gradient Corporation. Lynch's coauthor Barbara Beck represented EPA Region I on the 1986–1987 EPA Risk Assessment Forum and is currently with Gradient, which provides environmental consulting services to the Atlantic Richfield Company. ARCO owns the Anaconda Minerals Superfund Site in Montana (a former mining site where arsenic is a key contaminant) and has played a key role in supporting research and analysis on the risk of ingested arsenic.

[29] The current SAB Drinking Water Committee was formerly a subcommittee of the SAB Environmental Health Committee.

[30] As discussed below, Smith suggests that ongoing epidemiological studies in Argentina and Chile serve to strengthen the evidence of an association between arsenic ingestion and internal cancers (Allan Smith, Professor of Public Health, University of California, Berkeley, personal communication).

[31] Methyl is the organic molecule CH_3. Methylated arsenic is excreted in urine more rapidly than inorganic forms of arsenic, and as more methyl groups are added to arsenic during metabolism, arsenical compounds become less toxic (U.S. EPA/RAF 1988). A number of short-term animal experiments (reviewed in U.S. EPA/OW 1992) indicate 70%–95% urinary excretion of soluble forms of arsenic.

[32] The amount of a substance consumed in drinking water represents an "administered" dose that may differ from the "internal" dose that is absorbed and available for biological interaction because ingested substances may be excreted from the body or metabolized into different forms that are more or less toxic than the original substance.

[33] Trivalent arsenic is more acutely toxic than pentavalent arsenic (U.S. EPA/RAF 1988). There has been some confusion in the literature regarding the statistical analysis used in U.S. EPA/RAF 1988. Various sources have inaccurately stated that EPA employed the linearized multistage (LMS) model. Under the 1986 guidelines, this is the default model EPA applies to the results of animal carcinogen studies. When EPA reports point estimates for cancer risks from animal studies using the LMS, it uses an upper-limit estimate (that is, the 95th percentile upper confidence limit) of the dose-response slope. However, because the Taiwan

data were from an epidemiological study, the model used in U.S. EPA/RAF 1988 differs in some important respects from the standard animal study assessment practices. The linear quadratic model differs from the simple linear regression model in that it includes both linear and quadratic (squared) terms (that is, $y = b_0 + b_1x + b_2x^2$). This model provided a better statistical fit to the Taiwanese data than if only a linear term were used. Consistent with the 1986 guidelines, however, the model is linear in the low-dose region. The model also differs from the simplest model (which considers only dose levels) by incorporating duration of exposure information. U.S. EPA/RAF 1988 also based its skin cancer risk estimate on the predicted slope of the curve in the low-dose region (that is, the maximum likelihood estimate or MLE). The MLE is less than the upper limit slope estimate (that is, the MLE is a less conservative risk estimate).

[34] See U.S. EPA/RAF 1988 (Appendix E) and U.S. EPA/OW 1992 for a more complete discussion of the metabolism of inorganic arsenic.

[35] Mushak and Crocetti (1995) also state that blood arsenic is a poor indicator of chronic or prior exposures, with urine being a more stable indicator of chronic exposure and hair being the best measure of cumulative exposure. An EPA official explains that standard blood analyses make no distinction between forms of arsenic (organic or inorganic). "You can spike your arsenic blood with seafood" [containing organic arsenic]. Beck et al. (1995), however, argues that the ratio of methylated urinary metabolites of arsenic could be a more sensitive indicator of methylation saturation than the percentage of inorganic arsenic in urine. See Slayton et al. (1996) and Mushak and Crocetti (1996) for further debate over the evidence for nonlinearities in the arsenic dose-response curve.

[36] For example, in 1991, EPA reduced its risk estimate for methylene chloride by an order of magnitude on the basis of research on the pathways through which the substance is metabolized. After the agency's reevaluation, however, research began to focus attention on parameter uncertainties in the pharmacokinetic modeling that had not been considered during the reassessment. At least one analysis suggested that according to the new information, EPA should have raised, rather than lowered, its original risk estimate (NRC 1994).

[37] Guo has conducted work on arsenic as a postdoctoral researcher at the University of Cincinnati and at the Washington, D.C.–based firm RegNet Environmental Services. Guo's study of arsenic exposure and urinary cancers was done in his capacity as a researcher for RegNet and was supported by ARCO and the International Lead Zinc Research Organization (ILZRO) (*Risk Policy Report*, September 16, 1994, 11).

[38] Exposure may be poorly related to well ownership because owners of multiple wells may supply water to others who do not own wells.

[39] Because the cancer registry only requires hospitals with fifty beds or more to report, cancer cases may not be reported from smaller hospitals and pathology centers.

[40] See, however, the discussion below about a possible trade-off between protection against acute arsenic toxicity and chronic cancer.

[41] For example, NRC (1993) concluded that "even if an agent's mechanisms of [carcinogenic] action are well understood, it will still be very difficult to determine its dose-response relationship accurately enough to predict doses that correspond to [cancer] risk as low as one in a million" (p. 10). NRC 1993 (206–210) also discusses the considerable interindividual variability in such factors as detoxification that affect susceptibility to toxic substances. Such variability contributes to the inherent imprecision of risk estimates.

[42] This figure assumed direct water consumption of 3.5 l/day and indirect water consumption of 1 l/day for cooking rice.

[43] According to an EPA official, the water consumption rates assumed by the agency were based primarily on a discussion between Herman Gibb and some farmers in a Taiwanese medical center. The farmers showed Gibb water bottles that they carried into the field and said that they would generally drink one bottle in the morning and another in the afternoon. Gibb measured the volume of the bottles, and agency analysts made some additional assumptions about morning and evening drinking water consumption.

[44] As an EPA research official observed, this degree of sensitivity of risk estimates to changes in exposure assumptions is not surprising. As Mushak and Crocetti (1996) note, however, any individual component of uncertainty in risk estimates should also be considered in the context of other potential sources of variability and uncertainty. See Mushak and Crocetti (1995), Slayton et al. (1996), and Mushak and Crocetti (1996) for further debate regarding assumed drinking water consumption rates for chronically heat- and humidity-stressed, active rural populations of Taiwan.

[45] Mushak and Crocetti (1996) estimate that a combined half-dozen food samples were taken. The statistical confidence limits around such a small sample would tend to be large.

[46] Mushak and Crocetti (1995) argue that because the methods used by Yost et al. (1994) to analyze the Taiwanese food samples involved strong acid treatment to produce satisfactory recoveries of total arsenic, the measured levels may have been generated from organic forms present in the food as an artifact of the laboratory analytical methods. See Slayton et al. (1996) and Mushak and Crocetti (1996) for further details on the debate.

[47] *In vitro* ("in glass") refers to a variety of laboratory-based procedures not involving whole animal testing.

[48] A clastogen is a chemical that is able to cause structural damages in chromosomes, primarily breaks. Chromosomal and DNA damage are two possible cancer mechanisms. Rudel et al. (1996) suggest that arsenic's carcinogenic mechanism may result in a nonlinear dose-response relationship.

[49] Marc Mass is with EPA's Health Effects Research Laboratory.

[50] The harmful and beneficial chemical forms generally differ. For example, hexavalent chromium is carcinogenic, but trivalent chromium is essential (Mushak 1994).

[51] See the ethylene dibromide case study (Appendix E) for a discussion of the consequences of an episode in which an EPA policymaker during the early years of the Reagan administration attempted to depart from standard risk assessment procedures in pursuit of a regulatory relief policy.

[52] The NRC is the primary operational arm of the National Academy of Sciences.

[53] The ambient water quality criteria documents developed for the surface water program would include both human health and environmental effects information. Only a subset of the toxic pollutants for the surface water program would be germane to the drinking water program.

[54] According to this source, the draft criteria document for arsenic suggested an ambient water quality standard below the practical detection limits of analytical technology.

[55] Prior to a 1991 reorganization, the EPA Office of Drinking Water (now the Office of Ground Water and Drinking Water [OGWDW]) conducted its own risk assessments. As part of the reorganization, the separate health and ecological risk analytic staffs from the surface water and drinking water programs were consolidated in the Office of Water in the Office of Science and Technology. See further discussion of the reorganization and its effects below.

[56] For example, committee member Michael Gallo, a toxicologist and currently director of the National Institute of Environmental Health Sciences (NIEHS) Center of Excellence at New Jersey's Robert Wood Johnson Medical Center, has been described as a "leader of the campaign" to replace risk assessment assumptions, such as the linear dose-response model, with methods based on a biological understanding at the molecular level (Stone 1993).

[57] The Washington, D.C.–based Office of Health and Environmental Assessment produced the HAD. This office was successor to the Cancer Assessment Group in ORD and predecessor to the current ORD National Center for Environmental Assessment.

[58] Shortly thereafter, ODW and the regulated drinking water suppliers criticized ORD's assessment for using an epidemiological study to quantitatively estimate the risks of low-level arsenic ingestion, treating the Taiwanese study population as comparable to the U.S. population when the study population had a diet lower in protein, discounting the evidence from animal studies of arsenic's nutritional essentiality, making no distinction between Taiwanese and U.S. drinking water consumption rates (using 2 l/day for both), and using a simple linear dose-response model (least-squares linear regression). See the discussion of these scientific issues above.

[59] The 1986 SDWA Amendments required EPA to request comments from SAB "prior to proposal of a maximum contaminant level goal and national primary drinking water regulation," virtually eliminating the role of NRC in evaluating specific drinking water contaminants. However, the law allows SAB to respond "as it deems appropriate," and makes clear that SAB review must be conducted within the statutory timetable for promulgation of drinking water stan-

dards. In practice, the SAB Drinking Water Committee (formerly a subcommittee of the Environmental Health Committee) has selectively commented on the more controversial or high-stakes proposed drinking water standards, including arsenic.

[60] The House report on the 1974 SDWA bill stated that if there is no safe threshold for a contaminant, the health-based goal should be set at zero (*Federal Register* 56: 26460).

[61] Marcus (1988) produced a draft risk assessment of arsenic for the Office of Drinking Water.

[62] RTI is a contract research organization in Research Triangle Park, North Carolina.

[63] New York University's Norton Nelson was appointed SAB chairman during the post-Gorsuch years and refused to allow the board to be drawn too closely into the policy sphere on a number of occasions (see discussion in Jasanoff 1990).

[64] The SAB Secretariat is housed in the Administrator's office and staffed by EPA employees.

[65] See, for example, the discussion by Valentine et al. (1979) and Buchet et al. (1981) above. Marcus and Rispin's report was included as an appendix to U.S. EPA/RAF 1988.

[66] According to sources from EPA and industry, the Bull Run Coalition is simply an Oregon lawyer who takes advantage of statutory provisions that his fees are paid in cases in which the agency fails to meet its legal obligations. Bill Carpenter is the attorney for the group (*Environment Reporter*, August 25, 1995, 805). Such provisions were enacted to permit citizen groups to play a watchdog function over regulatory agencies.

[67] Chronic exposure to arsenic can lead to neurological, dermatological, vascular, hepatic, and other signs of noncancer toxicity (U.S. EPA/OW 1992).

[68] Note that using no uncertainty factor would result in a DWEL point estimate of 28 ppb, still nearly half the current MCL of 50 ppb.

[69] The range provided in IRIS is 0.1 to 0.8 µg/kg/day (U.S. EPA/OHEA 1996). By comparison, a DWEL of 50 ppb (the MCL for arsenic) implies an RfD of 1.43 µg/kg/day, exceeding the upper range of the RfD for arsenic provided in IRIS. Strictly speaking, there are other substances for which there are multiple IRIS values (fluoride and zinc), but in those cases, there has been a distinction made between the RfDs for children and adults. No such distinction was made for arsenic. It is worth noting that a 1994 Office of Water document (U.S. EPA/OW 1994) summarizing EPA drinking water standards has no RfD listed for arsenic but lists RfDs for other contaminants.

[70] The draft criteria document was prepared under contract by Life Systems, Inc.

[71] The Office of Water's nutritional essentiality argument was seriously weakened after SAB concluded that there was a lack of human evidence (U.S. EPA/SAB 1989). According to a former EPA drinking water official, water program risk

assessors continued to make the essentiality argument during the 1990s but not as prominently as the methylation argument. According to press reports, the EPA Office of Air and Radiation (which regulates arsenic as a hazardous air pollutant) was especially strong in its opposition to the essentiality argument (*Inside EPA*, October 7, 1988, 5–6).

[72] A formal analytic approach would allocate scarce research funds such that they are most likely to achieve the greatest reduction in uncertainty for decision-making. However, there is no consensus regarding the source or magnitude of the uncertainties or the likelihood or extent that various research proposals would reduce them.

[73] Task force members Barbara Beck, Ken Brown, and Warner North, Senior Vice President of Decision Focus, Inc., work for consulting firms that do work for the water, mining, and electric utility industries, but they have also done work for EPA, as employees and contractors. North has served as a member and consultant to committees of SAB. (See further discussion of North's role below.) According to a task force member, when the group was established, SEGH tried to give it a balanced membership. According to this source, member T. A. Tsongas of Washington State University has a background in public health and environmental activism. Paul Mushak was a member of the task force as it was originally constituted but resigned after the group's 1993 meeting. Rebecca Calderon and David Thomas, EPA/ORD scientists with HERL, are task force members. Former SAB staffer Richard Cothern has also been an active member.

[74] The Comprehensive Environmental Response, Compensation and Liability Act of 1980 established ATSDR. Funded by Superfund, ATSDR develops toxicological profiles for pollutants on contaminated sites.

[75] A former EPA drinking water official says that Guo's work was never presented to the drinking water program management.

[76] Huggett designated Peter Preuss, director of the ORD Office of Science, Planning and Regulatory Evaluation and chair of RAF, which reviewed the 1988 arsenic reassessment, to represent ORD at the workshop.

[77] The abstract reports a small difference in the prevalence of a biomarker (micronucleated bladder cells) between subjects in northern Chile exposed to high (600 µg/l) and low (20 µg/l) arsenic levels in drinking water. An ecological epidemiological study in the province of Cordoba, Argentina, found a relative risk ratio for bladder cancer of approximately two for both males and females, using all of Argentina as a referent population (Smith et al. 1994). A rule of thumb applied by some epidemiologists is that for a single study to be persuasive by itself, a relative risk ratio of three or four is the lower limit (see Taubes 1995).

[78] The intent here is to simply map out who presented scientific information to the decisionmaker. As a former EPA official noted, such access to EPA policymakers by advocates on both sides of the issue is not unusual.

[79] Goldman is a physician with a master's degree in public health.

[80] EPA's cost estimates may have changed since this analysis was conducted. According to an environmental lawyer, lower-cost water treatment technologies have been identified.

[81] The 1996 SDWA Amendments now make the issue moot; however, a former EPA official contends that EPA had sufficient flexibility prior to the amendments to adopt a midrange standard for arsenic. As evidence of this flexibility, the former agency official cites EPA's prior use of practical quantitation limits (PQLs, that is, setting standards based on technology-based analytical detection limits) to "come up with a midrange" for other drinking water contaminants. Another EPA official disagrees, however, stating that although the SAB Drinking Water Committee questioned the reliability of detection at 2 ppb, the PQL for arsenic was low enough that it would not provide a means of justifying a midrange standard. Therefore, in this case, disagreement regarding whether EPA had sufficient flexibility under SDWA may turn on differing judgments about the PQL for arsenic (which would involve a judgment about acceptable rates of false positives and false negatives) and the affordability of a standard based on the PQL for small systems.

[82] The agency based this interpretation on 1974 House report language and a 1986 senatorial statement in the Congressional Record (*Federal Register* 54: 22093–22094). This interpretation of feasibility was reinforced by the Senate report (S. Rept. 104-69, Discussion of Section 6) for the 1996 SDWA Amendments: "Feasible means the level that can be reached using the best available treatment technology that is affordable for large, regional drinking water systems."

[83] A former EPA drinking water official differs with this interpretation, suggesting that the decision never reached the point of seriously considering options such as waivers because the scientific uncertainties were a threshold issue.

[84] According to an independent risk analyst, EPA's decision not to regulate gasoline vapors on the basis of liver tumors in male rats provides a counterexample in which then Assistant Administrator for Pesticides and Toxic Substances John Moore explicitly stated what experimental evidence he would consider sufficiently compelling to depart from the agency's default assumption that humans are at least as sensitive as the most sensitive animal species. In the gasoline vapor case, the question of "how much science is enough" was formulated as a testable hypothesis, the studies were done, and the cancer mechanism in rats was found to be inapplicable to humans.

[85] The 1996 SDWA Amendments required EPA to consult with the National Academy of Sciences, other federal agencies, and interested public and private entities in conducting the arsenic study. EPA's December 1996 solicitation was reportedly made in response to a complaint filed by the Natural Resources Defense Council that there had been no opportunity for public input into the research plan.

REFERENCES

Abernathy, C., W. Marcus, and C. Chen. 1989. Report on Arsenic (As) Work Group Meetings (internal memorandum to P. Cook and P. Preuss). Washington, D.C.: U.S. EPA, Office of Drinking Water.

Abernathy, C., and E. Ohanian. 1992. Non-Carcinogenic Effects of Inorganic Arsenic. *Environmental Geochemistry and Health* 14: 35–41.

Beck, B., P. Boardman, G. Hook, et al. 1995. Response to Smith et al. (letter). *Environmental Health Perspectives* 102: 15–17.

Brown, K., and C. Abernathy. 1995. The Taiwan Skin Cancer Risk Analysis of Arsenic Ingestion: Effect of Water Consumption Rates and Food Arsenic Levels (Abstract). Presentation at the Society of Environmental Geochemistry and Health Second International Conference on Arsenic Exposure and Health Effects, San Diego, Calif., June 12–14.

Brown, K., and C. J. Chen. 1993. Assessment of the Dose-Response for Internal Cancers and Arsenic in Drinking Water in the Blackfoot Disease Endemic Region of Taiwan. Presentation at the Society of Environmental Geochemistry and Health International Conference on Arsenic Exposure and Health Effects, New Orleans, La., July 28–30.

Brown, K., H-R. Guo, and H. Greene. 1997a. Uncertainty in Cancer Risk at Low Doses of Inorganic Arsenic. *Human and Ecological Risk Assessment* (forthcoming).

Brown, K., H-R. Guo, T-L. Kuo, and H. Greene. 1997b. Skin Cancer and Inorganic Arsenic: Uncertainty-Status of Risk. *Risk Analysis* (forthcoming).

Buchet, J., R. Lauwerys, and H. Roels. 1981. Urinary Excretion of Inorganic Arsenic and Its Metabolites after Repeated Ingestion of Sodium Metaarsenite by Volunteers. *Int. Arch. Occup. Environ. Health* 48: 111–118.

Carlson-Lynch, H., B. Beck, and P. Boardman. 1994. Arsenic Risk Assessment. *Environmental Health Perspectives* 102: 354–356.

CAST (Council for Agricultural Science and Technology). 1976. *Application of Sewage Sludge to Cropland: Appraisal of Potential Hazards of the Heavy Metals to Plants and Animals.* Report No. 64. Prepared at request of Office of Water Program Operations, U.S. EPA. November 15. Ames, Iowa.

Chen, C. J., Y. Chuang, T. Lin, and H-Y. Wu. 1985. Malignant Neoplasms among Residents of a Blackfoot Disease-Endemic Area in Taiwan: High-Arsenic Artesian Well Water and Cancers. *Cancer Research* 45: 5895–5899.

Chen, C. J., Y. Chuang, V. You, et al. 1986. A Retrospective Study on Malignant Neoplasms of Bladder and Liver in Blackfoot Disease–Endemic Areas in Taiwan. *British Journal of Cancer* 53: 399–405.

Chen, C. J., T. Kuo, and H. Wu. 1988. Arsenic and Cancers (letter). *Lancet* 1: 414–415.

Dower, R. 1990. Hazardous Wastes. In *Public Policies for Environmental Protection,* edited by P. Portney. Washington, D.C.: Resources for the Future, 151–194.

Fowle, J. 1992. Health Effects of Arsenic in Drinking Water: Research Needs. *Environmental Geochemistry and Health* 14: 63–68.

Guo, H-R. 1993. Arsenic in Drinking Water and Urinary Cancers. Presentation at the Society of Environmental Geochemistry and Health International Conference on Arsenic Exposure and Health Effects, New Orleans, La., July 28–30.

Hopenhayn-Rich, C., A. Smith, and H. Goeden. 1993. Epidemiological Studies Do Not Support the Methylation Threshold Hypothesis for the Toxicity of Inorganic Arsenic. *Environmental Research* 60: 161–177.

Hopenhayn-Rich, C., M. L. Biggs, and A. H. Smith. 1998. Lung and Kidney Cancer and Mortality Associated with Arsenic in Drinking Water in Cordoba, Argentina. *International Journal of Epidemiology.* 27: 561–569.

IARC (International Agency for Research Against Cancer). 1980. *Monograph on the Evaluation of the Carcinogenic Risk of Chemicals to Humans. Some Metals and Metallic Compounds,* Vol. 23. Lyon, France: IARC, 39–141.

Jasanoff, S. 1990. *The Fifth Branch: Science Advisors as Policymakers.* Harvard University Press: Cambridge, Mass.

Landy, M., M. Roberts, and S. Thomas. 1994. *The Environmental Protection Agency.* New York: Oxford University Press.

Marcus, W. 1988. *Quantitative Toxicological Evaluation of Ingested Arsenic.* Draft. Prepared for the Office of Drinking Water. Washington, D.C.: U.S. EPA.

Mass, M. 1992. Human Carcinogenesis by Arsenic. *Environmental Geochemistry and Health* 14: 49–54.

Mushak, P. 1994. Persisting Scientific Issues: Arsenic and Human Health. In *Proceedings of the International Conference on Arsenic Exposure and Health Effects, New Orleans, La., July 28–30, 1993,* edited by W. Chappell, C. Abernathy, and C. Cothern. *Environmental Geochemistry and Health* 16 (Special Issue): 305–318.

Mushak, P., and A. Crocetti. 1995. Risk and Revisionism in Arsenic Cancer Risk Assessment. *Environmental Health Perspectives* 103: 684–689.

———. 1996. Response: Accuracy, Arsenic, and Cancer [letter]. *Environmental Health Perspectives* 104: 1014–1018.

NRC (National Research Council). 1977. *Drinking Water and Health,* Vol. 1. Report of the Safe Drinking Water Committee. Washington, D.C.: National Academy Press.

———. 1980. *Drinking Water and Health,* Vol. 3. Report of the Safe Drinking Water Committee. Washington, D.C.: National Academy Press.

———. 1983. *Drinking Water and Health,* Vol. 5. Report of the Safe Drinking Water Committee. Washington, D.C.: National Academy Press.

———. 1986. *Drinking Water and Health,* Vol. 6. Report of the Safe Drinking Water Committee. Washington, D.C.: National Academy Press.

———. 1987. *Drinking Water and Health: Pharmacokinetics in Risk Assessment,* Vol. 8. Report of the Safe Drinking Water Committee. Washington, D.C.: National Academy Press.

———. 1993. *Issues in Risk Assessment.* Washington, D.C.: National Academy Press.

———. 1994. *Science and Judgment in Risk Assessment.* Washington, D.C.: National Academy Press.

———. 1999. *Arsenic in Drinking Water.* Washington, D.C.: National Academy Press.

Rudel, R., T. Slayton, and B. Beck. 1996. Implications of Arsenic Genotoxicity for Dose-Response of Carcinogenic Effects. *Regulatory Toxicology and Pharmacology* 23: 87–105.

SEGH (Society of Environmental Geochemistry and Health). 1994. *Proceedings of the SEGH Conference on Arsenic Exposure and Health Effects, New Orleans, La., July 28–30, 1993.*

Slayton, T., B. Beck, K. Reynolds, et al. 1996. Issues in Arsenic Cancer Risk Assessment [letter]. *Environmental Health Perspectives* 104: 1012–1013.

Smith, A., C. Hopenhayn-Rich, M. Bates, et al. 1992. Cancer Risks from Arsenic in Drinking Water. *Environmental Health Perspectives* 97: 259–267.

Smith, A., C. Hopenhayn-Rich, L. Moore, M. Biggs, and R. Barroga. 1994. Epidemiological and Biomarker Findings Concerning Arsenic and Bladder Cancer in Chile and Argentina. *Workshop on Arsenic: Epidemiology and PBPK Modeling,* 2.

Smith, A., M. Biggs, C. Hopenhayn-Rich, and D. Kalman. 1995. Arsenic Risk Assessment. *Environmental Health Perspectives* 103: 13–17. (Correspondence).

Smith, A. H., M. Goycolea, R. Haque, and M. L. Biggs. 1998. Marked Increase in Bladder and Lung Cancer Mortality in a Region of Northern Chile Due to Arsenic in Drinking Water. *American Journal of Epidemiology.* 147: 660–669.

Southwick, J., A. Western, M. Beck, T. Whitley, R. Isaacs, J. Petajan, and C. Hansen. 1981. *Community Health Associated with Arsenic in Drinking Water in Millard County, Utah.* EPA-60011-81-064. Report to the Health Effects Research Laboratory, U.S. EPA. Cincinnati, Ohio: Health Effects Research Laboratory.

Stone, R. 1993. Toxicology Goes Molecular. *Science* 259: 1394–1399.

Taubes, G. 1995. Epidemiology Faces Its Limits. *Science* 269: 164–169.

Thomas, L. 1988. Memorandum from the Administrator of EPA, June 21. Included in the Preface of U.S. EPA/RAF 1988.

TNCC (The National Coal Council). 1995. *A Critical Review of Efficient and Environmentally Sound Coal Utilization Technology.* Arlington, Va.: TNCC, May.

Tseng, W-P., H. Chu, S. How, J. Fong, C. Lin, and S. Yen. 1968. Prevalence of Skin Cancer in an Endemic Area of Chronic Arsenicism in Taiwan. *Journal of the National Cancer Institute* 40(3): 453–463.

U.S. EPA (U.S. Environmental Protection Agency). 1986. Guidelines for Carcinogen Risk Assessment. *Federal Register* 51: 33992–34003.

———. 1992. Guidelines for Exposure Assessment. *Federal Register* 57: 22888.

U.S. EPA/OHEA (U.S. EPA Office of Health and Environmental Assessment). 1996. *Integrated Risk Information System (IRIS) Online.* Cincinnati, Ohio: National Center for Environmental Assessment, U.S. OHEA.

U.S. EPA/ORD (U.S. EPA Office of Research and Development). 1984. *Health Assessment Document for Inorganic Arsenic.* Final Report. EPA-600/8-83-021F. Research Triangle Park, N.C.: Office of Research and Development, Environmental Criteria and Assessment Office.

U.S. EPA/OW (U.S. EPA Office of Water). 1980. *Ambient Water Quality Criteria for Arsenic.* Washington, D.C.: Office of Water Regulations and Standards, U.S. EPA.

————. 1992. *Draft Drinking Water Criteria Document on Arsenic*. Washington, D.C.: Human Risk Assessment Branch, Office of Science and Technology, Office of Water, U.S. EPA (Prepared by Life Systems, Inc.).

————. 1994. *Drinking Water Regulations and Health Advisories*. Washington, D.C.: U.S. EPA, November.

U.S. EPA/RAF (U.S. EPA Risk Assessment Forum). 1988. *Special Report on Ingested Inorganic Arsenic: Skin Cancer; Nutritional Essentiality*. EPA/625/3-87/013. Washington, D.C.: U.S. EPA.

U.S. EPA/SAB (U.S. EPA Science Advisory Board). 1989. *Science Advisory Board's Review of the Arsenic Issues Relating to the Phase II Proposed Regulations from the Office of Drinking Water*. EPA-SAB-EHC-89-038. Washington, D.C.: U.S. EPA, September 28.

U.S. EPA/SAB/DWC (U.S. EPA Science Advisory Board Drinking Water Subcommittee). 1993. *Review of the Draft Drinking Water Criteria Document on Inorganic Arsenic*. EPA-SAB-DWC-94-004. Washington, D.C.: U.S. EPA.

U.S. GAO (U.S. General Accounting Office). 1996. *Superfund: How States Establish and Apply Environmental Standards When Cleaning Up Sites*. GAO/RCED-96-70FS. Washington, D.C.: U.S. GAO.

Uthus, E. 1992. Evidence for Arsenic Essentiality. *Environmental Geochemistry and Health* 14: 55–58.

Valentine, J. H. Kang, and G. Spivey. 1979. Arsenic Levels in Human Blood, Urine and Hair in Response to Exposure Via Drinking Water. *Environmental Research* 20: 24–32.

Walker, K., M. Sadowitz, and J. Graham. 1995. Confronting Superfund Mythology: The Case of Risk Assessment and Management. In *Analyzing Superfund: Economics, Science and Law*, edited by R. Revesz and R. Stewart. Washington, D.C.: Resources for the Future, 25–54.

Warrick, J. 1999. New Arsenic Limits Proposed for Water. *The Washington Post*. March 24, p. A3.

Yosie, T. 1988. The EPA Science Advisory Board: A Ten-Year Retrospective View. *Risk Analysis* 8(2): 167–168.

Yost, L., R. Schoof, H-R. Guo, et al. 1994. Recalculation of the Oral Toxicity Values for Arsenic Correcting for Dietary Arsenic Intake. Presentation at the Society for Environmental Geochemistry and Health Rocky Mountain Conference, Salt Lake City, Utah, July 18–19.

Appendix C

The 1987 Revision of the National Ambient Air Quality Standard for Particulate Matter

BACKGROUND

In October 1948, a dense fog blanketed Donora, Pennsylvania. Approximately 40% of the population of 10,000 suffered some ill symptoms. Twenty people, mostly adults with preexisting cardiopulmonary conditions, died during or shortly after the fog. A series of similar winter episodes occurred in London between 1948 and 1962 (U.S. EPA 1982a). In retrospect, the cause of these public health episodes—air pollution trapped by thermal inversions—seems painfully obvious. However, several years passed before scientists excluded alternative potential causes and identified air pollution as the culprit. Beginning in 1963 and culminating in 1990, Congress enacted an escalating series of statutes to address air pollution on a national basis.

The 1963 Clean Air Act (CAA) directed the Department of Health, Education, and Welfare (HEW) to prepare criteria documents (CDs) (hence the term "criteria" air pollutants) summarizing scientific information on widespread air pollutants for use by state and local public health agencies. The 1967 Air Quality Act specified the role of CDs in the development of state air quality standards. A year later, the Surgeon General established the National Air Quality Criteria Advisory Committee to review and advise on CDs. In 1969, HEW produced the original CD for particulate matter (PM). On the basis of HEW's CD, and in response to the 1970 CAA Amendments, EPA promulgated the first national ambient air quality standards (NAAQS) for PM in April 1971 (*Federal Register* 36: 8186).[1]

The 1970 CAA also required that the NAAQS be based on air quality criteria that reflect the latest scientific information and that both the criteria and standards be periodically reviewed and, if appropriate, revised. In 1976, as a result of internal EPA review and recommendations of the EPA Science Advisory Board (SAB),[2] the agency decided to revisit all of the criteria air pollutants, with work scheduled to begin on a combined PM/sulfur oxides (SO_x) criteria document in 1979.[3] Dissatisfied with EPA's progress in reevaluating the original NAAQS, Congress directed the agency in the 1977 CAA Amendments to revise or reauthorize all of the NAAQS by December 31, 1980, and every five years thereafter. The 1977 amendments also required the Clean Air Science Advisory Committee (CASAC) to review the air quality criteria *and* standards and report directly to the Administrator *and* to Congress (Section 109, emphasis added).[4]

Litigation brought against EPA by the Iron and Steel Institute in 1978 was designed to keep the particulate review "in limbo," according to a former EPA official. But in October 1979, the agency announced that it was in the process of revising the criteria document for PM and SO_x and reviewing the existing NAAQS for possible revisions (*Federal Register* 44: 56731). In April 1980, EPA released for external review the first of three drafts of the revised PM/SO_x criteria document, prepared by EPA Office of Research and Development Environmental Criteria and Assessment Office (ECAO) (*Federal Register* 45: 24913). From August 1980 to January 1981, EPA convened six public meetings of CASAC to review the CD, the second draft of which was released late in January (*Federal Register* 46: 9746). In the spring of 1981, the EPA air program Office of Air Quality Planning and Standards (OAQPS) released the first draft of the staff paper (SP), a document informed by the criteria document and additional analysis that presents air program office staff recommendations regarding the NAAQS. In July and November of 1981, CASAC reviewed the second and third drafts of ECAO's criteria and the first and second drafts of OAQPS's staff analysis and recommendations.

In January 1982, a little over two years since EPA's initial announcement that the review was under way, CASAC sent a closure memorandum to the Administrator, Anne Gorsuch, indicating satisfaction with the final drafts of the CD and SP. Meanwhile, in order to make it clear that the agency would not use a particulates revision as a backdoor route to institute a regulatory acid rain program, PM and SO_x were decoupled administratively.[5] Reducing the scope of the review might have been expected to facilitate a timely decision; however, to comply with a 1981 executive order issued by the Reagan administration (Executive Order 12291), EPA was now required to prepare an economic regulatory impact analysis (RIA) on proposed major regulations for submission to the White House Office of Management and Budget.[6] In addition, no administrative action was politically feasible during the tumultuous period surrounding Gorsuch's 1983 resignation.

In February 1984, EPA released an RIA for the particulates NAAQS, the monetized health benefits of which were based to a significant extent on epidemiological morbidity analyses.[7] A month later, the agency issued its notice of proposed rulemaking (*Federal Register* 49: 10408). On the same day, ORD's ECAO released a newly revised criteria document. Given the lag in EPA's review of the NAAQS for particulates occasioned by the change in administrations, the agency was appropriately concerned that its rulemaking would be susceptible to legal challenges that it did not "reflect that latest scientific knowledge."[8] Already two years had passed since CASAC signed off on the agency's scientific analysis. However, the criteria document also reinforced the notion that the epidemiological basis of the RIA was viewed with great skepticism by EPA's science analysts.

Some respondents suggested that the mistrust of regulatory analysis based on epidemiology was, at least in part, a manifestation of "CHESS syndrome." The CHESS (Community Health and Environmental Surveil-lance System) research program began in 1967 under HEW's National Air Pollution Control Administration, and EPA inherited the program when the agency was created in 1970. In the public health tradition, CHESS collected ambient measurements and conducted epidemiological studies for criteria air pollutants. Problems with the CHESS program came to light when questions arose concerning a 1974 EPA study of SO_x. A 1975 SAB review of the study criticized both the epidemiological methods and pre-sentation of the findings, and a series of 1976 *Los Angeles Times* articles alleged that EPA staff had deliberately distorted data to support their regu-latory position. The news was followed by hearings and investigations by the House Committee on Science and Technology, which concluded that the CHESS program at best served only to confirm previous scientific find-ings and that, as a result of inadequate peer review, the SO_x study was useless for regulatory purposes (Jasanoff 1990, 84–86). As an EPA official recalls, "The Brown report [Rep. George Brown (D-Calif.), then chair of the House Science Committee] on the CHESS studies set the in-house 'epi' program back and tarnished the studies that were to be used in the early eighties. It clouded 'epi' studies in general, and the agency lost eight years of 'epi' work."[9]

In December 1985, CASAC concluded that relevant new studies had emerged over the preceding four years and recommended that the agency prepare addenda to the criteria document and staff paper. During 1986, addenda prepared by EPA analysts in ECAO and OAQPS were drafted, reviewed internally and externally, approved by CASAC, revised, and released (U.S. EPA 1986a, 1986b). In July 1987, nearly eight years after the agency announced the revision was under way, Administrator Lee Thomas signed the final rule (*Federal Register* 52: 24634). Table C-1 provides background for the development of national regulation and policy for ambient air particulates. Table C-2 provides a time line of the 1987 NAAQS revision.

The 1971 primary NAAQS for particulates, based on total suspended particulates (TSP), included a daily (24-hour arithmetic) average of 260

Table C-1. Background on U.S. Regulation of Particulate Air Pollution

1948	Air pollution episode in Donora, Pa., affects 40% of the town's population. Similar episodes occur in London and the U.S. eastern seaboard over the next decades.
1963	Clean Air Act directs the Department of Health, Education, and Welfare (HEW) to prepare criteria documents (CDs) summarizing science on widespread air pollutants for use by state and local public health agencies.
1967	Air Quality Act specifies role of CDs in development of state air quality standards. National Air Pollution Control Administration initiates Community Health and Environmental Surveillance System (CHESS) research program.
1969	Surgeon General establishes National Air Quality Criteria Advisory Committee. Original Criteria document for particulate matter (PM), measured in total suspended particulates (TSP), produced by HEW.
1970	Clean Air Act Amendments call for national ambient air quality standards (NAAQS). EPA inherits national air pollution control authority and begins setting original NAAQS on the basis of HEW CDs.
1971	EPA promulgates original NAAQS for PM, setting a 24-hour level of 260 $\mu g/m^3$ and an annual level of 75 $\mu g/m^3$, measured in TSP.
1970s	Experimental findings indicate that TSP includes nonrespirable, coarse particles.
1976	EPA decides to revisit all of the NAAQS, with work scheduled to begin on a combined PM/sulfur oxides criteria document in 1979.
1977	Clean Air Act Amendments direct EPA to revise or reauthorize all of the NAAQS by December 31, 1980, and every five years thereafter. The amendments also establish the Clean Air Science Advisory Committee.
1987	EPA promulgates revised NAAQS for PM, setting a primary standard based on a 24-hour level of 150 $\mu g/m^3$ and an annual level of 50 $\mu g/m^3$ using the PM-10 indicator, which measures finer-sized particles than TSP.
1994	The U.S. district court of Arizona orders EPA to review and, if necessary, revise the current NAAQS for particulates by January 31, 1997.
1996	EPA proposes to revise the current annual and daily PM-10 standards by adding new annual and daily standards based on PM-2.5.

$\mu g/m^3$ and an annual (geometric)[10] average of 75 $\mu g/m^3$. Experimental studies conducted by academics in the late 1960s and 1970s demonstrated that the fraction of PM deposited deep in the pulmonary system (where particles may have adverse health effects) consisted of the smaller-diameter components. EPA researchers recognized the implications: by regulating on the basis of TSP in 1971, "the agency realized it was missing the boat by measuring the wrong size category and having industry focus on the wrong size category," according to an EPA official. The 1984 proposal generated a range of alternative standards based on PM-10, a PM indicator focused on particles with a diameter of 10 microns or less: 150–250 $\mu g/m^3$ for the daily (arithmetic) average and 50–65 $\mu g/m^3$ for the annual (arithmetic) average. The 1986 staff paper addendum included lower bounds for the daily and annual PM-10 ranges, 140 $\mu g/m^3$ and 40 $\mu g/m^3$, respectively (U.S. EPA 1986b). In 1987, the final rule established the lower bound of the proposed ranges for both the daily (150 $\mu g/m^3$) and the annual (50 $\mu g/m^3$) levels of PM-10. New scientific findings, in the form of the particle deposition experiments, had penetrated the agency and initiated the long, complex sequence of revision.

Ultimately, the 1987 revisions do not appear to have appreciably tightened or relaxed the standard levels. Instead the result was a more refined targeting of pollution control and monitoring efforts on the finer-sized fraction of PM. In October 1994, seven years after the 1987 revision, in response to a suit brought by the American Lung Association, a federal district court ordered EPA to review and, if necessary, revise the current NAAQS for particulates by January 31, 1997 (*ALA v. Browner*, DC AZ, October 6, 1994). In November 1996, EPA proposed to revise the national ambient air quality standards for particulate matter on the basis of a slowly accumulated mass of epidemiological work (*Federal Register* 27: 1721). (See also Friedlander and Lippman 1994.)

SCIENTIFIC ISSUES

Interviewees unanimously responded that, overall, there was adequate scientific information available in 1987 to inform the decision to revise the particulate NAAQS.[11] However, respondents identified a number of areas in which there was considerable scientific uncertainty, including the specific constituents of particulate matter that were responsible for causing the observed effects, extrapolation between different ambient particle measures (that is, British smoke, TSP, PM-10) and from acute (high-level, short-duration) to chronic (low-level, long-duration) exposures, and interpretation of epidemiological studies using subjective reported morbidity response data (for example, restricted activity days). Respondents' opinions varied markedly about the overall magnitude of uncertainty, although it seems that the most certain scientific issue was the implications of the studies demonstrating that smaller particles are deposited more deeply in the lungs. An EPA air program official commented that "there were many uncertainties. It's hard to know how they would combine quantitatively." The discussion below focuses on the determination of the appropriate par-

Table C-2. Time Line of the 1987 Particulate Matter NAAQS Revision

1974	EPA air program begins investigating fine-particle standard.
1979	EPA announces that it is in the process of revising the criteria document for particulate matter (and sulfur oxides), and reviewing existing air quality standards for possible revisions.
1980	April 11: Draft 1 of revised particulate matter/sulfur oxides criteria document (CD), prepared by the Environmental Criteria and Assessment Office (ECAO), made available for external review
	August 20–22: Clean Air Science Advisory Committee (CASAC) public meeting to review Draft 1 of CD. The committee recommends five additional public meetings be held at EPA to discuss the draft. Meetings held: November 7, 1980; November 20, 1980; November 25, 1980; December 4, 1980; and January 7, 1981.
1981	January 29: CD Draft 2 released.
	Spring: Draft 1 of staff paper (SP) prepared by the Office of Air Quality Planning and Standards.
	July 7–9: CASAC public meeting to review Draft 2 of CD, Draft 1 of SP.
	October 28: CD Draft 3 released.
	November 16–18: CASAC public meeting to review Draft 3 of CD and Draft 2 of SP.
	December: Final draft of CD completed.
1982	January: CASAC closure memorandum endorses CD and SP. Final draft of SP released.
	December: Final draft of CD released.

ticle size for the standard and on the consideration of the ambient particulate levels at which adverse health effects may occur.

Particle Size

Ambient particulate matter represents a broad class of chemically and physically diverse substances, including liquid droplets or solids ranging in size from 0.005 to 100 microns.[12] On the basis of the 1969 HEW criteria document, EPA set the 1971 NAAQS using the TSP indicator of particulate matter. Aerometric TSP samplers vacuum suspended particles from the air and collect them in sizes up to 25–45 microns on filters (U.S. EPA 1982b).

According to an EPA official, air researchers recognized early on that TSP was a crude indicator because a variety of experimental dosimetry[13] results published in the late 1960s and 1970s indicated that only particles much finer in diameter (on the order of ten microns or less) were likely to deposit in the thoracic region, deep in the lungs (see, for example, Albert

Table C-2. Time Line of the 1987 Particulate Matter NAAQS Revision *(Continued)*

1983	EPA Administrator Anne Gorsuch-Burford resigns.
1984	February: EPA releases regulatory impact analysis.
	March: Environmental Criteria and Assessment Office issues revised CD.
	March 20: Notice of proposed rulemaking (NPRM) proposes replacing TSP with PM-10, and a primary standard with a 24-hour level in the range of 150 to 250 $\mu g/m^3$ and an annual level in the range of 50 to 65 $\mu g/m^3$.
1985	December 16–17: CASAC meets to discuss relevance of new scientific studies on health effects of PM that emerged since the committee completed its review of the CD and SP in January 1982. The committee recommends that the agency prepare separate addenda to the CD and SP.
1986	May 22–23: Peer-review workshop at EPA (Research Triangle Park) to review first draft of CD addendum.
	July 3: External review draft of CD addendum available.
	September 16: External review draft of SP addendum available.
	October 15–16: CASAC public meeting to review draft CD and SP addenda. The SP and CASAC float reduced lower ends of the proposed ranges for the daily and annual levels (140 $\mu g/m^3$ and 40 $\mu g/m^3$, respectively).
	December: CASAC sends closure memorandum on CD and SP addenda to Administrator. Final CD and SP addenda released.
1987	July 1: EPA promulgates final rule replacing TSP with PM-10 and setting primary NAAQS levels at 150 $\mu g/m^3$ (daily) and 50 $\mu g/m^3$ (annual).

et al. 1973, Lippman 1977).[14] Larger particles tend to be blocked or expelled from higher regions of the oronasal passages and thus would have minimal impact on lung function; respiratory ailments, such as bronchitis; respiratory and cardiovascular disease; or mortality. In addition, atmospheric particles were shown to occur generally in a bimodal size distribution, with a fine fraction and a coarse fraction conventionally separated by a particle diameter of 2.5 microns (for example, Whitby 1975).[15] On the weight basis used to set the particulates NAAQS in 1971, the large, nonrespirable particles collected by TSP samplers dominated the smaller, respirable particles collected by the instruments.

Beginning in 1978, additional particulate depositional work was done at EPA by ORD's Health Effects Research Laboratory (HERL). Beyond synthesizing the accumulated literature, researchers at HERL conducted some original dosimetry work and, perhaps more importantly, connected the depositional results with the performance of aerometric sampling instruments to develop a recommended "cut value" for a revised particulates standard. A nominal, or 50%, cut point refers to the particle diameter for which the efficiency of particle collection by sampling instruments is 50%. Larger particles are not excluded but are collected with decreasing efficiency (that is, the larger the diameter of the suspended particle, the lower the likelihood of collection). A 1979 paper by HERL staff concluded that a cut value of fifteen microns or less would ensure that the NAAQS would focus on the inhalable fine fraction of suspended particulates (Miller et al. 1979).

Projections of areas that would fall into nonattainment for the particulates NAAQS without further controls were sensitive to the selection of the cut point, and the mining industry, whose emissions are predominately in the coarse fraction, argued for smaller cut points (for example, PM-6). However, the issue was largely put to bed when, in 1981, the International Standards Organization adopted a nominal ten-micron cut point for particles that could penetrate the thoracic region (ISO 1981). Although technical support for the ten-micron level was based on the operating efficiencies of sampling instruments, rather than on any health basis, a former EPA official suggests that the argument for international harmonization of standards provided political cover for what was essentially a judgment call within the range below fifteen microns. In 1984, the agency proposed to replace TSP with PM-10.[16]

Mortality and Morbidity

Having determined the particle size of concern, the next issue to be dealt with, as required under the Clean Air Act, was to determine the ambient concentration of PM-10 at which no reasonably expected adverse health effects would occur, including an allowance for an adequate margin of safety, and taking into consideration sensitive members of the population. However, this legislative framing of the issue presumes that there exists an identifiable threshold level below which no adverse effects occur. "This is the problem with ambient air quality standards," concludes a former EPA researcher. "What you're trying to do is ludicrous—set the level below which the most sensitive person in the population will have no adverse health effects. In fact, it is impossible to come up with a scientifically justifiable number. The only intelligent way is cost-benefit or cost-effectiveness analysis, but we don't do that." Saddled with a scientifically unsound congressional mandate, EPA undertook to determine the lowest levels of PM-10 at which mortality and morbidity effects could be detected or reasonably inferred.

The EPA HERL and other laboratories conduct clinical studies of humans in environmentally controlled exposure chambers, and numerous chamber studies were conducted for PM/SO_x prior to 1987 (see U.S. EPA 1982a, 1986a). However, the experimental conditions lack generalizability, in particular to chronic exposures, and the subjects employed are often healthy adults not representative of sensitive populations at risk: the elderly, those with preexisting conditions, and children.[17] According to some respondents, toxicological experiments with laboratory animals provided very clear mechanistic support for the adverse effects of SO_x aerosols in animals. Furthermore, dosimetry work helped researchers bridge the gap between the toxicological findings in animals and the chamber study results in humans. However, unlike most pollutants, PM is defined by its size; it is not chemically specified or even associated with a particular industrial process.[18] The effects of toxicological experiments with inhaled particulates depend largely on their chemical composition, which varies considerably in the environment. As a result, "the toxicological basis for particulates is weaker than for other criteria air pollutants," according to an EPA air program official.

Given the limits of experimental data in determining "safe" particulate levels, analysts turned to epidemiology to establish associations between observed mortality and morbidity in populations exposed to measured ambient particulate levels. Three sets of epidemiological studies are noteworthy: studies of London mortality data, which were relevant to acute exposures; the Harvard Six City Study, which was primarily relevant to chronic exposures;[19] and a final set of studies (which were questioned by health scientists but included in the economic RIA) relating chronic particulate exposures to reported morbidity measures.

For the purposes of setting regulatory levels of PM-10, most attention was focused on studies that would influence the daily standard because it was generally considered to drive the annual standard. Considering the 24-hour and annual standards together, for example, EPA (1986b) projected that attaining a 24-hour standard of 150 $\mu g/m^3$ would substantially reduce annual average levels in a number of areas to below 50 $\mu g/m^3$. Analyses of London mortality data had been accumulating in the literature for more than twenty years at the time of the PM review (for example, Martin and Bradley 1960), but transfer of the results to the United States was frustrated by the incomparability of British and U.S. air pollution measures. The British smokeshade (BS) indicator analyzes light reflected from a stain formed by particulate matter collected on paper. The collection equipment typically employed had a cut point of approximately 4.5 microns (U.S. EPA 1982a). Unlike the TSP or PM-10 measures, the BS method did not permit direct measurement of the weight or chemical composition of collected particles. Instead, it reflected the density of the stain and the optical properties of the collected materials. As a result, the BS method was a better measure for sooty dark particles, such as elemental carbon, than for aerosols, such as sulfates.

In the 1980s, academic epidemiologists, EPA policy analysts, and others reanalyzed the London data (for example, Mazumdar et al. 1981; Ostro 1984; Schwartz and Marcus 1986). The reanalyses suggested the lack of a threshold in the relationship between "British smoke" levels and mortality. The conundrum of comparing incomparable measurements was bypassed by reanalyzing the London data on their own merits. "Nobody knew how to convert from BS to PM-10," allows a former EPA analyst, "but the lack of a threshold was clear, or if there was one, it clearly went down to very low levels."

There remains some uncertainty in exporting the results, however, as a consequence of the differences in composition of particulate matter between London, circa 1958–1972 (with soot from coal use) and contemporary U.S. cities where particulates of concern are predominately aerosols and fine dusts. Mortality studies conducted solely on the basis of epidemiological data in contemporary U.S. cities, however, suffer from an inability to tease the potentially independent effects of particulates apart from the potential contributions of other air pollutants that may be correlated with PM, such as ozone and SO_x. In addition, it is uncertain to what extent the relationship between mortality and acute air pollution exposures reflects the death of previously ill people whose lives may be shortened by a matter of days or weeks, the ghoulishly dubbed "harvesting effect."

The Harvard Six City Study (Ware et al. 1986) found that increased rates of bronchitis and lower respiratory illness in preadolescent children were associated with annual average ambient fine particulate concentrations. In addition to their focus on a biologically sensitive—and politically salient—subpopulation, the relevance of the Harvard study results was that the effects were neither associated with historic episodes of gross air pollution nor observed in Europe, but were observed at concentrations experienced in eastern and midwestern U.S. cities between 1974 and 1977.[20]

A series of epidemiological studies by economists and policy analysts developed quantitative relationships between chronic ambient particulate exposures and morbidity effects for which monetary values could be readily attributed (for example, Lave and Seskin 1977; Ostro, 1983). According to a number of respondents, these studies tended to suggest net economic benefits at levels of particulates lower than current standards. However, experimental scientists had numerous objections to relying on these studies for quantitative assessments of the health impacts of different ambient PM-10 concentrations. For example, the studies used air-monitoring data available from networks established for determining areawide attainment of NAAQS. Such monitoring data are often poorly correlated with actual human exposures within the area, and, in addition, most of the PM data were measured in terms of TSP rather than PM-10. While concurrent TSP and PM-10 measurement was initiated in the 1980s on some monitoring sites in some cities to facilitate eventual conversion between the indicators, the relationship between TSP and PM-10 can vary markedly among sites. For the majority of monitoring sites without concurrent TSP and PM-10 measurements, any estimated relationship between "backcasted" PM-10 exposures and reported morbidity effects was, therefore, highly uncertain.

In addition, these studies used indicators of morbidity that were economically relevant and available over a broad range of areas and exposure levels, such as responses from the National Health Interview Survey.[21] Such surveys include questions like, "Did you go to work today?" Without some formal calibration (which could be achieved by conducting a followup, independent confirmation of the accuracy of responses for a subsample of the survey), the reliability of survey data is hard to assess. Evaluating subjective "ordinal" responses from different individuals is also problematic (for example, the same response to "Is your cough severe or mild?" from different people can mask substantial differences in the severity of symptoms). Experimental scientists are generally skeptical of data they regard as "subjective." But, by relying on overt "clinical" symptoms associated with acute exposures, analysts may not capture important health effects resulting from "real-world" exposure levels and patterns. A senior EPA careerist stated that the epidemiological analysis conducted by economists and included in the RIA "was not sufficiently developed or broadly accepted." ECAO's criteria document gave the Harvard study a generally favorable review but concluded that few of the studies on morbidity effects associated with chronic exposures to airborne particles provided useful results by which to derive quantitative conclusions concerning exposure-effect relationships (U.S. EPA 1986a).

Regarding the epidemiological evidence overall, responses from the interviewees were mixed. Some respondents characterized the epidemiology available for the 1987 PM revision as good to very good, but a former EPA researcher judged that "the 'epi' was only fair. The methods were underdeveloped, even for their time. Very few of the studies controlled for confounding variables."[22] According to a statistical analyst, "Epi, dose-response studies are simply hard to carry out. There's always going to be different interpretations with these types of studies. Some of it points this way; some of it points that way. There are not enough scientists, time, and money to come up with definitive answers." As a result, controversies over such studies generally boil down to differing judgments about what constitutes sufficient evidence in the context of a specific decision. These judgments will invariably be conditioned by underlying values.

THE PROCESS WITHIN EPA

Setting the Agenda

The legislative mandate for revisiting NAAQS on a five-year cycle was the factor most frequently (six) mentioned by respondents as being responsible for getting the PM revision on the agency's agenda. However, activity within the agency began at least three years prior to the 1977 CAA Amendments. One air program official "came into the agency in 1974 to look into a fine particle standard." Despite the legislative requirement for periodic review, in the view of a former agency political appointee, "What got the decision on the agency's agenda was the depositional studies." Therefore, new scientific information played a key role in setting the regulatory agenda.

Most of the early activity within EPA occurred in Research Triangle Park, North Carolina, where air program offices (OAQPS) and research offices (ECAO and HERL) are colocated. A 1974 paper by John Bachmann in the air program's OAQPS addressed the question of whether sulfates should be regulated separately from the rest of suspended particulates. Four years later, OAQPS requested that ORD's ECAO begin to address the issue of particle size and the appropriate cut value for monitoring equipment. In 1979, Fred Miller, who had previously been conducting academic research on ozone dosimetry, and others at HERL made the recommendation to set the nominal cut point at fifteen microns or less.

The Criteria Document–Staff Paper Process

As detailed in the background section above, ECAO and OAQPS spent much of the next seven years drafting, vetting, and revising criteria documents and staff papers for the PM revision. Much of the writing of the criteria documents for PM was contracted out to academics and independent researchers. But, with the exception of some of the exposure assessment work that was contracted, the staff papers, according to an air program official, "were basically a four-person effort" conducted by OAQPS staff. A number of respondents commented that the practice of relying on contracted experts to contribute to criteria documents can result in authoritative but unsynthesized analysis. But, ECAO resources are insufficient to maintain a cadre of full-time staff dedicated to each criteria air pollutant.[23]

While EPA drafted and revised CDs and SPs, the agency's stated intention to replace the TSP measure motivated others outside the agency to conduct additional dosimetry work (which, according to an EPA air program official, confirmed earlier findings) and ambient air monitor manufacturers to develop alternative sampling technology. During the course of the ongoing scientific review, the agency also initiated simultaneous monitoring of PM-10 and TSP in several areas. An academic remarked, "Give the agency credit for gathering that data prior to finishing the review. The agency set about to create research networks in numerous cities to make scientifically valid conversion factors, investing in the resources to get the necessary data. It was started too late if they were going to [finalize the rule] by 1985, but it was in time for 1987."

Early in the review process, EPA staff also approached Harvard investigators about conducting the Six City Study. The agency was motivated to support the research because prior epidemiological studies came from the field of occupational health that did not cover potentially sensitive or highly exposed populations, such as children, the elderly, or outdoor laborers. According to an EPA air program official, the Harvard study represented an early model of environmental regulators "posing questions that the scientists had not previously asked." Combining support from EPA, the National Institute of Environmental Health Sciences (NIEHS), and the Electric Power Research Institute (EPRI), the Harvard group began issuing a series of reports on or about 1981 that provided input into the PM revision (for example, Ware et al. 1980). Due to the length of the PM review period, the Harvard group was able to produce some answers to regulators' questions prior to the decision, a rarity for multiyear, complex epidemiological studies.

According to an EPA official, OAQPS requested that Joel Schwartz, an analyst in the agency's Office of Policy Analysis (OPA), conduct a reanalysis of the British mortality data. Previous reanalyses had been conducted, including one by OPA's Bart Ostro (Ostro 1984), but Schwartz and Marcus (1986) addressed a number of methodological questions raised by reviewers. This reanalysis carried considerable weight. According to an EPA official, "Schwartz had clout because he was smarter than everybody else." (Schwartz had also established some credibility with CASAC on the strength of his work on the effects of lead air pollution. See Appendix A.) Access to the raw British mortality data was also a key factor that permitted Schwartz and Marcus to address the reviewers' questions. In the majority of cases, EPA regulatory analysts do not have access to raw data. Currently a biostatistician in ECAO, Marcus "was pleasantly surprised that they were able to get the raw data. . . . It made me optimistic that in-house researchers could do just as good a job as PIs [principal investigators] and extract the information that was relevant to the agency. Now, I find that's not always the case." Even when the agency is able to gain access to the raw data, says Marcus, "we generally don't have the time and resources to redo the analysis to make sure that [the original investigators] did it right."

The proximity of ECAO and OAQPS to each other, the frequent circulation of personnel between the offices, and their detachment from headquarters offices in Washington, D.C., have resulted in a level of interoffice coordination that, although not without tensions, is uncommon within the agency. On the downside, according to a senior careerist, the close arrangement can sometimes "discourage debate." As an illustration, this EPA official points to the acquiescence to ECAO's critique of much of the epidemiological evidence. Another factor that may have contributed to curtailing substantive debate was tension between the air program headquarters and OAQPS. According to an EPA air program official, early in the 1980s, headquarters was complaining about the political flak it was catching for failing to make revisions more quickly. Reflecting a typical face-off between careerists and politicos, "D.C. felt that OAQPS was hard-headed, not dealing with D.C. as a client. RTP [Research Triangle Park] said, 'We know what we're doing.'" According to respondents, headquarters dispatched Gerald Emison to be the new director of OAQPS in 1984 "to make sure the trains run on time."

Respondents all characterized the agency's treatment of the scientific information available at the time of the decision to revise the PM standard as good to very good. However, an academic provided a qualified endorsement. "It was very good in terms of the best that can be done, and very poor in terms of not having enough data. EPA doesn't support any research to speak of, so they have to find information available from studies with other objectives. They're skillful in taking what information is out there and drawing inferences from it, but the data are very weak. The agency has never put enough effort into generating sufficient data."

Communicating the Science to Agency Leadership

Respondents rated the communication of the scientific information to agency decisionmakers as good to very good. EPA staff briefed three successive Administrators. Staff briefed Administrator Anne Gorsuch prior to her resignation, according to an air program official, and the benefits analysis included in the RIA reportedly prevented Gorsuch from relaxing the standard. Administrator Ruckelshaus was initially briefed in 1983 and later received a two-day briefing. However, agency leaders believed that data at that point were inadequate and questioned why a research strategy had not been laid out and conducted in advance to provide a better database for the eventual review of the standard. Why, for example, had sensitive populations not been addressed? According to a former EPA official, Administrator Ruckelshaus's concern that CASAC had not endorsed a specific level for the PM-10 standard led to the unprecedented decision to propose a range for a NAAQS.

As is generally the case, the majority of the communication of scientific information to agency decisionmakers was conducted orally or through brief memos. However, Al Alm, Ruckelshaus's Deputy Administrator, who was described as "a voracious reader," was supplied with a stream of materials prior to the proposal in 1984. Alm and Assistant Administrator for Air Joseph Cannon were also briefed on the benefits analysis, and, according to an air program official, "it helped justify going to the lower bound of the proposed range in their minds." Finally, Lee Thomas was briefed prior to issuing the final rule in 1987.

There were a number of key agency staff involved in the briefings, including Bruce Jordan, the OAQPS Ambient Air Standards branch chief and ECAO's Les Grant. During his tenure in the Ruckelshaus administration as Assistant Administrator for ORD, Bernard Goldstein was also influential in translating and interpreting the science for agency decisionmakers. But in each case, John Bachmann, the lead author of the OAQPS staff papers, was the principal staffer responsible for communicating the science to the leadership. In his role as lead author of the SP, Bachmann served as the bridge between ECAO and OAQPS. Bachmann was also responsible for preparations for the lengthy series of CASAC meetings. Respondents remarked that Bachmann was an effective communicator because straddling the science-policy divide, he grasped the relevance of the scientific findings, and he could talk about the science to decisionmakers in simple terms without jargon.

Regarding the PM revision, an air program official noted, "The process took a long time from proposal to finalization. In terms of key players, it was like a conveyor belt." Over the entire life of the review, there was not only a changing of the guard in the Administrator's Office but also turnover at senior political and administrative levels. Between the proposal and final rulemaking, there was turnover in the air program at the Assistant Administrator level, from Joseph Cannon to Craig Potter, and the directorship of OAQPS changed hands from Walter Barber to Gerald Emison. But careerists, such as Bachmann, Grant, and Jordan, were constants. While agency careerists may retain a regulatory issue in their portfolio for several years, if not their entire careers, the tenure of senior political appointees in the federal government is generally on the order of two or three years. As a result, policymakers are often dependent on careerists for information, and careerists often have to make allowances for bringing new political appointees up to speed.

The Role of External Scientists in the Process

All respondents characterized the role of nonagency scientists in influencing the PM revision process as significant to very significant, and there was a consensus that nonagency scientists also played a significant role in legitimizing the agency's decision. However, the influence was not entirely positive. A former EPA researcher argued that the scientific review process was "protracted by a conscious decision to have the adversaries present" at agency working sessions. Some of this input may have served little more than to strengthen the agency's record of decision. On the other hand, some agency officials believed that industry analysts, from the electric utility sector, for example, helped "sharpen the agency's analysis" and prompted the agency to "slay those dragons which can be slain." Since EPA chose to adopt the International Standards Organization's nominal cut value of ten microns for particles that could penetrate the thoracic region, it appears that industry input into the agency's analysis for setting the NAAQS levels may have contributed more than industry analysis regarding the determination of the particle size cut value. (Recall that the cut value selection had differential impacts across domestic industrial sectors. Once the cut value was fixed, setting the level of the standard would affect which areas were likely to fall into nonattainment.)

As detailed in the background discussion above, CASAC played an extensive role in the PM review process. There was considerable continuity between the 1982 and 1986 panels. One of the members served on both panels, and four of the 1986 members had served as consultants to the 1982 panel, including Morton Lippman, chair of the 1986 panel.[24] Lippman had also been a principal researcher involved in the particulate deposition studies responsible for getting the PM revision on the agency's agenda.

EPA analysts were confronted with a lack of official agency guidance for conducting NAAQS reviews, in particular, and for noncancer risk assessments, in general. As a result, respondents noted that the agency "sought CASAC endorsement of what they were doing" methodologically to ensure scientific credibility and acceptance. For example, a critical area in which there was no generally accepted scientific method was converting measurements between TSP and PM-10, so the agency relied on CASAC review and endorsement to ensure the credibility of its conversions. Apart from methodological review, CASAC also played a role regarding the NAAQS levels finally adopted by EPA. The entire range for the standard proposed by the agency in 1984 (150–250 $\mu g/m^3$) was intended, in the words of an EPA official, "to be protective of public health."

At its 1986 meeting, however, CASAC suggested that the agency adopt the lower end of the proposed range and even consider going below it in order to provide a greater margin of safety. Although the adequacy of the margin of safety is generally regarded as an issue in the policy domain, providing such advice seems well within CASAC's statutory authority to comment on standards as well as the criteria. According to one former EPA official, CASAC suggested that EPA consider a standard below the proposed range to provide the agency with cover to justify promulgating the standard at the lower bound of the proposed range (150 $\mu g/m^3$). If this account is accurate, one could reasonably argue that engaging in this type of strategic behavior exceeds CASAC's legislative authority to advise and comment on agency criteria and standards. However, it should be noted that the 1986 staff paper also included reduced lower bounds of the range. Another former EPA official says that CASAC Chair Lippman's view was that the upper end of the proposed range provided little, if any, margin of safety and the rest of the committee deferred to Lippman's expertise.

SCIENCE IN THE FINAL DECISION

In 1987, the final rule established the lower bound of the proposed ranges for both the daily (150 $\mu g/m^3$) and the annual (50 $\mu g/m^3$) averages for the PM-10 NAAQS. An EPA official commented, "Without CASAC endorsement, the agency would be concerned about the credibility and legal defensibility of the decision." However, CASAC's role should not be overstated. In issuing the proposal in 1984, Administrator Ruckelshaus acknowledged that he leaned toward primary standards from the lower portions of the proposed ranges. Respondents also suggested that OAQPS staff and ECAO's Lester Grant recommended setting the daily standard at 150 $\mu g/m^3$. Finally, according to an air program official, "One thing a decisionmaker doesn't want is to be surprised by studies that are going to come out [after the decision]." In the case of the PM revision, agency leadership "was told at the time that there would probably be studies forthcoming suggesting concern at even lower levels."

Respondents rated the level of consideration given to the scientific issues by agency decisionmakers as thorough to very thorough, and there was a consensus that the scientific information had a high impact on the ultimate decision. The minds of agency leaders were focused because of the economic stakes involved in the PM NAAQS. EPA estimates compliance costs of more than $500 million per year (U.S. EPA 1986c). (The threshold for a "major" regulation is $100 million per year.) A senior EPA careerist noted, Administrator Lee Thomas "put a lot of time into it," and an air program official recalled, "Lee Thomas had a lot of questions." According to an air program official, the reason the "communication [of the scientific information] was so good was because decisionmakers spent a lot of time trying to understand what the issues were." Despite focusing considerable attention on the scientific issues, according to a former EPA official, "Thomas was given a very narrow range of staff recommendations; he was given little wiggle room."

Respondents pointed to a scattering of factors that impeded the use of science in the PM revision, with no one factor being mentioned more than twice. However, grouping the problems as follows suggests that respondents believed that the major impediments were associated with the data and methods and the problems of matching science with policy:[25]

- Problems with the data and methods (seven): large scientific uncertainties (two); inadequate epidemiological and toxicological tools and methods (two); the large volume of information available; the lack of relevant information; the lack of EPA funding for research.
- Problems matching science with policy (five): the CAA's presumption that a pollutant concentration exists below which there are no adverse health effects; scientists lack understanding of policymakers' information needs; policymakers perceive staff scientific analysis as deficient due to the quality of EPA staff; EPA policymakers lack understanding of what science can contribute to decisionmaking; the adversarial rulemaking process.
- Problems with disciplinary science (two): disciplinary inertia prevents consideration of new scientific findings; disciplinary boundaries result in fragmented, narrow analyses.

Peer review (both internal and external) was the factor most frequently mentioned by respondents as facilitating the use of science in the PM revision (six). Other factors included the skill of EPA staff (two); the high economic stakes; the clear health basis of the NAAQS under the CAA; the legislative requirement for periodic revision of the NAAQS; and good communications between scientists and policymakers.

CONCLUDING OBSERVATIONS

More than ten years before EPA proposed to revise the NAAQS for particulates, the narrowly scoped results of new particle deposition experiments had penetrated deep into the agency, far removed from decisionmakers in Washington. The findings initiated a long, complex sequence of activity that was promoted and inhibited by multiple factors. A key air office staffer, John Bachmann, was primarily responsible for transferring the accumulating scientific information from the researchers through the air program and finally to politically accountable decisionmakers.

The primary source of support for the depositional studies conducted by NYU researchers was NIEHS. Agency staff in Research Triangle Park dedicated to surveying the scientific literature for new, relevant information were exposed to the information and grasped its significance. However, additional work was needed to transform the depositional findings into an operational cut value for use in the design of air pollution monitoring equipment and for setting the NAAQS. Much of this supplementary work was conducted within EPA by researchers at HERL. The mining industry attempted to influence the agency's use of the information to further reduce the cut size, but the International Standards Organization's position was operationally practical and provided political cover. EPA adopted the PM-10 indicator.

Due to the lengthy review period, Harvard researchers, supported by NIEHS, EPA, and EPRI were able to provide information prior to the final decision regarding the effects on sensitive populations of particulate matter not at levels associated with acute episodes in foreign lands, but at typical concentrations experienced in U.S. urban areas. This study was prompted by questions posed by environmental policymakers, not by the scientific community. A reanalysis of the British mortality data requested by OAQPS and conducted by an EPA policy official suggested the lack of a discernible threshold in the relationship between particulate levels and mortality. EPA was able to access the raw data due to international linkages established by ECAO. Both of these studies were evaluated favorably by EPA scientists and CASAC and provided scientific support for adopting levels of the PM-10 standards at the lower ends of the proposed ranges.

The epidemiological morbidity studies originating from the economic literature, however, experienced a slightly different fate. These studies were based on available monitoring and health survey data that enabled researchers to evaluate health effects in economically relevant terms. In this case, the internal receptor and promoter within the agency was the Office of Policy Analysis. ECAO and CASAC, acting in their science gatekeeper roles, rendered these studies somewhat impotent. Officially, EPA concluded that too many fundamental questions remained considering the RIA methodology, and, in any event, the CAA did not permit the consideration of costs in setting the NAAQS. Unofficially, according to respondents, the RIA's projection of net benefits at levels below the proposed range for PM-10 helped justify the final decision by the Thomas administration to select the lowest end of the range and prevented the Gorsuch administration from relaxing the NAAQS. Therefore, the economic analysis appears to have indirectly influenced the outcome toward the same end as scientific analysis. Regardless of the impact the information had on the decision, this case study illustrates how, in the context of a single decision, different organizations and disciplines within EPA use different sources of scientific information for different purposes and judge the information by different criteria. Figure C-1 illustrates the fate and transport dynamics of these key sources of scientific information in the 1987 PM revision.

In the course of discussing the 1987 PM review, a number of respondents expressed frustration with the criteria document–staff paper

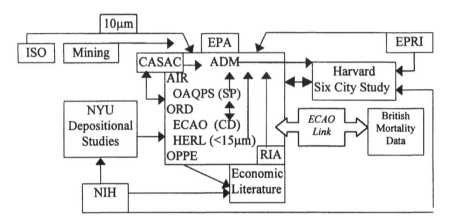

Figure C-1. Fate and transport of science in the 1987 particulate matter revision

process, in general. The most frequently mentioned source of the problem was ECAO's practice of developing voluminous CDs. However, respondents also noted that while CASAC has called for more succinct CDs, individual panel members display a tendency to strive to ensure that the documents cover their own disciplines in exhaustive detail. Whether, in any particular case, this is "just academics being academics," as one respondent put it, or strategic behavior intended to keep their discipline high on the environmental research agenda cannot be determined.

Another criticism of the process was that ECAO and the air program appear reluctant to work on more than one criteria pollutant at a time. Again, a number of factors may be at play. According to an academic researcher, ORD management has not made criteria document development an organizational priority, and ECAO Director Les Grant's direct involvement as a principal author may also limit the number of criteria pollutants that can be addressed at any given time. The air program, in turn, is reluctant to focus on more than one pollutant at a time because doing so would increase the burdens on state officials and EPA program staff in terms of developing and reviewing state implementation plans for NAAQS attainment. There may also be a mismatch in terms of ECAO staffing. Although NAAQS are increasingly driven by epidemiological findings, most agency scientific staff have backgrounds in toxicological and other experimental sciences. Relatively fewer have backgrounds in epidemiology and statistics.

This case study provides an illustration of nonagency scientists operating in multiple, overlapping roles. Three out of seven of the contributors to the 1982 criteria document epidemiological chapter were from the Harvard study group (Benjamin Ferris, Frank Speizer, and James Ware). Ware later served as a member of CASAC in 1986. It is not uncommon for scientists to provide evidence as academic researchers, contribute to assessments as contractors, and evaluate results from a position on an independent advisory committee. In part, this reflects the small number of expert practitioners in highly specialized areas. On the other hand, it poses the risks associated with insularity and inbreeding.

EPILOGUE

Regarding the most recent (1996–1997) PM review, an air program official noted that the criteria document and staff paper process, which extended over many years and occurred sequentially in the 1987 revision, has been compressed and run in parallel under the court-ordered deadline. Although the CD has been widely faulted for being too long and unfocused, this official suggested that it is sometimes hard to predict what the OAQPS staff and CASAC will regard as critical. In prior reviews, the first draft of the CD was revised while the SP was being developed, providing a feedback loop in the process of reviewing and interpreting the scientific information. The court-ordered deadline has the effect of eliminating or crimping this feedback loop. CASAC also bridled at the compressed review schedule that permitted the committee only a single opportunity to review the CD and SP in draft. (After EPA proposed to revise the NAAQS for PM in November 1996, the agency announced that it would convene a meeting of CASAC to enable the committee to comment on the proposals [*Inside EPA*, December 20, 1996, 10].)

In March 1996, CASAC Chair George Wolff sent Administrator Carol Browner a consensus closure letter regarding the new criteria document for particulate matter. In the letter, Wolff reported that "about half of the Panel members expressed concern that the case made in the Criteria Document for PM-2.5 being the best available surrogate for the principal causative agent in PM-10 may be overstated, and that EPA has not adequately justified its rejection of other alternative explanations. . . ." However, former CASAC Chair Morton Lippman and three other CASAC members took the rare step of sending Browner a "supplementary" letter contending that there is sufficient evidence of a causal relationship between PM exposure and excess mortality and morbidity; PM-2.5 is better than any alternative measure; and because some adverse health effects may be linked to coarser particles, separate fine (that is, PM-2.5) and coarse (that is, PM-10) standards should be considered (*Risk Policy Report*, April 19, 1996). Shortly after the elections of November 1996, EPA proposed to supplement the current primary PM-10 standards by adding two new primary PM-2.5 standards set at 15 $\mu g/m^3$, annual mean, and 50 $\mu g/m^3$, 24-hour average (*Federal Register* 27: 1721).

Despite CASAC's importance in legitimizing EPA's NAAQS decision-making, the agency would revise the standard without the panel's unanimous endorsement. In June 1997, President Clinton endorsed EPA's final plans for promulgating the new NAAQS for particulates, and on July 18, 1997, EPA promulgated the final rule. The primary PM standards were revised in several respects (*Federal Register* 62: 38651–38701). Two new PM-2.5 standards were added, set at 15 $\mu g/m^3$, based on the three-year average of annual arithmetic mean PM-2.5 concentrations, and 65 $\mu g/m^3$, based on the three-year average of the ninety-eighth percentile of 24-hour PM-2.5. The current 24-hour PM-10 standard was also revised to be based on the ninety-ninth percentile of 24-hour PM-10 concentrations. The secondary standards were revised by making them identical to the new suite of primary standards.

Under the final regulations, implementation of the standards was to be phased in over several years. On May 14, 1999, however, the U.S. Court of Appeals for the District of Columbia remanded the revised NAAQS for particulate matter and ozone to EPA. In its *American Trucking Association, Inc. v. EPA* decision, the court focused on Section 109 of the Clean Air Act, which requires that EPA set ambient air quality standards at the level "requisite to protect the public health" with an "adequate margin of safety." Noting that EPA considers both PM and ozone to be nonthreshold pollutants, the court reasoned that because zero (or no exposure) is the only risk-free concentration for these pollutants, for EPA to pick any non-zero level, it must explain the degree of harm permitted and it must justify lower standards as providing additional health protections. Without a specific scientific justification for any particular standard, the court was unwilling to uphold the agency's decision. At the same time, the court reaffirmed that Section 109(b) of the Clean Air Act does not allow EPA to consider the cost of implementing NAAQS (see http://www emi.org/public_policy/clean_air_epa_setback.htm).

The legal core of the ruling was the judges' resuscitation of the "nondelegation" doctrine, under which the court found that the executive branch agency had acted on legal assumptions that amounted to "unconstitutional delegations of legislative power" (Warrick and McCallister 1999). The appellate court's decision was a marked departure from the Supreme Court's landmark 1984 decision in *Chevron U.S.A., Inc. v. Natural Resources Defence Council, Inc.* (467 U.S. 837). In this decision, the Supreme Court noted that the legislature commonly delegates to an executive branch agency the power to administer a congressionally created program. When Congress leaves gaps in the program, either explicitly by authorizing the agency to adopt implementing regulations or implicitly by enacting an ambiguously worded provision that the agency must interpret, the Supreme Court argued that Congress has explicitly or implicitly delegated to the agency the power to fill those gaps. By reviving the nondelegation doctrine in its 1999 ruling, the appellate court was essentially saying that setting ambient air quality standards was too important for Congress to delegate to EPA.

ENDNOTES

[1] Primary NAAQS are intended to protect the public from known or reasonably anticipated adverse health effects with an adequate margin of safety. Secondary NAAQS are intended to protect public welfare from known or reasonably anticipated adverse effects on environmental and economic resources and personal well-being. However, this case study does not discuss the secondary NAAQS for particulates.

[2] SAB was created by an administrative order in 1974 (Jasanoff 1990).

[3] The criteria air pollutants now include carbon monoxide, lead, nitrogen oxides, ozone, particulates, and sulfur oxides. According to an EPA official, the agency elected to schedule a revision of the ozone criteria first because the states "hadn't worked hard to implement" that NAAQS. SO_x was included with PM because sulfate aerosols constitute a substantial portion of fine suspended particulate matter.

[4] The 1977 Clean Air Act Amendments authorized the Clean Air Science Advisory Committee to review all criteria documents prior to proposal and promulgation of national ambient air quality standards. The Environmental Research, Development and Demonstration Act of 1978 gave statutory footing to the EPA Science Advisory Board and also required that SAB review the criteria and the NAAQS. Consequently, CASAC is administered by SAB but has a distinct charter and independent status. In a provision that provides CASAC a broader scope of legitimate commentary than the rest of the agency's official science advisory committees, the 1977 amendments authorized the committee to advise the Administrator of "any adverse public health, welfare, social, economic, or energy effects which may result from various strategies for attainment and maintenance of such national ambient air quality standards." Not surprisingly, CASAC's exercise of its authority to comment on extrascientific matters has been an episodic source of friction between the committee and EPA.

[5] According to an EPA official, to do so, the agency would have had to institute a PM "cut value" of 3.5 microns, and indirectly implementing an acid rain program via the particulate standard would have been much more costly than a direct approach. In 1996, EPA proposed to revise the PM standard by adding PM-2.5 standards to the current PM-10 standards (*Federal Register* 27: 1721).

[6] An estimated annual cost of $100 million was the threshold for "major" regulations under Executive Order 12291.

[7] The principal economic analysis was prepared in 1983 by Mathtech for the EPA Economic Analysis Branch, OAQPS.

[8] American Iron and Steel filed suit in the U.S. District Court for the District of Columbia alleging that EPA did not use the latest science in the revised criteria document (*Inside EPA*, May 11, 1984, 5).

[9] Although epidemiology has regained some credibility within the agency, several respondents characterized the agency's scientific culture as being dominated by experimental sciences, such as toxicology; with observational sciences, such as epidemiology and economics, being subordinate. Note the debate in the arsenic in drinking water case study (Appendix B) about the sufficiency of epidemiological evidence in the absence of supporting toxicological evidence.

[10] Because the geometric mean is less sensitive to extreme values than is the arithmetic mean, it is commonly used as an indicator of central tendency for asymmetric distributions such as the lognormal. The annual "geometric average" is lower than the arithmetic average and is less dependent on days with the highest recorded particulate levels.

[11] The unanimity of respondents, of course, should not be interpreted to mean that there were no dissenters in the scientific community. For example, some British scientists criticized EPA for using London mortality data associated with air pollution episodes to estimate human health effects at low particulate concentrations (for example, Holland et al. 1979, cited in Friedlander and Lippman 1994). Some scientists and analysts affiliated with U.S. industry (notably the electric power and steel sectors) also suggested that the uncertainties associated with quantitative risk assessments of particulates were too great to make them useful for regulatory purposes (for example, Roth et al. 1986).

[12] 1 micron $= 1\ \mu\text{m} = 10^{-6}$ meters $= 1$ micrometer.

[13] Dosimetry refers to measurements of the dose of a chemical substance delivered to a specific organ in the body. The delivered dose can differ markedly from ambient concentrations to which people are exposed in rate, concentration, and kind (for example, size fraction, metabolites).

[14] According to an academic researcher, the original depositional studies were conducted primarily at New York University (NYU) and Frankfurt, Germany. The NYU studies received early support from the National Institute for Occupational Safety and Health and the American Medical Association but were primarily funded by National Institute of Environmental Health Sciences.

[15] Although the reference for Whitby (1975) reports EPA grant support, according to the EPA official, this work was funded by both EPA and the National Science Foundation.

[16] In a letter sent to EPA Administrator Carol Browner after reviewing the first draft of EPA's 1995 staff paper for PM, CASAC Chairman George Wolff reported that several CASAC members "felt that the selection of a 2.5 [micron] cutpoint was arbitrary" (*Risk Policy Report*, January 19, 1996, 5–6).

[17] Some sulfur dioxide chamber studies were conducted with asthmatics, however.

[18] An illustration of the latter would be whole effluent testing, which applies aquatic toxicology to an unspecified waste stream originating from a specified process, such as leather tanning.

[19] In addition to the chronic exposure study discussed below, the Harvard Six City Study also included investigations of reversible declines in lung function in children associated with acute exposures (Dockery et al. 1982). Only the acute

study was longitudinal in design, meaning that the study population was observed over time. The chronic study was cross-sectional in design, meaning that the health status of the study population observed at one particular time was related to their estimated chronic exposure levels.

[20] The six cities included in the study were Steubenville, Ohio; Watertown, Massachusetts; Portage, Wisconsin; Topeka, Kansas; St. Louis, Missouri; and Harriman, Tennessee. The Harvard study received support from the National Institute of Environmental Health Sciences, EPA, and the Electric Power Research Institute.

[21] See Adams et al. (1984) for a discussion of the trade-offs for environmental decisionmaking between replications at particular exposure levels and coverage over the range of exposures. In sum, replication contributes to the precision necessary to discriminate the rank order of the distribution of net benefits under a limited set of policy alternatives, but without adequate coverage to reduce model uncertainty, the magnitude of the net benefits remains uncertain.

[22] Confounding variables are alternative, uncontrolled factors that may account for variation in the observed effects—for example, the potential effects of ozone and SO_x discussed above.

[23] According to a former EPA air program official, the cost to EPA of developing a NAAQS is in the $4–$8 million range. This estimate may not reflect, however, the cost associated with EPA research (in house and extramural) that contributes to the foundation of the standard. ECAO and OAQPS also have increasing demands on their resources for conducting risk analyses for hazardous air pollutants as a result of the 1990 Clean Air Act Amendments.

[24] The member who served on both panels was Mary Amdur of the Massachusetts Institute of Technology. Edward Crandall of Cornell, Timothy Larson of the University of Washington, and Roger McClellan of the Lovelace Research Institute served as consultants in 1982 and members in 1986.

[25] Numbers in parentheses refer to the frequency with which a factor was mentioned by respondents as impeding or facilitating the use of science.

REFERENCES

Adams, R., T. Crocker, and R. Katz. 1984. Assessing the Adequacy of Natural Science Information: A Bayesian Approach. *The Review of Economics and Statistics.* 66: 568–575.

Albert, R., M. Lippman, H. Peterson, J. Berger, K. Sanborn, and D. Bohning. 1973. Bronchial Deposition and Clearance of Aerosols. *Archives of Internal Medicine* 131: 115–127.

Friedlander, S., and M. Lippman. 1994. Revising the Particulate Ambient Air Quality Standard: Scientific and Economic Dilemmas. *Environ. Sci. Technol.* 28(3): 148–150.

ISO (International Standards Organization). 1981. Size Definitions for Particle Sampling: Recommendations of *Ad Hoc* Working Group Appointed by TC 146 of the International Standards Organization. *Am. Ind. Hyg. Assoc. J.* 42: A64–A68.

Jasanoff, S. 1990. *The Fifth Branch: Science Advisers as Policymakers.* Cambridge, Mass.: Harvard University Press.

Lave, L., and E. Seskin. 1977. *Air Pollution and Human Health.* Baltimore, Md.: Johns Hopkins University Press.

Lippman, M. 1977. Regional Deposition of Particles in the Human Respiratory Tract. In *Handbook of Physiology, Section 9*, edited by D. Lee, H. Falk, and S. Murphy. Bethesda, Md.: American Physiological Society, 213–232.

Martin, A., and W. Bradley. 1960. Mortality, Fog and Atmospheric Pollution—An Investigation during the Winter of 1958–59. *Mon. Bull. Minist. Health, Public Health Lab. Serv.* 19: 56–72.

Mazumdar, S., H. Schimmel, and I. Higgins. 1981. Daily Mortality, Smoke, and SO_2 in London, England, 1959–72. In *Proceedings of Am. Pol. Control Ass. Conference on the Proposed SO_2 and Particulate Standard, Pittsburgh, Pa., September 16–18*, 219–239.

Miller, F., D. Gardner, J. Graham, R. Lee, W. Wilson, and J. Bachmann. 1979. Size Considerations for Establishing a Standard for Inhalable Particles. *J. Air Pollut. Control Assoc.* 29: 610–615.

Ostro, B. 1983. The Effects of Air Pollution on Work Loss and Morbidity. *J. Environ. Econom. Mgt.* 10: 371–382.

———. 1984. A Search for a Threshold in the Relationship of Air Pollution to Mortality: A Reanalysis of Data on London Winters. *Environmental Health Perspectives* 58: 397–399.

Schwartz, J., and A. Marcus. 1986. Statistical Reanalyses of Data Relating Mortality to Air Pollution During London Winters 1958–72. Appendix to U.S. EPA 1986b.

U.S. EPA (U.S. Environmental Protection Agency). 1982a. *Air Quality Criteria for Particulate Matter and Sulfur Oxides*. Research Triangle Park, N.C.: Environmental Criteria and Assessment Office, Office of Research and Development, December.

———. 1982b. *Review of the National Ambient Air Quality Standards for Particulate Matter: Assessment of Scientific and Technical Information*. OAQPS Staff Paper. Research Triangle Park, N.C.: Office of Air Quality Planning and Standards, January.

———. 1986a. *Second Addendum to the Air Quality Criteria for Particulate Matter and Sulfur Oxides (1982)*. Research Triangle Park, N.C.: Environmental Criteria and Assessment Office, Office of Research and Development, December.

———. 1986b. *Review of the National Ambient Air Quality Standards for Particulate Matter: Updated Assessment of Scientific and Technical Information*. Addendum to the 1982 OAQPS Staff Paper. Research Triangle Park, N.C.: Office of Air Quality Planning and Standards, December.

———. 1986c. *Regulatory Impact Analysis of the National Ambient Air Quality Standards for Particulate Matter*. Second Addendum. Research Triangle Park, N.C.: Office of Air Quality Planning and Standards, December.

Ware, J., B. Feris, D. Dockery, J. Spengler, D. Stram, and F. Speizer. 1986. Effects of Ambient Sulfur Oxides and Suspended Particles on Respiratory Health of Pre-Adolescent Children. *Am. Rev. Respir. Dis.* 133: 834–842.

Warrick, J., and B. McCallister. 1999. Court Voids EPA Air-Pollution Rules. *Washington Post* May 15: A1.

Whitby, K. 1975. *Modeling of Atmospheric Aerosol Particle Size Distributions: A Progress Report on EPA Research Grant No. R800971, "Sampling and Analysis of Atmospheric Aerosols."* Minneapolis: University of Minnesota.

Appendix D

The 1993 Decision Not To Revise the National Ambient Air Quality Standard for Ozone

BACKGROUND

Health effects from ambient levels of tropospheric ozone were first reported in 1967 in terms of lowered performance by high school athletes in California on high-exposure days (Lippman 1991). Since then, regulation of tropospheric ozone has unfolded slowly, despite a legislative requirement for periodic review, and has been characterized by discord and litigation. The role of science and scientists in the ozone standard-setting process has increased and has become more formal over time. But this has not diminished the controversy, in large part, because the Clean Air Act is based on the false scientific premise that a threshold level exists below which health effects from ubiquitous air pollutants will not be observed. As a consequence of this mistaken legislative presumption, new scientific developments inevitably point toward ever more stringent ambient ozone standards and preordain—in principle—the outcome of periodic reviews of the scientific basis of ozone regulation. In practice, EPA's response has been to delay the inevitable.

The national ambient air quality standard (NAAQS) for all photochemical oxidants was originally set in 1971 at a maximum hourly average of 0.08 ppm (parts per million), not to be exceeded more than once per year.[1] According to a former EPA political appointee, science played a little role in the initial ozone NAAQS.[2] "[Former Senator Edmund] Muskie [D-Maine] wanted a 95% reduction—that was his goal." The principal scientific rationale for the 1971 standard was one study that reported that asthmatics had more attacks on days when ozone levels were above 0.10 ppm. A subsequent reassessment of the study revealed that the actual exposure levels were considerably higher (Landy et al. 1994). In 1976, the American Petroleum Institute (API) and the city of Houston petitioned EPA to revise the standard (Marraro 1982). Conse-

267

quently, EPA initiated a review of the NAAQS for photochemical oxidants in 1977 (McKee 1994).

In 1978, EPA finalized the criteria document (CD, an assessment of the available scientific information regarding health and welfare effects) for photochemical oxidants and proposed to revise the NAAQS. The agency proposed to (1) raise the primary NAAQS to 0.10 ppm and retain the secondary NAAQS of 0.08 ppm,[3] (2) measure and enforce the standard using tropospheric ozone as the chemical indicator species for the family of photochemical oxidants,[4] and (3) change the form of the standard from deterministic to statistical to allow for interannual variation in weather conditions.[5] In 1979, after President Carter's Regulatory Analysis Review Group (operating primarily out of the Office of Management and Budget), the White House Office of Science and Technology Policy, the Council of Economic Advisors, and API critically evaluated EPA's proposal, the agency set both the primary and secondary standards for ozone at 0.12 ppm. Subsequently, API, the Natural Resources Defense Council, and several other groups petitioned the courts to review the standard. The D.C. circuit court of appeals held in favor of EPA in its 1981 *API v. Costle* ruling. At the center of the controversy were the roles of official and ad hoc scientific advisory groups and of the Executive Office of the President in the agency's decisionmaking process and EPA's use of a controversial evaluation procedure.[6] An outgrowth of the controversial 1979 ozone revision was the institutional formalization of the NAAQS review process. (For more on the historical development of the NAAQS process, see Jasanoff 1990, Chapter 6).

By 1983, the EPA Office of Research and Development Environmental Criteria and Assessment Office (ORD/ECAO) had initiated a review of the ozone CD in response to the statutory requirement to review the NAAQS by 1985.[7] It was not until 1985, however, that the Clean Air Act Science Advisory Committee (CASAC) reviewed the first draft of the CD. After CASAC's initial review, ORD/ECAO began revisions, and the EPA Office of Air and Radiation Office of Air Quality Planning and Standards (OAR/OAQPS) began drafting an ozone staff paper (SP, an analysis forming the basis of policy options and staff recommendations for NAAQS). The following year, CASAC reviewed the first draft of the SP and indicated its satisfaction with the second draft of the CD in a "closure letter" to the Administrator.

During its 1987 review of the second draft of the SP, however, CASAC argued that EPA should capture in its review of the ozone NAAQS new, emerging data on the health effects and agricultural crop damages from longer exposures (six to eight hours and seasonal exposures). In response, ORD/ECAO drafted a supplement to the 1986 CD while OAR/OAQPS revised the SP again. CASAC reviewed the documents in 1988. Although the committee urged EPA to complete its review as rapidly as possible (it was now five years overdue), CASAC's 1989 closure letter to EPA Administrator William Reilly revealed an internal division. The committee indicated that the supplement to the 1986 CD and the SP "provide an adequate scientific basis for the EPA to *retain or revise*" the NAAQS for ozone (McKee 1994, emphasis added). CASAC Chairman Morton Lippman reported that the committee did reach consensus on the definition of adverse health effects and sensitive populations groups, but only half of the members of CASAC believed the current ozone standard was adequately protective of human health (Lippman 1989). In particular, according to an ORD official, "CASAC members had a split view on the multihour effects."

While ECAO, OAQPS, and CASAC were busily reviewing ozone science, the Bush administration and the Democrat-controlled Congress were immersed in legislative negotiations to revise the Clean Air Act (CAA). For the first time since 1977, it appeared that the political planets were aligned to enable a reauthorization of the act. According to McKee (1994), "Following numerous discussions and briefings within EPA on the need for modified and possibly more stringent [ozone] standards, it was determined that any such changes might be disruptive of the ongoing Clean Air Act negotiations and that such action should be delayed." According to an environmental lawyer, there was a meeting late in 1989 during which senior EPA officials made the political decision not to revise the NAAQS for ozone "unless they were forced to." Fearing that it would alienate stakeholders involved in the legislative negotiations, however, the agency was unwilling to announce its decision not to act. According to an OAQPS official, "There was a policy decision that the time was not ripe to take either action," that is, go public with a decision that revisions are not appropriate or revise the standard on the basis of the existing record of decision. "The agency," said this source, "didn't want to rock the apple cart."

After the 1990 Clean Air Act amendments were enacted, however, the American Lung Association (ALA) and other plaintiffs filed suit in October 1991 to force EPA to complete its review of the NAAQS for ozone. The U.S. District Court for the Eastern District of New York ordered EPA to announce its proposed decision on whether to revise the NAAQS for ozone by August 1992 and to announce its final decision by March 1993 (McKee 1994). In August 1992, then Administrator William Reilly proposed that revisions to the ozone standards were not appropriate at that time. The notice did not take into account more recent studies that had not been assessed in the 1986 criteria document or its supplement or reviewed by CASAC. EPA estimated that two to three years would be needed to assess and review the new information.

After the defeat of George Bush in the election of November 1992, President Clinton appointed Carol Browner as Administrator of EPA. In compliance with the court order, Administrator Browner finalized the decision not to revise the ozone NAAQS in March 1993. Table D-1 provides a summarized background of the 1993 decision not to revise the NAAQS for ozone.

Shortly after President Clinton was reelected in November 1996, EPA reversed positions and proposed to revise the NAAQS for ozone. The case study focuses on the earlier decision not to revise the standard.

SCIENTIFIC ISSUES

Ozone is a highly reactive gas composed of three atoms of oxygen (O_3). In the upper atmosphere (the stratosphere), ozone filters out hazardous ultraviolet radiation from the sun. Stratospheric ozone is being depleted by chlorofluorocarbons and other chemical compounds. Closer to home in the troposphere, ozone is an atmospheric pollutant and the most important of a class of pollutants called photochemical oxidants. Ozone is described as a "secondary" air pollutant, because it is not emitted directly. Instead, it is formed by atmospheric chemical reactions involving nitrogen oxides (NO_x), volatile organic compounds (VOCs), and oxygen (O_2) in the presence of sunlight. High ambient levels of ozone generally occur only when temperatures exceed eighty to ninety degrees Fahrenheit.

Four out of five interviewees responded that the scientific information available in 1993 was adequate for regulatory decisionmaking. (However, much of the information had not been through the formal NAAQS review process.) There is very little—if any—disagreement among respondents concerning whether the principal scientific studies demonstrate biological effects in humans, animals, and plants from ozone at levels below the current standard of 0.12 ppm for exposure periods exceeding one hour. It is generally agreed that the effects of long-term exposure to low concentrations of ozone can produce greater effects than short-term exposure at peak concentrations, and that the effects become progressively more significant as the duration of exposure increases. "That the observations exist is incontrovertible," notes an industry scientist. "There is," however, "a raging debate about what is harmful and what is not." The question hinges on what constitutes an adverse effect in a continuum of measurable biological changes occurring down to background levels of ozone.

Health Effects

Respiratory effects have been the primary focus of study and concern for health effects of ozone. These include decreased lung function; respiratory symptoms, such as coughing and shortness of breath;

Table D-1. Background on the 1993 Ozone NAAQS Decision

1971	EPA sets NAAQS for photochemical oxidants at 0.08 ppm not to be exceeded more than one hour per year.
1977	EPA initiates review of air quality criteria document for photochemical oxidants.
1978	EPA publishes final criteria document for ozone and other photochemical oxidants.
	EPA proposes to raise the primary NAAQS to 0.10 ppm, retain the 0.08 secondary standard, base the photochemical standard on ozone, and change to standards with a statistical form (that is, expected exceedances).
1979	After public comment and Executive Office of the President review, EPA revises the NAAQS to 0.12 ppm (primary and secondary).
1981	U.S. district court, D.C., upholds EPA's decision in *API v. Costle*.
1982	EPA/ORD/ECAO initiates review of the ozone criteria document.
1985	CASAC reviews draft criteria document. EPA/ORD/ECAO prepares a new draft.
	EPA/OAR/OAQPS begins drafting ozone staff paper.
1986	CASAC reviews first draft of staff paper, second draft of criteria document, sends EPA Administrator closure letter indicating satisfaction with final criteria document.
1987	CASAC reviews second draft of staff paper. CASAC recommends incorporating new scientific information.
1988	CASAC reviews draft supplement to the 1986 criteria document, third draft of staff paper.
1989	CASAC sends EPA Administrator closure letter indicating that the supplement and staff paper "provide an adequate scientific basis for the EPA to retain or revise" the NAAQS for ozone and suggests that EPA complete its review as rapidly as possible.
	EPA delays changing the standard to avoid disrupting ongoing Clean Air Act negotiations.
1990	Clean Air Act Amendments passed.
1991	American Lung Association sues to compel EPA to complete the ozone NAAQS review.
1992	Under court order, EPA Administrator Reilly proposes not to revise the NAAQS for ozone. EPA/ORD/ECAO initiates update of the CD.
1993	Administrator Browner announces that EPA would not modify the NAAQS for ozone.

decreased exercise performance; increased airway sensitivity to allergens (for example, contributing to asthma attacks); aggravation of existing respiratory diseases, such as chronic bronchitis and emphysema; pulmonary inflammation and morphological effects (that is, structural changes) in the lungs; and decreased ability to defend against respiratory infections. Ozone also causes a variety of detectable biological changes outside the respiratory system (for example, in red blood cell morphology and enzyme activity) and appears to be one of many contributing factors to overall morbidity and mortality (Horvath and McKee 1994).[8]

Evidence of these effects comes from both human and animal studies. The best documented evidence of health effects of ozone exposure are temporary decreases in lung function (for example, reduced lung capacity for a period of days).[9] Evidence of decreased lung function comes from controlled human exposure, field, epidemiology, and animal toxicology studies.[10] Clinical studies of human subjects under controlled exposure conditions (that is, in exposure chambers) provide the most reliable quantitative human exposure-response data (in the 0.12- to 0.24-ppm range), but are limited to short-term (one to eight hours) exposures and, generally, to healthy subjects.[11] According to an industry scientist, the "seminal" clinical ozone studies originated from the EPA Human Studies Division of the National Health and Environmental Effects Research Laboratory (NHEERL).[12] Heavily exercising, healthy, young (mostly male) subjects have experienced measurable (that is, small, but statistically significant) decreases in lung function during controlled exposures of ≥0.12-ppm ozone for one to two hours.[13] Measurable decreases in lung function have also been observed in intermittently exercising subjects exposed to concentrations as low as 0.08 ppm when exposures last six to eight hours. The level of exercise and individual responsiveness (that is, sensitivity) to ozone play a large role in the extent of lung function decrease at a given ozone concentration and exposure duration (Horvath and McKee 1994).[14]

Studies have also observed variability among individuals in their ability to tolerate ozone. That is, after successive ozone exposures, the pulmonary function response attenuates to some degree. According to an EPA research official, this has been described as a positive response (evidence that the effects are not long lasting) or a negative response (indicating that the lungs' defenses are being circumvented). An industry scientist says, "With ozone tolerance, we don't know what's going on."

Epidemiology and field studies provide evidence of measurable decreases in lung function from ambient ozone exposures ranging from 0.08 to 0.12 ppm, depending upon length of exposure. However, it may be difficult to separate the independent effects of ozone from those of other air pollutants (for example, particulates and acid aerosols) and other environmental variables (for example, high temperature and humidity) in these studies. The observed effects may be an additive combination of factors, or alternatively, ozone may interact with these variables to produce decreases in lung function "greater than the sum of the parts."

The first of several epidemiological studies of children engaged in normal activity at summer camps in the Northeast and California appeared in 1983 (Lippman et al. 1983).[15] These studies reported an association between decreased lung function in children and short-term exposure to ambient ozone concentrations. One of the summer camp studies (Spektor et al. 1988a) reported measurable decreases in lung function in children exposed to ozone levels that did not exceed the current one-hour standard of 0.12 ppm. In some cases, the lung function decreases persisted for up to a week (Horvath and McKee 1994). An industry scientist finds the epidemiological studies with children in summer camps to be the most reliable data on ozone health effects because children at play during the summer are a highly exposed, susceptible population and because their level of activity was not under artificial experimental control. The first of a series of epidemiological studies of healthy adults engaged in outdoor exercise appeared in 1988 (Spektor et al. 1988b). These studies also reported measurable decreases in lung function following short-term exposure at ozone concentrations below 0.12 ppm.[16] Thus, a variety of sources indicate measurable lung function decrements in different populations engaged in varying levels of activity in response to ozone concentrations at or below the current standard. However, these results beg the question of the health significance of measurable decreases in pulmonary function.

The summer camp studies also provided indications of persistent pulmonary inflammation and lung morphological changes in the children. According to Horvath and McKee (1994), these effects represent a potentially more important ozone health response than the more transient effects reported in most controlled exposure studies. Some have suggested that the responses could contribute to maldevelopment of the lungs in children. A number of studies provide evidence that short-term exposures to relatively low levels of ozone can inflame lung tissue. Researchers in EPA's Human Studies Division have reported pulmonary tissue inflammation in intermittently exercising subjects following exposure to 0.08- to 0.10-ppm ozone for more than six hours (for example, Koren et al. 1991). While a limited number of such exposures would not cause permanent damage in most individuals because of the body's self-repair capacity, many long-term (for example, months or years) animal studies have shown that repeated ozone exposures causing lung inflammation eventually result in morphological changes in the lungs and accelerated permanent loss of lung function (that is, "lung aging"). An autopsy study of cadavers in Los Angeles County (where ozone levels are typically the highest in the country) also revealed a higher-than-expected number of lung lesions, although researchers were unable to determine the contribution from smoking (Horvath and McKee 1994).

Some scientists dispute the significance of lesions observed in the lungs of animals and of cadavers from Los Angeles associated with chronic exposures to ambient ozone levels. Former CASAC panel chair Roger McClellan of the Chemical Industry Institute of Toxicology testified, for example, that he found the results of a study with rodents exposed five days per week to ozone at levels of 0.12 ppm and higher for two years suggested that long-term exposure at current ambient concentrations of ozone was unlikely to produce serious, irreversible changes in the lungs. The findings reduced McClellan's concern for the long-term impact from brief exposures that produce reversible effects (http:www.ciit.org/ACT97/ActivitiesFeb97/McClellan.html). The reasoning is that lungs have excess capacity or a reserve capability and that the lesions are not manifest clinical symptoms. According to an OAQPS official, however, the lung lesions might impinge on quality of life by causing shortness of breath from mild exertion with age rather than causing or contributing to mortality, as some have suggested. According to an industry scientist, there is considerable uncertainty about the reversibility of the pulmonary lesions observed in the long-term animal studies. "The problem is to determine what is a homeostatic response, a normal reversible change associated with defense, as opposed to what changes are associated with permanent, long-term damage." An ORD official also points to the uncertainties in extrapolating the results of the chronic exposures from animals to humans. "We have made some progress on dosimetric models for extrapolating from animals to humans. The hope was that they would be ready for this round of review, but we're not there yet."

Nonhealth Effects

The effects of ozone exposure on vegetation have been the primary focus of the secondary standard. There is general acknowledgment that vegetative effects occur at levels of exposure below 0.12 ppm and that certain plant species are more sensitive to ozone than are humans. According to an OAQPS official, "It is certain that ecological effects are occurring." Similar to the situation of ozone health effects, however, there is no collective certainty about what to make of the observed changes. Many early "phytotoxicity" studies assessed the effects of ozone on plants in terms of foliar injury (for example, leaf necrosis—dead leaves—or leaf drop); however, the extent of foliar injury (in terms of percentage of the crop exhibiting symptoms) can be much greater than that of crop yield loss. Conversely, significant yield losses can occur with little or no foliar injury (Tingey et al. 1994). An ORD scientist hypothesizes that tropospheric ozone may have its greatest ecological impact not on plant productivity but by exerting selection pressure that ultimately alters species composition within ecological communities and genetic variability within species. For example, the fact that many trees in the eastern United States are less sensitive to ozone than their western cousins may be an adaptive evolutionary response to higher ozone levels. Whether such changes are to be regarded as "important" is another issue.

The chief concern for the secondary effects of ozone has been on agricultural and timber crop yields due to their economic value. Using data collected by a multisite experimental program initiated in 1980 (the National Crop Loss Assessment Network, NCLAN), EPA analysts estimate that current ambient ozone exposures result in a 14% yield loss for major U.S. crops and that yield losses would continue to occur even if all U.S. sites would attain the current secondary NAAQS of 0.12 ppm (Tingey et al. 1994). However, reviewers (for example, Adams et al. 1984) have criticized NCLAN's experimental design, use of sensitive crop cultivars that maximize plant response to ozone, and failure to incorporate agricultural practices (for example, increased fertilization) that could offset the yield reductions caused by ozone exposure.[17] A factor that has seriously impeded the assessment of the effects of ozone on crops and natural resources is the lack of rural ambient air monitoring sites.

The selection of the ozone exposure indicator related to plant yields has been a difficult problem in formulating a secondary standard. The indexes that best predict yields are measures of cumulative exposure that give greater weight to peak exposure days (for example, a weighted sum over the growing season). There remain questions, however, regarding the practicality of this form of the indicator for regulatory purposes (Tingey et al. 1994).

The Effects of Weather

Hot, stagnant weather increases ambient ozone levels. Ozone and its precursors (VOCs and NO_x) are also transported considerable distances by air masses. As a result, ozone levels—and the number of exceedances of the NAAQS level—are sensitive to variations in meteorological conditions (Yosie et al. 1994). In addition, high heat and humidity appear to intensify the health effects of ozone (Horvath and McKee 1994).

THE PROCESS WITHIN EPA

Setting the Agenda

The 1991 ALA lawsuit against EPA was the factor most frequently cited by respondents as being responsible for getting the ozone NAAQS on the agency's agenda. According to an independent policy analyst, "They [EPA] were going to delay as long as they could until ALA forced the issue." Of course, legislative provisions requiring periodic review of the NAAQS and permitting citizen lawsuits enabled ALA's agenda-setting action.

According to an air program official, something that contributed to elevating ozone on the agency's agenda was that prior to the 1989 decision not to revise the standard, "Vocal scientists, particularly [former Assistant Administrator for ORD] Bernard Goldstein and [CASAC Chairman] Mort Lippman, met with the Administrator and pounded the table that they had studies that . . . suggested that [adverse health effects] were happening on a longer time frame, six to eight hours," than the form of the current standard (one hour).

Another factor that may have contributed indirectly to forcing ozone higher on the EPA agenda was the scientific information that agency researchers had generated since the 1988 supplement to the criteria document. These data demonstrated that one could induce measurable biological changes in humans, animals, and plants at ever-lower ozone concentrations. From 1988 to 1992, clinical studies reported by EPA's Human Studies Division incrementally lowered the bar by increasing the subjects' level of activity and duration of exposure. Similarly, NCLAN researchers designed experiments to detect plant responses at the lowest levels of ozone. In both cases, the pattern of searching for the lowest levels of ozone at which biological effects could be detected was consistent with the misguided presumption under the Clean Air Act of a threshold concentration below which no adverse effects would occur. According to an ORD scientist, the EPA air program framed the research question as, "At what level can you detect a change?" This source describes the situation as one in which "analysts, not policymakers, were involved in discussing what effects to measure." Absent these and other studies documenting measurable responses to ozone concentrations below the existing standard, ALA presumably would have had little institutional incentive to compel the agency to complete its review.

Assessing the State of the Science

According to an environmental lawyer, the most important studies on the effects of six- to eight-hour low-level ozone exposures were reported in

1989–1993, after the supplement to the 1986 criteria document was closed. An ORD official says the agency was overwhelmed by attempting to keep up with the science. "There had been over 1,000 new scientific papers in the literature since the last ozone decision, and the agency is expected to analyze each one." No formal NAAQS review of the science was conducted prior to the 1993 decision not to revise the ozone standard, but EPA staff scientists who tracked the expanding literature believed that the information that had accumulated since the 1988 CD supplement "pointed in the direction of multihour exposures having effects on pulmonary responses at lower concentrations than the current standard," according to another ORD official.

Communicating the Science to Agency Leadership

Sometime during the winter of 1992–1993, between unpacking boxes and "reinventing" EPA, members of the EPA transition team and the new administration were briefed on the ozone NAAQS by OAQPS staff. The staff indicated the "direction" in which the new science was pointing, according to an EPA official. But since ECAO was in the earliest stages of preparing a new criteria document at the time of the briefing,[18] it was clear that there would be insufficient time to conduct a full-blown NAAQS review prior to the court-ordered deadline. According to a former EPA air program official, this sort of communication, in which decisionmakers are briefed about the "directionality" of the evolving science and the studies currently under way that could have a major impact on the rulemaking, is typical of a decision not to revise a standard.

The Role of External Scientists in the Process

Although external scientists apparently had no direct involvement in EPA's 1993 decision not to revise the NAAQS for ozone, CASAC's historical involvement played a significant role. In contrast to the position of the official scientific advisory panel that reviewed EPA's 1979 revision, CASAC adopted a more precautionary approach regarding what constitutes an adverse health effect in the 1980s review of the criteria document for ozone (Jasanoff 1990, 115–116). By defining reversible changes in lung function and nonclinical symptoms as adverse health effects, CASAC set a precedent that narrowed policy options and paved the way for the lack of consensus on a policy recommendation. After setting this precedent, according to an independent policy analyst, the committee was "unwilling to redefine reversible health effects resulting from short-term exposures as 'not adverse.'" While the committee justified their negotiated definition of reversible changes as adverse on the basis of the understood effects of short-term studies, the concern of some members may have been focused more on the poorly understood effects of chronic exposures. As this source suggests, "In part, this [the committee's precautionary definition of adverse effects] may be due to a recognition by the committee of the inability of the available observational methods to capture potentially irreversible effects resulting from chronic exposures."

Respondents stated that the significance of the reported effects from exposures below 0.12 ppm for periods of six to eight hours was the primary source of division within CASAC that led it to recommend that EPA *"retain or revise"* the existing NAAQS for ozone. These were the same effects that the committee had, by "consensus" defined as adverse. Apparently, this consensus was a fragile one. According to EPA officials, there was a strong consensus within CASAC that the standard should not be relaxed, but some committee members felt that the level of the one-hour standard should be lowered or the form of the standard should be changed to an eight-hour average, while others believed that the database was not sufficiently "ripe" to recommend any change. According to an industry scientist, Morton Lippman, CASAC chair during the 1980s review, "firmly believes that his studies in the children's camps demonstrate detrimental effects at ambient levels."

SCIENCE IN THE FINAL DECISION

According to a former senior EPA official, the Reilly administration proposed not to revise the ozone NAAQS in 1992 "because CASAC was divided and the 1990 CAA Amendments had recently been passed." Regarding the role of science in the decision not to revise, an OAQPS official says, "Based on record up through 1988, there was not a compelling case to revise the standard. And, if there was a compelling case, it would have never gotten through the agency."

There was apparently some informal calculus that considered the science in terms of the incremental benefits arising from a marked change in the form, if not in the overall stringency, of the standard. (Recall that such a change was enacted in the 1987 PM-10 revision. See accompanying case study on particulate matter.) An EPA air program official noted the resultant "disruption that would occur in the ozone control program." Changing the form of the standard from one hour to eight hours would require "new models [and] new monitoring procedures and would essentially bring the control program to a stop for two to three years." This raised the question in the minds of air program officials of whether the incremental protection that the agency would gain by simply adopting a longer averaging time was worth the disruption in the program. "We concluded that it was not ready. It was more strategic than tactical," says this source.

According to respondents, the focus of decisionmaking by the Browner administration early in 1993 was on negotiating the ozone NAAQS review schedule with the plaintiff (ALA). Although the role of science in the final decisionmaking was not substantive, the scientific review *process* was considered. In the view of an ORD official, the 1993 decision was "science driven" in that EPA needed sufficient time to adequately analyze the vast amount of new information that had accumulated on ozone since 1988. Of course, the Browner administration had no control over the early stage at which it inherited the ozone NAAQS review process from the Reilly administration. Furthermore, despite the members' familiarity with the available scientific literature, the lack of consensus within CASAC ruled out the possibility that the committee's review would be rapid or perfunctory. An OAQPS official sympathizes with the Administrator's position in finalizing the agency's decision not to revise the standard. "In fairness to Browner, she had no choice but to sign it."

CONCLUDING OBSERVATIONS

In terms of a fate and transport analogy, this case illustrates that for criteria air pollutants, a full-blown NAAQS review is required to make science "available for uptake" by EPA decisionmakers. Information generated inside and outside EPA may be released into the public domain for several years and subjected to peer review. It may permeate throughout the agency, with its implications fully understood by staff. But, the information is not actionable until the NAAQS procedural requirements are satisfied. Furthermore, the lack of a consensus recommendation from CASAC to revise the standard provides the Administrator with an "out" to justify not revising the NAAQS on the basis of "scientific uncertainty." Therefore, the scientific evidence of effects at levels below the existing standard that is available to the decisionmaker can be neutralized by policy disagreements within CASAC. This is illustrated schematically in Figure D-1.

EPILOGUE

After EPA's 1993 decision not to revise the NAAQS for ozone, the American Lung Association sought judicial review of the EPA decision, citing the agency's failure to consider all relevant, available scientific information. The decision was voluntarily remanded to the agency, but in 1994,

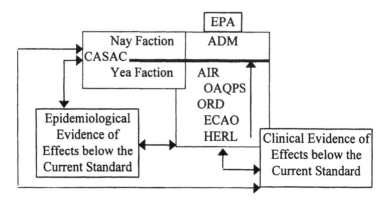

Figure D-1. Fate and transport of science in the 1993 decision not to revise the ozone standard

ALA again sued (unsuccessfully) to force EPA to complete the review by the end of 1995 (*Environment Reporter*, October 14, 1994). In 1995 (ten years after the statutory deadline for completing the review), OAQPS released a staff paper that recommended changing the standard to 0.07 ppm to 0.09 ppm measured over eight hours. Noting the continuum of biological effects observed down to background ozone levels, CASAC endorsed replacing the one-hour primary standard with an eight-hour standard but did not recommend any specific level. In CASAC's 1995 closure letter to Administrator Browner, the committee stated that "ozone may elicit a continuum of biological responses down to background concentrations. This means that the paradigm of selecting a standard at the lowest-observable-effects level and then providing an 'adequate margin of safety' is no longer possible" (*Environment Reporter*, January 19, 1996, 1756–1757).

Shortly after the elections in November 1996, EPA proposed to revise the NAAQS for ozone, basing the primary standard on eight-hour averages, and setting it at a level of 0.07–0.09 ppm. The agency also proposed to replace the current secondary standard with one of two alternative standards: one identical to the new primary standard or, alternatively, a new seasonal standard expressed as a sum of hourly ozone concentrations greater than or equal to 0.06 ppm, cumulated over twelve hours per day during the ozone season (*Federal Register* 27: 1672).

During the 1996–1997 revision of the ambient air quality standards for ozone, CASAC members came to general agreement that, within the range of standards under consideration, there was no "bright line" to distinguish any of them as being significantly more protective of public health than the others. Setting the standard, the panel agreed, was, therefore, a policy choice about the desired margin of precaution. Because the CASAC panel is authorized to offer policy advice, however, more than half the panel went on to offer EPA various and conflicting personal judgments as to the appropriate level and form of the standard. Policymakers and stakeholders seized whichever of these judgments they agreed with to fortify their own policy positions and to attack those of their adversaries with opinion masked as "sound science." Although much of the discussion blurred the line between science and policy, there were moments of honest debate about whether the benefits of the rule warranted the costs.

Ultimately, President Clinton endorsed EPA's plans for promulgating the new NAAQS for ozone in June 1997. On July 18, 1997, EPA promulgated the final rule (*Federal Register* 62: 38855–38896). The agency replaced the existing one-hour primary standard with an eight-hour standard at a level of 0.08 ppm with a form based on the three-year average of the annual fourth highest daily maximum eight-hour average ozone concentrations. The one-hour secondary standard was replaced by an eight-hour standard identical to the new primary standard.

Like the accompanying revised NAAQS for particulate matter, implementation of the new ozone standards was to be phased in over several years. On May 14, 1999, however, a U.S. Court of Appeals found that EPA had acted on legal assumptions that amounted to "unconstitutional delegations of legislative power" and remanded the regulations to EPA (Warrick and McCallister 1999).

ENDNOTES

[1] Photochemical oxidants are a class of highly reactive gases that includes ozone. They occur in the troposphere (ground-level atmosphere) and are formed in chemical reactions that occur in the presence of sunlight. Stratospheric ozone, on the other hand, occurs higher in the atmosphere and filters ultraviolet radiation.

[2] The 1970 Clean Air Act gave EPA ninety days to propose six NAAQS.

[3] Primary NAAQS are intended to protect the public from known or reasonably anticipated adverse health effects with an adequate margin of safety. Secondary NAAQS are intended to protect public welfare from known or reasonably anticipated adverse effects on environmental and economic resources and personal well-being.

[4] Of the major photochemical oxidants in ambient air, only ozone occurs at sufficiently high concentrations to be a significant health concern. Other photochemical oxidants such as hydrogen peroxide (H_2O_2) and peroxyacetyl nitrate produce effects only at higher concentrations than those found in most cities (Horvath and McKee 1994).

[5] A deterministic form defines areawide attainment of the standard as no more than x (for example, one) exceedances of the standard in any given year. A statistical form defines areawide attainment in terms of an expected number of exceedances over a period (for example, an average of no more than three exceedances over three consecutive years).

[6] The assessment procedure involved an expert judgment elicitation process known as subjective probability encoding. See Marraro (1982) and Nazar (1994) for case studies of the 1979 revision.

[7] The 1977 Clean Air Act Amendments required EPA to revise or reauthorize all of the NAAQS by December 31, 1980, and every five years thereafter.

[8] At the time of the 1993 decision, relatively few studies of the effects of ozone on mortality had been conducted. However, a recent draft study reported by the independent Health Effects Institute indicates that ozone has an effect on mortality that is independent of the effects of ambient particulates, sulfur oxides, nitrogen oxides, and carbon monoxide. As a group, the other pollutants also appear to contribute to mortality, but their independent, individual effects could not be identified (*Risk Policy Report*, April 19, 1996, 9).

[9] Clinical, field, and epidemiological studies have also reported associations between ozone levels as low as 0.10–0.12 ppm and a variety of mild to moderate respiratory symptoms (for example, throat irritation and coughing), but these data are regarded as less reliable than the lung function measurements because the symptoms are inherently more subjective (Horvath and McKee 1994).

[10] Field studies contain elements of both controlled human exposures and epidemiological studies.

[11] In some cases, asthmatic subjects have been tested.

[12] NHEERL is a division of the Office of Research and Development. The EPA facility is located at University of North Carolina, Chapel Hill. There are a limited number of facilities capable of conducting clinical ozone exposure studies. According to an ORD official, other such facilities are located at the University of California, Santa Barbara; Rancho Los Amigos Medical Center, Downey, California; and University of Michigan, Ann Arbor. According to this source, the University of California facility receives federal research support from the National Institute of Environmental Health Sciences (NIEHS) and EPA. Rancho Los Amigos is supported by the Electric Power Research Institute, General Motors (GM), and other industry sources. The University of Michigan facility is supported by GM.

[13] Average lung function decreases by approximately 5% for very heavily exercising human subjects exposed to 0.2-ppm ozone for two hours. However, there is considerable interindividual variability in response. Controlled exposures to 0.12-ppm ozone during very heavy exercise have resulted in individual lung function decreases up to 16% for adults and up to 22% for children (Horvath and McKee 1994).

[14] Controlled studies of resting human subjects (conducted mainly in the 1970s) found little or no change in lung function. An increased level of exercise is

associated with increased respiration rates, deeper breathing, and oral breathing. For a given ambient ozone concentration, the respiratory system receives an increased ozone dosage with elevated respiration rates. Oral breathing also results in greater penetration of ozone into the respiratory system.

[15] New York University Medical Center's Lippman and colleagues conducted studies in New York, New Jersey, and Connecticut with support from NIEHS and EPA. University of Michigan at Ann Arbor's Higgins and colleagues conducted another youth summer camp study in the San Bernardino mountains of southern California.

[16] In addition to the studies by Spektor and colleagues, a group from the Harvard School of Public Health also conducted a study of adult hikers on Mt. Washington, New Hampshire, that found an association between ambient ozone levels and decreased lung function (Korrick et al. 1992).

[17] The primary criticism of the experimental design has to do with the narrow set of ozone concentrations that were used. A large number of experimental replications over a few concentration levels increase the chances that the study would detect effects even if they are relatively small and may facilitate a mechanistic understanding of the effect of ozone on plant growth. However, Adams et al. (1984) observe that spreading the experimental trials over a larger number of concentration levels provides better input into an exposure-response analysis.

[18] In 1992, ECAO released a draft supplement to the air quality criteria for ozone summarizing selected new information on the effects of ozone on health and vegetation (Horvath and McKee 1994).

REFERENCES

Adams, R., T. Crocker, and R. Katz. 1984. Assessing the Adequacy of Natural Science Information: A Bayesian Approach. *The Review of Economics and Statistics.* 62: 568–575.

Horvath, S., and D. McKee. 1994. Acute and Chronic Health Effects of Ozone. In *Tropospheric Ozone: Human Health and Agricultural Impacts,* edited by D. McKee. Boca Raton, Fla.: Lewis Publishers, 39–83.

Jasanoff, S. 1990. *The Fifth Branch: Science Advisors as Policymakers.* Cambridge, Mass.: Harvard University Press.

Koren, H., W. McDonnell, S. Becker, and R. Devlin. 1991. Inflammation of the Lung Induced by Extended Exposure of Humans to 0.08 to 0.10 ppm Ozone. In *Tropospheric Ozone and the Environment: Papers from the First International Symposium on Tropospheric Ozone,* edited by R. Berglung, D. Lawson, and D. McKee. AWMA Transaction Series No. TR-19. Pittsburgh, Pa.: Air and Waste Management Association, 71–82.

Korrick, S., D. Gold, D. Dockery, et al. 1992. Pulmonary Effects of Ozone Exposures in Hikers on Mount Washington. *American Review of Respiratory Disease* 145: A94.

Landy, M., M. Roberts, and S. Thomas. 1994. *The Environmental Protection Agency*. New York: Oxford University Press.

Lippman, M. 1989. Health Effects of Ozone: A Critical Review. *Journal of the Air Pollution Control Association* 39: 672–695.

———. 1991. Health Effects of Tropospheric Ozone. *Environmental Science and Technology* 25: 1954–1962.

Lippman, M., P. Lioy, G. Leikauf, et al. 1983. Effects of Ozone on the Pulmonary Function of Children. In *International Symposium on Biomedical Effects of Ozone and Related Photochemical Oxidants. Advances in Modern Technology*, Vol. 5, edited by S. Lee, M. Mustafa, and M. Mehlman. Princeton, N.J.: Princeton Scientifica Publishers, 423–446.

Marraro, C. 1982. Revising the Ozone Standard. In *Quantitative Risk Assessment in Regulation*, edited by L. Lave. Washington, D.C.: Brookings Institution, 55–98.

McKee, D. 1994. Legislative Requirements and Ozone NAAQS Review. In *Tropospheric Ozone: Human Health and Agricultural Impacts*, edited by D. McKee. Boca Raton, Fla.: Lewis Publishers, 9–18.

Nazar, V. 1994. Revising the Ozone Standard. In *The Environmental Protection Agency: Asking the Wrong Questions, From Nixon to Clinton*, edited by M. Landy, M. Roberts, and S. Thomas. Oxford, England: Oxford University Press, 49–88.

Spektor, D., M. Lippman, and P. Lioy. 1988a. Effects of Ambient Ozone on Respiratory Function in Active Normal Children. *American Review of Respiratory Disease* 137: 313–320.

Spektor, D., M. Lippman, G. Thurston, et al. 1988b. Effects of Ambient Ozone on Respiratory Function in Healthy Adults Exercising Outdoors. *American Review of Respiratory Disease* 138: 821–828.

Tingey, D., D. Olszyk, A. Herstron, and E. Lee. 1994. Effects of Ozone on Crops. In *Tropospheric Ozone: Human Health and Agricultural Impacts*, edited by D. McKee. Boca Raton, Fla.: Lewis Publishers, 175–206.

Warrick, J., and B. McCallister. 1999. Court Voids EPA Air-Pollution Rules. *Washington Post* May 15: A1.

Yosie, T., H. Feldman, T. Hogan, and W. Ollison. 1994. Ozone Policy from a Petroleum Industry Perspective: Lessons Learned, Future Directions. In *Tropospheric Ozone: Human Health and Agricultural Impacts*, edited by D. McKee. Boca Raton, Fla.: Lewis Publishers, 301–322.

Appendix E

The 1983–1984 Suspensions of Ethylene Dibromide under the Federal Insecticide, Fungicide and Rodenticide Act

BACKGROUND

Ethylene dibromide (EDB) was once widely used as an agricultural pesticide and as an additive to leaded gasoline to remove lead deposits from engines.[1] As a highly toxic pesticide detected in wells and consumer grain products, EDB became front-page news in the winter of 1983–1984. However, the EDB story began, in fact, many years earlier, and it continues today as substitutes for EDB are themselves being phased out. In the case of one substitute, methyl bromide, the reason for concern is an environmental effect that was completely off EPA's radar screen when it announced emergency suspensions of EDB in 1983–1984—stratospheric ozone depletion.

EDB was an economically important pesticide used on food and non-food crops. More than 90% of EDB's pesticide usage was as a soil fumigant. The pesticide was injected directly into the soil before planting crops to destroy nematodes, small parasitic worms that attack the roots of plants (U.S. EPA 1983). EDB-treated fields grew more than thirty fruit and vegetable crops in California, Hawaii, and the southern states (Jasanoff 1990). EDB was also used as a fumigant to control insects and molds on grain milling machinery and flour mills and on stored grain products. An estimated 75% of the nation's grain mills used EDB as a fumigant. Another important use of EDB was as a postharvest fumigant for grains, fruits and vegetables (*Federal Register* 42: 63134–63161; *Federal Register* 49: 4452–4457).[2] Agricultural imports and exports to and from the United States were commonly required to be fumigated with EDB to prevent the international spread of agricultural pests and pathogens.

Farmers began using EDB in the 1920s, and the first reports of adverse effects of EDB on test animals surfaced in 1927 (Jasanoff 1990). EDB was registered under the Federal Insecticide, Fungicide and Rodenticide Act (FIFRA) in 1948 when the U.S. Department of Agriculture (USDA) was responsible for administering pesticide registrations. Under FIFRA, pesticides are registered separately for particular uses, and between 1949 and 1956, EDB was registered as a fumigant for soil, fruits and vegetables, and stored grain. At the same time, the Food and Drug Administration (FDA) was responsible for setting tolerance levels for pesticide residues and food additives under the Federal Food, Drug and Cosmetics Act. Based on data originally submitted by pesticide manufacturers, FDA set tolerances in 1955 for organic bromine compounds and inorganic bromine residues in food resulting from the use of EDB. In 1956, FDA granted Dow Chemical an exemption from a tolerance for inorganic bromine residues when EDB was used as postharvest fumigant for raw grain (Krimsky and Plough 1988).

The federal approach to regulating pesticides began to change slowly in response to public concern about pesticides and the environment sparked by Rachel Carson's landmark 1962 book, *Silent Spring*. In 1970, the administrative reorganization establishing EPA shifted the pesticides registration and tolerance setting divisions from USDA and FDA to the new agency. Congress amended FIFRA two years later, propelling government into the active regulation of the harmful environmental and health effects arising from pesticide use. FIFRA requires the registration and labeling of all pesticide products (Section 3) and authorizes EPA to suspend, cancel, or restrict an existing pesticide registration if it presents "any unreasonable risk to man or the environment, taking into account the economic, social, and environmental costs and benefits of the use of any pesticide" (Section 2). The suspension provisions of FIFRA (Section 6) authorize the Administrator, on a finding of "imminent hazard," to take regulatory action immediately (announce a suspension) to avoid the lengthy time required for cancellation proceedings, but do allow registrants five days to request an expedited hearing. The Administrator is authorized to issue an "emergency suspension," which does not permit a hearing, upon findings of an "imminent hazard" and an "emergency."

As a consequence of the "unreasonable risk" standard, FIFRA is considered a "balancing" statute requiring the Administrator to weigh the costs and benefits of pesticide use. Because FIFRA places the burden of proving that a pesticide product satisfies the criteria for registration on the registrant, FIFRA is a licensing program that permits EPA to require pesticide manufacturers to provide chemical testing and other data needed for regulatory decisionmaking. FIFRA also requires external review of EPA proposals to suspend or cancel pesticide registrations by USDA and an official Scientific Advisory Panel (SAP).

In order to expedite the risk-benefit review of some 50,000 pesticide registrations that had been approved during the previous thirty years,

EPA in 1975 instituted a modified review process called rebuttable presumption against registration (RPAR). Pesticides flagged as posing a substantial question of safety by a screening process using a series of risk criteria entered the RPAR process, which began with an informal comment period during which stakeholders could either rebut the presumption of unreasonable risk or seek to document the hazard. Despite EPA's hopes that RPAR would streamline the review of existing pesticide registrations, three to seven years were ordinarily required for a full RPAR review (Jasanoff 1990).

By 1973, a number of scientific studies had reported that EDB caused mutations and reproductive damage in animals. In 1975, the National Cancer Institute (NCI) reported that EDB should be considered a human carcinogen. The findings of cancer and noncancer effects in animals, the discovery of EDB traces in food, and a petition by the Environmental Defense Fund (EDF) led EPA to issue an RPAR notice for EDB in 1977 (*Federal Register* 42: 63134–63161). Following the notice, EPA was peppered with rebuttal comments from industry. In December 1980, the outgoing Carter administration issued a preliminary notice of its proposal to cancel EDB's uses as a fumigant for grain and grain milling machinery and as a citrus fumigant (*Federal Register* 45: 81516–81524).

During its first years, the Reagan administration sat on the proposal, and EDB became "a symbol of congressional concern" about the delay in pesticide reviews under FIFRA, in general, and EPA pesticide policy under Administrator Anne Gorsuch-Burford, in particular (Krimsky and Plough 1988). The new administration had sharply reduced the pesticides program staff and in particular, the staff responsible for reviewing registered pesticides. Meanwhile, EDB took on complicated interstate and international dimensions. In 1981, California agricultural authorities embarked on a massive fumigation program using EDB in an effort to control the Mediterranean fruit fly (Medfly). In response, the state's Occupational Safety and Health Administration (Cal OSHA) proposed a stringent workplace standard for EDB, and Japanese dockworkers refused to unload fruit arriving in their country until the Cal OSHA exposure standard was adopted. (Grapefruit exports to Japan were worth $65 million in 1982.) The Japanese government, on the other hand, was more concerned with preventing the introduction of the Medfly and insisted on fumigation of all produce arriving from California. California supermarkets, in the meantime, started boycotting EDB-fumigated fruit from Texas and Florida to meet Cal OSHA standards (Jasanoff 1990).

The process within EPA was given a prod when EDB groundwater contamination was detected between 1982 and 1983 in California, Florida, Georgia, and Hawaii. While EPA had expressed concerns about EDB's use in soil fumigation in the 1980 notice, groundwater contamination was not thought to be a problem due to EDB's volatility. It was, after all, a *fumigant*. According to Rick Johnson, the EPA EDB review team leader, the discovery of EDB in groundwater "was the straw that broke the camel's back" (Sun 1984). Shortly after Administrator Anne Gorsuch-Burford, John Todhunter, Assistant Administrator for the Office of Pesticides and Toxic Substances (OPTS), and most other senior officials left EPA in a 1983 housecleaning, Administrator William Ruckelshaus and OPTS acting Assistant Administrator Don Clay acted quickly, issuing an emergency suspension of EDB's soil fumigant registration in October 1983 (*Federal Register* 48: 46228–46248).

Shortly thereafter, public alarm about EDB in grain products mushroomed. FDA had begun monitoring EDB residues in food in 1978 after EPA first placed EDB in the RPAR process. In December 1983, the Florida Agricultural Commission reported that EDB residues had been detected in numerous grain products and started a product embargo. According to a former senior EPA official, when the state of Florida started pulling bread off grocery shelves because of EDB, it created a public panic. Other states responded with similar emergency measures against grain products. An EPA official relates a story revealing the level of hysteria: "I had a mother call me on the phone, and she would not let her children eat bread, orange juice, drinking water. She said, 'I have not fed my babies in two days because I don't know what to feed them.'" What seemed to scare the public most was that EDB was reported to be potentially capable of causing infertility and genetic mutations. Such dreaded effects that could damage future generations were perceived as being "riskier" than cancer. Shortly after being reinstated at the helm of EPA, Administrator Ruckelshaus gave a major press conference at which he announced that the agency was conducting an accelerated study to determine the risks of EDB in an attempt to calm public fears.

According to a former senior EPA official, a Florida state official sought to divert attention to the use of EDB as a grain fumigant to cover the state's negligence in permitting Florida citrus growers to use EDB as a soil fumigant greatly in excess of standards, which had led to high levels of groundwater contamination. Added to the alarm of domestic consumers were international concerns about the safety of U.S. grain exports. To allay public fears about the nation's food supply, EPA issued an emergency suspension of EDB as a fumigant for grain and grain milling machinery in February 1984 (*Federal Register* 49: 4452–4457).

Since the emergency suspensions of EDB in 1983–1984, EPA has announced the phaseout of methyl bromide, a substitute for some EDB uses, due to its contribution to stratospheric ozone depletion, and California suspended, then restricted the use of Telone, another EDB substitute. The suspensions and restrictions of EDB substitutes is particularly ironic since

EDB itself was an approved replacement for another suspended fumigant, DBCP. Table E-1 provides a summary chronology on EDB regulation.

SCIENTIFIC ISSUES

EDB is a synthetic halogenated organic compound (1,2-dibromoethane). (Halogens include chlorine, bromine, fluorine, and iodine.) While chlorinated organic pesticides (for example, DDT) show a strong tendency to accumulate in the environment and food chain, EDB was originally believed not to accumulate because of its relatively high volatility. Improvements in analytical measurement techniques (capable of detecting EDB levels at one part per billion [ppb]), however, revealed traces of EDB in food and animal feed in the 1960s and 1970s, with EDB levels in some wheat flour samples reaching four parts per million (*Federal Register* 42: 63138–63139). During 1982–1983, EDB was also detected in groundwater in several states as a result of soil fumigation, with levels varying from 0.02 ppb to more than 100 ppb (*Federal Register* 48: 46230). According to an EPA official, "the biggest leap in scientific understanding" regarding the risks of soil fumigants was their discovery in groundwater, a discovery that started with DBCP.

Between 1972 and 1974, an NCI-sponsored cancer study was conducted that concluded that EDB was very carcinogenic in rats and mice. Moreover, the tumors were observed to begin after a short time compared with the results of tests of other chemical carcinogens. Critics of the study, however, pointed out what they viewed as its inadequacies. First, the cancers were formed in the forestomach of the rodents, and though humans have an esophagus, they do not have a forestomach. Second, the EDB was administered to the rodents by intubation into the stomach. To apply the results to exposures to farmworkers via inhalation and dermal absorption required extrapolating from one exposure route to another. However, EDB was also found to be a carcinogen in a variety of species of laboratory animals when the chemical was administered by inhalation, skin painting, and intubation into the stomach. A number of studies, including a 1977 National Institute for Occupational Safety and Health (NIOSH) study, also found EDB to be mutagenic, reinforcing the evidence of EDB's carcinogenic potential across species. Evidence showing that EDB is mutagenic to germ cells (for example, sperm), also suggested the possibility of human genetic damage. Animal husbandry researchers in the 1960s and 1970s found that ingested or injected EDB caused spermicidal and other infertility effects in rams and bulls, and in 1978, an EPA contractor produced a report demonstrating that inhaled EDB resulted in reduced sperm counts in rats (U.S. EPA 1983).

Table E-1. Summary of Background on EDB Regulation

1925	EDB is first used as a pesticide.
1927	First reports of EDB's toxicity in exposed animals.
1947	Congress passes the Federal Insecticide, Fungicide and Rodenticide Act (FIFRA).
1948	Dow Chemical Co. registers EDB as a pesticide with the U.S. Department of Agriculture (USDA).
1949	EDB is registered for use as a soil fumigant.
1956	EDB is registered as a fumigant for stored grain, fruits, and vegetables.
1956–1958	Food and Drug Administration (FDA) sets tolerances and exemptions from tolerances for inorganic bromide residues resulting from use of EDB on a variety of fruits, vegetables, and grains.
1970	EPA is created.
1972	FIFRA is amended.
1973	Reports show EDB causing mutations and reproductive damage in animals.
1974	National Cancer Institute (NCI) issues a "memorandum of alert," identifying EDB as a potential carcinogen.
1975	July: EDF petitions EPA under FIFRA to investigate the carcinogenic potential of EDB pesticides and to either suspend or cancel their registrations.
	NCI reports that EDB should be considered a human carcinogen based on animal study results.
1977	EPA issues position document (PD) 1, stating the agency's rationale for the rebuttable presumption against registration of EDB.
	National Institute of Occupational Safety and Health (NIOSH) study concludes EDB is mutagenic.
	Occupational Safety and Health Administration (OSHA) issues guidelines to limit occupational exposures to EDB.
1978	FDA begins to monitor EDB residues in food.
1979	NIOSH recommends that OSHA reduce permissible workplace exposures from 20 ppm to 130 ppb.
	EPA suspends another fumigant, DBCP, for all uses except Hawaiian pineapple.
1980	To replace DBCP, EPA grants emergency approval for soil fumigation with EDB for soybeans.
	Carter administration EPA issues PD 2/3, a proposal to end EDB's uses as a fumigant for grain and grain milling machinery. Use as a citrus fumigant would end on July 1, 1983. Other uses would be continued but on a restricted basis.

Table E-1. Summary of Background on EDB Regulation *(Continued)*

1981	Levels in some USDA/EPA wheat samples reached 4200 ppb.
	FIFRA Science Advisory Panel (SAP) concurs with EPA's proposal to cancel EDB uses.
	Mediterranean fruit fly crisis.
	Reagan White House Office of Science and Technology Policy chairs interagency task force to review EDB problem.
1982	Reagan appointee John Todhunter, Assistant Administrator for OPTS, sits on EDB proposal.
	Groundwater residues of EDB, from soil injection in citrus fields, detected in Georgia, California, and Hawaii.
1983	March: Administrator Anne Gorsuch-Burford, Todhunter, and other high-ranking EPA officials resign.
	Florida and California suspend use of EDB as soil fumigant.
	September: Congressional hearing held on EDB.
	September: EPA releases PD 4, declaring an emergency suspension of EDB as a soil fumigant for agricultural crops and a cancellation proceeding against all other pesticide uses of EDB.
	October: EPA issues *Federal Register* notice revealing its intent to cancel registrations of pesticide products containing EDB, except use as a postharvest citrus fumigant, which may continue until September 1, 1984, to allow time for alternatives to be found.
	November: SAP meeting with Office of Pesticide Programs representatives to discuss alternatives to EDB.
	December: Florida Agricultural Commission reports that residues of EDB are in numerous grain-based products and starts a product embargo.
1984	February: EPA announces emergency suspension of EDB as a fumigant for grain and grain milling machinery.
	June: SAP meets to review the emergency suspensions of grain fumigation uses of EDB and discusses EPA's lack of attention to the hazards of probable alternatives to EDB.
1990	Stratospheric ozone depletion potential of methyl bromide recognized.
	California suspends Telone, a soil fumigant containing 1,3-dichloropropane (1,3-D) after the state's Air Resources Board detects 1,3-D "levels of concern" in ambient air.
1993	EPA announces phaseout of methyl bromide under the Clean Air Act.
1994	California allows restricted use of Telone.

There were no strong epidemiological studies of any adverse health effects of EDB in humans. But, to give the reader an indication of EDB's cancer potency estimated by EPA from the animal studies, EPA calculated that the workers most highly exposed to the pesticide (warehouse workers who over forty-year working lives handle agricultural commodities that are fumigated postharvest for pest quarantine) could have a lifetime risk of cancer as high as two in ten (2×10^{-1}) (*Federal Register* 48: 46237).

THE PROCESS WITHIN EPA

Setting the Agenda

The NCI animal study was sufficient to push EDB into the RPAR process, and the EDF petitions may have forced EPA to issue its regulatory proposals in 1980. But the suspension of DBCP (for which EDB was a substitute), the Medfly infestation in California, and federal pesticides policy during the first years of the Reagan administration combined to prevent further action on EDB. According to current and former EPA officials, the detection of EDB groundwater contamination in 1982–1983 provided the impetus for proceeding with the suspension of soil fumigation uses, and Todhunter's resignation in March 1983 removed the principal internal obstacle. The scope of the problems associated with soil fumigation were geographically limited, however. The safety of the nation's food supply was placed onto the regulatory agenda by Florida's actions regarding EDB residues in grain products, which ignited a public panic.

Assessing the Science

Six out of seven interviewees responded that there was adequate science to inform EPA's decision to suspend EDB registrations in 1983–1984. The data on human exposure to EDB were considered particularly abundant. According to a former SAP member, it was one of the more "cut-and-dried" pesticides reviews in terms of the availability of scientific information. Respondents characterized the scientific uncertainty at the time of the 1983–1984 suspensions as moderate to very small. Areas of scientific uncertainty identified by respondents were an incomplete understanding of noncancer effects and of the risks of substitutes for EDB. Interviewees agreed that EPA's final treatment of the available science was good to very good.

Under the rebuttable presumption against registration process, EPA laid out its scientific and economic analyses in a series of position documents (PDs) prepared by a working group managed by the Office of Pesticide Programs (OPP) Office of Special Pesticide Reviews. PD 1, issued in December 1977, stated the agency's rationale for placing EDB in the RPAR process. PD 2/3, issued in December 1980, responded to information submitted in rebuttal, reported the review's findings, and proposed regulatory action. PD 4, issued in September 1983, summarized the data, public commentary, and scientific review and announced a final regulatory decision.

PD 1.   The primary basis for the EDB RPAR review was the NCI rodent study, although it was available in 1977 only in manuscript form. OPP was also able to take advantage of a 1977 NIOSH criteria document (produced by a contractor) assessing the available EDB health effects information. The EPA Office of Research and Development Cancer Assessment Group (CAG) conducted a quick review of summary data from the NCI study and concluded that EDB caused a significant increase in the incidence of stomach cancer in rodents and that the tumor rates appeared high. A review of the EDB mutagenicity studies was conducted by an OPP contractor and supplemented by the OPP Criteria and Evaluation Division (OPP/CED). The EDB working group performed a review of the EDB reproductive effects studies, and the EPA Office of Toxic Substances (OTS)[3] contracted for a 1976 study of the developmental toxicity of EDB inhaled by rodents. OPP/CED conducted some preliminary EDB exposure assessments based on available literature, which included studies by the National Institutes of Health and industry contractors. The exposure assessments looked at exposures to EDB applicators, residues on raw agricultural commodities, and residues in grain products, but did not consider the possibility of groundwater contamination.

PD 2/3.   In the 1980 PD 2/3, EPA responded to an accumulation of rebuttal comments questioning the validity of the NCI study, the mutagenicity studies, and EPA's exposure analyses. In addition to the information provided by the rebuttals, EPA was authorized under FIFRA to require registrants to provide much of the data used in the analysis. CAG analyzed the rebuttals regarding the carcinogenicity of EDB. Two studies issued after PD 1, one of which was by NCI, showed that inhaled EDB produced a significant increase in rodent tumors. The agency reviewed a number of new studies that supported the earlier findings of EDB mutagenicity. The agency rejected as technically flawed two human epidemiological studies submitted by the Ethyl Corporation in rebuttal to the PD 1 analysis of reproductive risks. The agency also sponsored an epidemiological reproductive study that found no effects but considered it of limited significance due to its small size. Using data submitted to the agency, OPP conducted a more extensive assessment of human exposures to EDB. Residues of EDB were detected in flour and bread, on fruits and vegetables, in air inhaled by exposed workers, and from skin contact with treated commodities.

EPA estimated that the range of cancer risks to people exposed occupationally or through dietary exposures was from moderate (one in one hundred thousand, 10^{-5}) for the general population to very high (one in ten, 10^{-1}) for some agricultural workers. The agency also concluded qualitatively that EDB posed mutagenic and reproductive risks to exposed humans. Weighing these health risks against the benefits of continued registered uses of EDB and the availability of alternative pesticides, the agency proposed to cancel immediately the use of EDB on stored grain, flour mill machinery, and felled logs; to cancel the use of EDB for postharvest fumigation of citrus and tropical fruits as of July 1983; and to allow continued but restricted use for soil fumigation and other uses (*Federal Register* 45: 81516).

SAP Review of PD 2/3.   The FIFRA SAP reviewed the EDB PD 2/3 in 1981. In addition to briefings from OPP's Special Pesticide Review Division (SPRD) staff and an EPA reproductive biologist, the panel received statements from representatives of USDA; the state of Florida, Department of Citrus; the Hawaii papaya industry; Ethyl Corporation; Dow Chemical Corporation; corporations recommending irradiation as an alternative to EDB for citrus fumigation; the National Association of Wheat Growers; the National Pest Control Association; academics; and others. SAP concurred with the agency's decision to cancel the registration for stored grain, flour mill machinery, and felled logs. However, the SAP considered the phaseout of citrus fumigation to be unnecessary, as well as impracticable, due to the uncertain feasibility of irradiation to control fruit flies on citrus. Instead, the panel recommended additional restrictions on this registered use, including increased protection for exposed workers. That panel also recommended that, prior to final rulemaking, EPA require additional data and analysis of exposures resulting from the use of EDB for flour mill fumigation and EDB residues in grain and grain products; monitoring of highly exposed workers in the grain, citrus, and minor use areas; and an additional rodent reproductive study. Finally, the panel also indicated its concern about the occurrence of EDB in groundwater and urged that groundwater be monitored closely in high use areas (*Federal Register* 48: 46242–46244.)

Todhunter's "Reassessment" of EDB.   John Todhunter served as Assistant Administrator of OPTS from 1981 to March 1983. Disinclined to proceed with the EDB regulatory actions proposed by the outgoing Carter administration, Todhunter produced his own "alternative" EDB risk assessment, which downplayed the risks of the pesticide. According to an

EPA official who served on the EDB working group, Todhunter's analysis consisted of "back of the envelope calculations." Edwin Johnson, Director of OPP, testified to Congress that Todhunter's reassessment of EDB was "an example of Todhunter playing scientist . . . [and he] used poor methods" (cited in Lash 1984). According to Lash (1984), Todhunter had also asked EPA statistician Anne Barton to alter her risk estimates for EDB using an obscure scientific theory she had never seen before (or after). Todhunter's EDB reassessment conveniently presumed that risk levels declined exponentially (instead of linearly) with decreasing exposure time (Walsh 1982, 1595). An environmentalist reports that Todhunter compared the risk of EDB to that of smoking one cigarette in a lifetime. In 1983, prior to his resignation, Todhunter asked SAP to rereview the EDB regulatory package (Jasanoff 1990).

Todhunter's penchant for conducting his own reanalysis was not limited to EDB. His highly controversial analysis of formaldehyde, which OTS identified in 1981 as a candidate for priority attention under the Toxic Substances Control Act, deviated from carcinogen risk analyses previously performed at EPA and rested on a number of questionable assumptions (Jasanoff 1990, 197–198). It was in the wake of these episodes that the National Research Council recommended that EPA develop formal risk assessment guidelines to ensure methodological consistency (NRC 1983).

PD 4. Citrus-growing states began reporting detection of EDB in groundwater in 1982, and the policy vacuum at EPA was filled by the states until Todhunter resigned in March 1983. During the spring and summer of 1983, under the new Ruckelshaus administration, the EDB working group worked feverishly to update the EDB regulatory package in light of the new data and to respond to comments on the EDB PD 2/3. The final EDB team, led by SPRD's Rick Johnson, included nineteen EPA staff with training in statistics, toxicology, chemistry, entomology, plant pathology, reproductive biology, law and economics (U.S. EPA 1983).

Communicating the Science to Agency Leadership

Three respondents characterized the communication of the scientific information to agency decisionmakers as very good. Due to the high stakes and public visibility, the decisionmakers were "remarkably and unusually interested in the details," says one EPA official. According to this source, Administrator Ruckelshaus was briefed on EDB by the review team manager two to three times a month during 1983 and 1984.

A former senior EPA official says that the issue was first raised to the new EPA leadership when Don Clay, the acting Assistant Administrator of OPTS, requested a meeting with the new Deputy Administrator, Alvin

Alm. At this 1983 meeting, OPP presented the toxicological evidence on EDB, saying that "they had never seen a substance that caused so many tumors so quickly" in laboratory animals. The leadership was also impressed by the hard data that the agency had assembled on exposure to EDB. The staff recommended that on the basis of the new data on EDB groundwater contamination, the agency should immediately suspend the use of EDB as a soil fumigant to control nematodes. Alm questioned whether the agency should also suspend EDB use as a grain fumigant, but Clay convinced Ruckelshaus and Alm not to do so due to the lack of data on dietary exposure from grain products.

SCIENCE IN THE FINAL DECISION

According to a former senior EPA official, the 1983 emergency suspension of EDB as a soil fumigant was driven by scientific information. "The science was adequate for the suspension—it was not a hard call," says this source. But the 1984 emergency suspension of EDB for use to treat stored grain and for treating grain milling equipment was dictated "purely by economics and public hysteria." The grain product bans and inconsistent EDB residue tolerance levels being enacted by the states were disrupting the market, causing economic losses, and creating concerns about the safety of U.S. grain exports. A former EPA political appointee recalls that the EDB grain suspension was "unlike any regulatory action [he] ever dealt with in that the people who normally opposed regulation were pushing hardest for regulation. . . . USDA, the Grocery Manufacturers Association, and the food industry all demanded EPA action. They felt that grain and other food supplies were threatened by public panic if EPA didn't take charge and calm people's fears. [There was] something like $3 billion worth of stored grain. The Russians threatened not to buy U.S. grain because of EDB" and the Japanese needed high-level assurances that U.S. grain was safe. Most of the states, according to this source, "were glad to be bailed out" by EPA's decision.

CONCLUDING OBSERVATIONS

The story of EDB suggests the power of new scientific data—the occurrence of EDB in groundwater—to break through a logjam erected by a reluctant regulator—Assistant Administrator Todhunter. In terms of a fate and transport analogy, the data produced outside EPA on the occurrence of EDB in groundwater and consumer grain products were released into the public domain beginning in 1982 prior to Todhunter's resignation.

EPA had no control over the data and lost control over interpretation of the data. The fact that the states got out ahead of EPA on the issue in reaction to the new data makes it seem likely that the agency would have been forced to take regulatory action even if there had been no turnover in EPA's political leadership. It should be underscored that the new findings could not have been completely unexpected, since SAP strongly urged EPA to require groundwater monitoring in 1981. EDB also illustrates the downside of EPA's losing control over the regulatory agenda. Cross (1989) concluded that while EPA eventually regulated the carcinogenic hazard from EDB, agency action was unnecessarily slow and final EPA restrictions may have been unnecessarily stringent.

A former senior EPA official believes that although Todhunter deserves a fair amount of criticism for his role in the EDB decision, the culpability extends beyond the former Assistant Administrator. "There was dawdling within the agency. The [EDB toxicity] data that the agency acted on were in the agency before [Todhunter took office]. They're giving one individual an awful lot of credit." Whether or not agency staff contributed to the slow pace of the EDB review, however, Todhunter was the politically accountable decisionmaker. Furthermore, to continue the fate and transport analogy, EPA's reservoir of public trust was contaminated by Todhunter's track record for cooking the data to provide regulatory relief. The agency had no public credibility to portray the dietary risks of EDB in an objective manner or call for more time to evaluate the public health risks (as Clay had counseled Alm and Ruckelshaus). The news media were handed a juicy story line that included unseemly official conduct, bumbling bureaucrats, and horrific public health threats contained in something as familiar as a bread wrapper. Those with competing agendas filled the vacuum, whipping the public into a frenzy and further eroding trust and confidence in public institutions. These dynamics are illustrated in Figure E-1.

Graham (1991) cites EDB as an example of the credibility problem exacerbated by political appointees in the Reagan administration who pursued a policy of "regulatory relief." Because early Reagan-era EPA policymakers could not be trusted to resist adjusting the science to meet their policy preferences, the agency spent the next decade developing formal guidelines for risk assessment that have now come to be viewed by some EPA critics as inflexible and overly simplistic. These episodes also fed the perception that there should be a fire wall between risk assessors and risk managers within the agency. With the growing recognition of the value of participation by decisionmakers and stakeholders to help guide judgments made in the risk analysis process (for example, regarding the scope and intended use of the analysis and what environmental or health effects are of concern), the virtue of a rigid distinction between risk

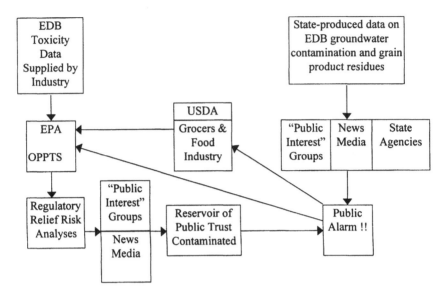

Figure E-1. Fate and transport of science in the 1983–1984 EDB suspensions

assessment and risk management is increasingly being called into question. (See, for example, NRC 1996.)

The case of EDB also indicates the unintended consequences of considering a pesticide and its alternatives independently and the need in regulating pesticides to thoroughly consider the risks of substitutes as well as other risks that may be created unintentionally by regulatory action. Because the 1983–1984 EDB suspensions were issued on an emergency basis, SAP did not review the decisions until after they were final. When EPA presented its decisions to the panel in 1983 and 1984, SAP members cautioned the agency about the lack of knowledge concerning the health effects of alternative nematocides (principally Telone) and grain fumigants (primarily methyl bromide). EDB, the panel noted, had itself replaced the canceled pesticide DBCP as a soil fumigant for soybeans in 1980. Would Telone be next? Would methyl bromide, which a Dutch study had found to be an animal carcinogen, cause the next spasm of public hysteria? (Jasanoff 1990). An EPA official explains what the SAP member recognized—that nematocides are simply "a nasty group" of pesticides.[4] "To be efficacious, they [nematocides] need to leach down into the soil, and they are very toxic as a group," says this official. After suspending permitted uses of Telone in 1990, California in 1994 allowed the use of Telone as a soil fumigant to continue on a restricted basis. According to a statement by James Wells, Director of the California EPA

Department of Pesticide Regulation, restricted uses of Telone were rein-stated because "farmers are trying to cope with a diminishing number of pest management tools . . . this problem is especially critical with soil fumigants . . . [and] limited use under strict conditions does not pose a significant risk and can be of benefit to the state's agricultural economy" (http://www.cdpr.ca.gov/docs/archives/pressrls/1994/94-42.arc).

However, the types of risk issues that SAP raised in 1983–1984 were the familiar risks (for example, cancer) of alternative pesticides. A substi-tute risk that EPA was aware of at the time regarding EDB's use as a grain fumigant was that of aflatoxin, a very potent animal carcinogen produced by a fungus on corn and peanuts that was controlled by EDB. (Gray and Graham [1995] state that EPA never considered formally the potential increase in risk due to aflatoxin, but Administrator Ruckelshaus was reportedly aware of the substitute risk.) EPA failed to foresee, however, the unconventional environmental effects that environmental releases of EDB substitutes could have. The agency, for example, responded to SAP's concerns about methyl bromide by indicating that since it was more volatile than EDB, it could be expected to pose fewer risks of contaminat-ing the food chain or the environment (Jasanoff 1990). However, an inde-pendent risk analyst notes that the higher volatility of methyl bromide results in greater transport of the halogen bromine into the upper atmos-phere, or stratosphere, than would result from the use of EDB. Therefore, methyl bromide poses a greater threat to stratospheric ozone than does EDB, resulting in a greater risk of environmental and health effects from unfiltered ultraviolet radiation.

The Molina and Rowland hypothesis of stratospheric ozone depletion by the halogen chlorine was first published in 1974 (Molina and Rowland 1974). After the Antarctic ozone hole was discovered in 1985 and grew unexpectedly in 1987, the Montreal Protocol was signed in September 1987. According to an EPA official, within three years, researchers had a good understanding of the contribution of methyl bromide to stratos-pheric ozone depletion. In December 1993, EPA announced that the pro-duction and importation of methyl bromide would be phased out by 2001.

ENDNOTES

[1] Of over 300 million pounds of EDB produced each year by 1984, around 20 million pounds were used as pesticide and 230 million pounds went into leaded gasoline (Jasanoff 1990, 130).

[2] Additional uses, described by EPA as minor, included control of mountain pine bark beetles in the western states by federal and state forestry agencies, con-trol of dry-wood and subterranean termites in structural pest operations, and control of wax moth in honeycombs (*Federal Register* 42: 63134–63161).

³ OPP and OTS are both under the Assistant Administrator for Pesticides and Toxic Substances.

⁴ Methyl bromide has also been used as a soil fumigant, but its greater volatility requires covering fields with tarps to ensure penetration sufficient to control nematodes.

REFERENCES

Cross, F. 1989. *Environmentally Induced Cancer and the Law: Risks, Regulation, and Victim Compensation.* Westport, Conn.: Greenwood Press, Inc.

Graham, J. 1991. Science in Environmental Regulation. In *Harnessing Science for Environmental Regulation,* edited by J. Graham. New York: Praeger, 1–9.

Gray, G., and J. Graham. 1995. Regulating Pesticides. In *Risk vs. Risk: Tradeoffs in Protecting Health and the Environment,* edited by J. Graham and J. Wiener. Cambridge, Mass.: Harvard University Press, 173–225.

Jasanoff, S. 1990. *The Fifth Branch: Science Advisors as Policymakers.* Cambridge, Mass.: Harvard University Press.

Krimsky, S., and A. Plough. 1988. Pesticide Residues in Food: The Case of EDB. In *Environmental Hazards: Communicating Risks as a Social Process.* New York: Auburn House.

Lash, J. 1984. *A Season of Spoils: The Story of the Reagan Administration's Attack on the Environment.* New York: Pantheon Books.

Molina, M., and Rowland, F. 1974. Stratospheric Sink for Chlorofluoromethanes: Chlorine Atom-Catalysed Destruction of Ozone. *Nature* 249: 810–814.

NRC (National Research Council). 1983. *Risk Assessment in the Federal Government: Managing the Process.* Washington, D.C.: National Academy Press.

———. 1996. *Understanding Risk: Informing Decisions in a Democratic Society.* Washington, D.C.: National Academy Press.

Sun, M. 1984. EDB Contamination Kindles Federal Action. *Science* 223: 464–466.

U.S. EPA (U.S. Environmental Protection Agency). 1983. *Ethylene Dibromide (EDB) Position Document 4.* Washington, D.C.: Office of Pesticide Programs, September 27.

Walsh, J. 1982. Spotlight on Pest Reflects on Pesticide. *Science* 215: 1592–1596.

Appendix F

The 1989 Asbestos Ban and Phaseout Rule under the Toxic Substances Control Act

BACKGROUND

Asbestos is a commercial term applied to a class of naturally occurring fibrous minerals. Asbestos has been called a "miracle fiber" due to its many useful properties (Peck 1989). Products containing asbestos are naturally flexible, heat resistant, noncombustible, chemically inert, and possess high tensile strength. Asbestos-containing products include asbestos cement pipe, flooring products (for example, tiles), paper products (for example, padding), friction materials (for example, brake linings and clutch facing), roofing products (for example, tiles), fire-retardant/fireproof insulation, and coating or patching compounds (NRC 1984). About 30 million tons of asbestos were used in the United States between 1900 and 1980 and, as of 1986, when EPA proposed an asbestos ban, hundreds of products were still being made with asbestos (EESI 1996).

Health effects from exposure to asbestos have been recognized in the United States since the turn of the century, and about 65,000 Americans are now estimated to suffer from asbestosis, a chronic lung disease that makes breathing progressively more difficult (EESI 1996).[1] The majority of asbestos-related disease incidence has been traced back to occupational exposures to miners, shipbuilders, insulators and factory workers (NRC 1984). Asbestos first gained broad notoriety in the U.S. in 1964 when Dr. Irving Selikoff of Mt. Sinai Hospital and colleagues issued a landmark paper on high rates of mesothelioma and other cancers in New York City insulation workers exposed to asbestos. The study, which was presented at the annual meeting of the New York Academy of Science, "blew the asbestos issue out of the quiet circles of scholarly study and into the public consciousness," according to the *Boston Globe* (January 18, 1983). During the 1970s, concern mounted among the general public about potential health effects resulting from nonoccupational exposures to asbestos. Production and use of asbestos in the United States has fallen dramatically since it peaked in the late 1960s and early 1970s.[2] Product liability ("toxic tort") litigation against domestic asbestos product manufacturers has played a significant role in the industry's decline since 1973.[3]

After five years of development and debate, Congress passed the Toxic Substances Control Act (TSCA) in 1976 to address nonpesticidal toxic chemicals in commerce.[4] According to a source involved in negotiations over the bill, TSCA evolved specifically with substances like asbestos in mind.[5] TSCA was designed to provide a "life-cycle" framework for management of chemical risks, taking into account risks from product manufacturing, use, and disposal and exposures associated with the air, water, and land (Shapiro 1990). TSCA requires premanufacture evaluation of most new chemicals (under Section 5)[6] and allows EPA to regulate existing chemical hazards (under Section 6). The risk management tools available to EPA for existing chemicals range from labeling standards to outright bans.

TSCA is unlike most other pollution control statutes (with the exception of the Federal Insecticide, Fungicide, and Rodenticide Act) in that it primarily regulates commercial products—not wastes—and, therefore, acknowledges the many benefits of commercial chemicals. Most environmental statutes require decisions to be based on precautionary margins of safety or the availability of pollution control technology and seek to limit explicit consideration of economic factors. In contrast, to enact regulatory controls under TSCA, EPA must have a "reasonable basis" for finding that a toxic substance poses an "unreasonable risk" to health or the environment and must adopt the "least burdensome" control option that the agency is authorized to require to limit the risk to an acceptable level. This "unreasonable risk" provision has been interpreted as requiring EPA to balance the costs and benefits of proposed regulatory decisions, taking into account the availability of substitutes and other adverse effects (for example, the risks posed by substitutes) that the regulation may have (Shapiro 1990).[7] TSCA is also unlike other environmental statutes in that EPA is subject to the "substantial evidence" standard of judicial review (under Section 19). By comparison, under other statutes, EPA must meet the less demanding "arbitrary and capricious" judicial review standard. Therefore, TSCA is formulated to require EPA to provide a considerable weight of evidence and regulatory analysis to document its regulatory decisions and defend them under judicial review.

In 1979, EPA formally announced its intent to pursue a complete ban of all uses, manufacturing, mining and importation of asbestos under TSCA Section 6 (*Federal Register* 44: 60061).[8] The international asbestos industry and, according to a former senior EPA official, the Canadian government lobbied vigorously against the proposal,[9] and there were years of conflict between EPA and the Office of Management and Budget (OMB) over the rule (EESI 1996). Considerable controversy was generated in 1985 when EPA tentatively referred asbestos to the Occupational Safety and Health Administration (OSHA) and the Consumer Product Safety Commission (CPSC) and later reversed its decision. (According to a former senior EPA official, OMB raised objections that the rule had not been coordinated with OSHA and then EPA Administrator Lee Thomas pulled the rule back from OMB.)[10] In 1986, seven years after EPA announced its initial intent, the agency proposed a rule to ban asbestos (*Federal Register* 51: 3738). It took another three years (during which EPA held informal rulemaking hearings and developed or updated several of the major supporting documents) before the rule was finalized in 1989 (*Federal Register* 54: 29467). The rulemaking prohibited (in three staged intervals) the future manufacture, importation, processing, and distribution in commerce of virtually all asbestos products by 1997.

EPA's 1989 asbestos ban and phaseout rule was based on one of the longest (dating back to the 1940s) and largest substance-specific human health databases ever amassed.[11] When OSHA established its occupational standard for asbestos in 1986, it stated, "OSHA is aware of no instance in which exposure to a toxic substance has more clearly demonstrated detrimental health effects on humans than has asbestos exposure" (*Federal Register* 51: 22612). EPA's rulemaking procedure was also one of the most extensive in the agency's history (Carnegie Commission 1993). According to an EPA official, the agency spent approximately $10 million on the regulation's cost-benefit analysis alone. In the judgment of EPA Administrators Lee Thomas and William Reilly and Assistant Administrator for Pesticides and Toxics John Moore, the health database and the agency's analysis provided substantial evidence that asbestos posed an unreasonable public health risk. In 1991, in deciding the case of *Corrosion-Proof Fittings v. EPA* (947 F.2d 1201 [5th Cir. 1991]), the U.S. Fifth Circuit Court of Appeals ruled otherwise, concluding that EPA had not presented sufficient evidence to justify banning, for all practical purposes, all commercial uses of asbestos.

The court justified its decision by finding that EPA had failed (1) to give public notice of its intended use of "analogous exposure" data to calculate expected regulatory benefits and (2) to promulgate the least burdensome, reasonable regulation required to provide adequate protection. Arguing that the agency bears a heavier burden when it seeks a ban of a substance than when it merely seeks to regulate that product, the court held that the rule was not promulgated on the basis of substantial evidence as required under TSCA. The court delved into substantive issues in which administrative agencies are traditionally granted judicial deference, discussing at considerable length what it regarded as shortcomings in EPA's cost-benefit analysis and in the agency's analysis of the health and safety risks posed by asbestos substitutes. The court also considered *amicus curiae* briefs relating to scientific details, such as toxicological differences in asbestos fiber types and sizes. The court held that EPA's failure to more thoroughly consider the risk of asbestos substitutes "deprives its order of a reasonable basis" because "EPA cannot say with any assurance that its regulation will increase workplace safety when it refuses to evaluate the harm that will result from the increased use of substitute products." The court concluded that "eager to douse the dangers of asbestos, the agency inadvertently actually may increase the risk of injury Americans face." But what ultimately seemed to drive the decision was that the judges felt that the estimated cost of $30–$40 million per life saved by the asbestos ban was unreasonable. The court stated, "The EPA, in its zeal to ban any and all asbestos products, basically ignored the cost side of the TSCA equation."

The Department of Justice elected not to appeal the decision. The following year, a *Federal Register* notice stated that, in a subsequent clarification of its decision, the court had allowed EPA's 1989 rule to continue to govern "new" uses of asbestos in products that were not being manufactured, imported, or processed on July 12, 1989 (*Federal Register* 57: 11364). In 1994, EPA issued an amended rule that bans "new uses" of asbestos. With a few exceptions, products containing asbestos originally manufactured before July 12, 1989, can still be made today, but any asbestos products made thereafter are now banned (EESI 1996). Prior to the *Corrosion-Proof Fittings* decision, Shapiro (1990) observed that TSCA had resulted in less regulatory activity than was expected when the law passed. Many observers now interpret the 1991 decision as paralyzing EPA's use of TSCA Section 6 regulatory powers (for example, Mazurek et al. 1995).[12]

Over the past fifty years, industry guidelines for "safe levels" of exposure to asbestos in the workplace have given way to federal regulation of asbestos in a variety of settings. In 1946, the American Council of Governmental and Industrial Hygienists established a nonregulatory occupational standard of 5 million particles/ft^3 (approximately 177 fibers per milliliter [f/ml]). In its first attempt to regulate a substance as a carcinogen, OSHA set the federal regulatory workplace standard for asbestos at 5 f/ml in 1972. The agency subsequently lowered the standard, most recently to 0.1 f/ml in 1994 (Bates 1994; U.S. ATSDR 1995).[13] One year later, in the first regulatory action addressing nonoccupational asbestos exposure, EPA listed asbestos as the first hazardous air pollutant under the Clean Air Act (Section 112). At the time, EPA rejected a proposed asbestos ban primarily for economic reasons.[14] In 1978, EPA banned all spray applications of asbestos (especially popular as a fire-retardant/fireproof insulation in the 1950s during the postwar building boom) (Bates 1994).

During the years leading up to *Corrosion Proof Fittings*, the public debate over EPA's ban and phaseout was overshadowed by the more politically visible controversy over the occurrence and removal of asbestos from schools and other public buildings. Of all forms of indoor air pollution, asbestos has probably received the greatest publicity. During the 1980s, a flurry of federal statutes and regulations were enacted to address the occurrence and removal of asbestos in schools and other public buildings (Bates 1994).[15] While there was initially much public outrage about children's exposure to asbestos from school building materials, the controversy soon expanded to include the question of local government financing of costly inspection and abatement procedures and the health risks posed by asbestos removal itself. Some skeptics viewed the political support for asbestos abatement as an effort to create work for construction firms who discovered a sideline during the building slump that began with the 1982–1983 economic recession. After gaining some experience with asbestos abatement, it became apparent that removal was warranted only when asbestos was exposed and friable (crumbly), causing fibers to be released into the air. In many cases, it seems, asbestos abatement was doing more harm than good, and existing asbestos materials could be encapsulated at lower cost *and* risk. The high cost and unintended consequences of asbestos abatement in schools and public buildings may have cast doubts—warranted or not—on the trustworthiness of EPA's judgment concerning asbestos regulation under TSCA Section 6.[16] Table F-1 provides a summarized chronology of asbestos regulation. Augustyniak (1997) provides additional background on asbestos regulation and focuses on EPA's development and use of economic analysis in the 1989 rulemaking.

SCIENTIFIC ISSUES

Asbestos is released into the environment from the weathering of natural deposits and the degradation and/or destruction of man-made asbestos products. Because asbestos fibers do not evaporate, dissolve, or break down into other compounds, they accumulate in the environment over time. Small asbestos fibers can remain suspended in the air for long periods, can be carried long distances by wind or water currents before settling, and can be resuspended in the air after settling out. Inhalation of fibers suspended in the air is the most common source of human exposure to asbestos. Asbestos fibers may also be present in drinking water from erosion of natural deposits or piles of waste asbestos, from asbestos-containing cement pipes, or from filtering through asbestos-containing filters. However, in this country, concentrations of asbestos in drinking

Table F-1. Summary Chronology of Asbestos Regulation

1907   Asbestosis recognized in asbestos workers.

1940s WW II shipbuilders and pipe fitters exposed to high levels of asbestos used for insulation.

1946   American Council of Governmental and Industrial Hygienists recommends occupational standard of 177 f/ml.

1955   Doll links asbestos exposure to lung cancer in *British Journal of Industrial Medicine.*

1964   Selikoff et al. report strong link between occupational asbestos exposure and lung cancer in insulators in *Journal of American Medical Association.* New York Academy of Sciences meeting produces "the first signal event," pronouncing asbestos as serious public health issue.

1971   Council on Environmental Quality calls for toxic substance legislation.

1972   Occupational Safety and Health Administration (OSHA) sets asbestos occupational standard of 5 f/ml.

1973   EPA lists asbestos as hazardous air pollutant under Clean Air Act.

       First damages awarded in court for asbestos-related disease.

1974   Court overturns OSHA occupational standard for asbestos.

1976   Congress passes the Toxic Substances Control Act (TSCA).

1977   Consumer Product Safety Commission (CPSC) bans asbestos in patching compounds and artificial fireplace logs.

1978   Department of Health, Education and Welfare (HEW) releases a report (the "Estimates Document") projecting 2 million deaths over the next three decades from asbestos exposure.

       EPA bans spray applications of asbestos.

1979   February: EPA releases its preliminary evaluation, *Exposure to Asbestos.*

       October: EPA is petitioned to prohibit the future use of asbestos cement pipe in water systems. The agency announces intent to ban most remaining uses of asbestos.

1981   Reagan administration Executive Order 12291 requires Office of Management and Budget (OMB) review of new regulations.

       Department of Education promulgates rule under the Asbestos School Hazard Detection and Control Act; no funding appropriated.

       Doll and Peto denounce HEW's asbestos Estimates Document as "clearly inflated" and a "confidence trick" in *Science.*

1982   EPA promulgates asbestos in schools identification and notification rule under TSCA.

       International Agency for Research on Cancer classifies asbestos as a human carcinogen.

       Largest U.S. asbestos producer declares bankruptcy due to toxic tort liability.

1983   EPA issues initial guidance for controlling asbestos in school buildings (the "Blue Book").

Table F-1. Summary Chronology of Asbestos Regulation *(Continued)*

1984	Congress passes Asbestos School Hazard Abatement Act.
	National Academy of Sciences issues *Asbestiform Fibers: Non-occupational Health Risks.*
	March: Asbestos Industry Association argues EPA should refer asbestos to OSHA.
	July: EPA Science Advisory Board reviews *Airborne Asbestos Health Assessment Update.*
	September: Natural Resources Defense Council petitions EPA to prohibit further use of asbestos in motor vehicle brakes.
	December: OMB finds EPA's proposed asbestos ban not cost-effective, calls for EPA to refer asbestos to OSHA.
1985	February: EPA refers asbestos to OSHA and CPSC.
	April: Rep. John Dingell (D-Mich.) holds oversight hearings on EPA's asbestos referral.
	June: EPA issues new guidance for controlling asbestos in school buildings (the "Purple Book"), emphasizing that asbestos can often be safely managed in place instead of being removed.
	August: EPA releases its first regulatory impact analysis (RIA) for asbestos ban.
1986	Congress passes Asbestos Hazard Emergency Response Act.
	January: OMB signs off; EPA proposes asbestos ban and phaseout.
	July and October: EPA holds informal administrative hearings on the proposed asbestos ban.
	OSHA reduces its permissible exposure level to 0.2 f/ml.
1988	March: EPA releases updated RIA, Asbestos Exposure Assessment.
	August: Congress earmarks $2 million of EPA's appropriations for Health Effects Institute research on asbestos in buildings.
	September: EPA conducts informal hearings on proposed asbestos ban.
	December: Symposium on Health Aspects of Exposure to Asbestos in Buildings held at Harvard's Energy and Environmental Policy Center. Findings reported suggest that children in schools with asbestos-containing materials are exposed to a 1-in-100,000 risk of death. This risk is characterized as far less than other indoor air health risks (EEPC 1989).
1989	Congress reauthorizes Asbestos School Hazard Abatement Act.
	EPA finalizes asbestos ban and phaseout rule.
1991	September: Health Effects Institute–Asbestos Research releases study suggesting low risk from buildings with asbestos containing materials.
	October: Fifth Circuit Court of Appeals overturns EPA's asbestos ban.
1992	Court clarifies its decision.
1993	EPA announces six asbestos products remain subject to the 1989 ban.
1994	OSHA tightens asbestos occupational standard to 0.1 f/ml.

water are generally less than 1 million fibers per liter (U.S. ATSDR 1995).[17] Asbestos most commonly affects the lungs, and the diseases primarily associated with asbestos are asbestosis, lung cancer, and mesothelioma.[18]

There is an unusual amount of epidemiological evidence (both domestic and international) on occupationally related cancers and disease arising from asbestos exposure. According to an EPA official, the link between asbestos and human health effects "was among the strongest ever seen." Evidence from animal studies confirms the carcinogenicity of asbestos. The International Agency for Research on Cancer classifies asbestos as a Group 1 carcinogen, meaning that there is sufficient evidence of carcinogenicity in humans and animals. EPA classifies asbestos as a Group A (known) human carcinogen. In 1986, EPA estimated that about 2,560 lung cancers and mesotheliomas and an undetermined number of mortalities from asbestosis and other cancers in the United States would result from production of asbestos products over fifteen years without EPA action under TSCA (*Federal Register* 51: 3738–3759).

There was virtually no doubt in the scientific community of the hazard posed by asbestos exposure at the time of EPA's ban and phaseout rulemaking. According to an EPA official, there were four principal scientific issues remaining: the differential toxicity of different asbestos fiber types; EPA's choice of a single, no-threshold dose-response model for quantitative cancer risk assessment; the accuracy and precision of EPA's exposure analysis; and the risk of asbestos substitutes.

Fiber Type

There are two major groups of asbestos minerals: serpentine (chrysotile) and amphibole (amosite, crocidolite, and tremolite). Chrysotile represents the majority (90%–95%) of asbestos used commercially in the United States, but products usually contain a mixture of chrysotile and other types of asbestos. According to an EPA official, there was considerable scientific uncertainty regarding the relationship between the length and type of asbestos fibers and their potency, with some evidence suggesting that chrysotile asbestos would be less potent than amphibole forms. In what has become known as the amphibole hypothesis, various commenters have suggested that chrysotile is far less carcinogenic than amphiboles, and that this distinction should be included in any asbestos risk assessment.[19]

A former senior EPA official, however, notes that epidemiological evidence from "North Carolina, where the overwhelming exposure of the population was almost exclusively chrysotile," supported the agency's position that all forms of asbestos increased relative health risks. Whereas previous British epidemiological studies had failed to find an effect from chrysotile, the Carolina study was larger and provided more power to detect effects of exposure.[20] On that basis, says this source, "You can't say asbestos is not a risk," despite differences that may be associated with fiber type. An EPA official says, "There are different forms of asbestos—nasty stuff and not-so-nasty stuff. Most of the studies were on the nasty stuff; there were fewer studies on the not-so-nasty stuff. The agency made a policy call to treat them the same." Another EPA official explains that the agency did not want to run the risk of waiting five or ten more years for scientific consensus on the relative risk of different fiber types saying, "We won't play with public health as an experimental group."

Five years prior to the ban, a committee of the National Academy of Sciences concluded that "difficulties in interpreting studies of various groups of workers do not allow for reliable assessments of the role of fiber type (chrysotile or crocidolite) in determining risk for developing either lung cancer or mesothelioma" (NRC 1984). According to a former senior EPA official, "The linchpin to treating all fibers equally was the Academy report." In a more recent review of the scientific evidence, Stayner et al. (1996) state, "Given the evidence of a significant lung cancer risk, the lack of conclusive evidence for the amphibole hypothesis, and the fact that workers are generally exposed to a mixture of fibers, we conclude that it is prudent to treat chrysotile with virtually the same level of concern as the amphibole forms of asbestos." A former senior EPA official reports being "skeptical" about treating all fibers as toxicologically equivalent but recognizes that if the agency had proceeded down the "slippery slope" of assigning different toxicity values to different forms of asbestos, any specific toxic equivalency factor (a ratio of the toxicity of one substance to that of a reference substance) would have been construed as "arbitrary" and subject to attack on that basis.[21] In the face of uncertainty and the lack of scientific consensus, therefore, EPA viewed equal treatment not as arbitrary (as some have argued) but rather as precautionary.

Dose-Response Analysis

Epidemiological evidence suggests that asbestos is a potent carcinogen at relatively high occupational exposure levels. (EPA estimates that the 1986 OSHA standard of 0.2 f/ml represented a lifetime cancer risk on the order of 1 in 1,000 [10^{-3}] [*Federal Register* 54: 29467].) There remains, however, more uncertainty about the potency of asbestos at lower-level ambient exposures. (As discussed above, the potency may differ by fiber type.) NRC (1984) observed that few epidemiological studies had established quantitative dose-response relationships between levels of asbestos and health effects and that there was considerable variability observed among the studies that had. A factor that complicates interpretation of the epidemiological studies is that the risk of lung cancer from asbestos exposure is greatly increased by smoking (that is, a synergistic effect), and a large proportion of the populations first identified as suffering from the effects of asbestos (for example, World War II shipyard workers) were smokers. Some scientists held that evidence of DNA repair and asbestos fiber elimination would suggest a sublinear dose-response form, an argument advanced by the Asbestos Industry Association (Gaheen 1995). (A sublinear curve would estimate lower risks at low exposure levels than would a linear dose-response form.) Two years after EPA finalized the ban and phaseout rule, however, the Health Effects Institute–Asbestos Research (HEI-AR) concluded that there was no experimental evidence for establishing the precise dose-response relationship and observed that occupational exposure levels have never been recorded with much accuracy, contributing to the uncertainty in the dose-response analysis (HEI-AR 1991).[22]

Consistent with its 1986 cancer risk assessment guidelines, EPA estimated dose-response relationships for asbestos-related lung cancer and mesothelioma using a linear, no-threshold model. The agency estimated the dose-response constants as the geometric means of the "best estimates" (maximum likelihood estimates) from a number of epidemiological studies of effects associated with occupational exposures.[23] Thus, EPA directly extrapolated from the observed effects at relatively high, imprecisely measured occupational exposure levels to predict effects that would occur at low, ambient exposure levels to which most of the general public would be exposed. An academic mineralogist observes that there is no direct data or scientific evidence to validate the choice of model for extrapolation and concludes that the matter "could not be resolved on scientific grounds."

Exposure Analysis

TSCA's life-cycle perspective called for EPA to use an innovative approach to exposure analysis. Traditionally, EPA evaluates emissions or releases of toxic substances at the "end of the pipe" (that is, point source emissions from production facilities).[24] EPA estimated that the majority of ambient asbestos exposures resulted from the use—not the manufacture—of automobile drum brake linings and disk brake pads (*Federal Register* 54: 29478). By banning asbestos, EPA sought to reduce or eliminate exposures from "cradle to grave" in mining, product fabrication, distribution, end use, and disposal.

However, as is the norm for any particular toxic substance (even one as intensely scrutinized as asbestos), national data were critically lacking on the number of people exposed to various levels of asbestos as a result of nonoccupational exposures. EPA's asbestos exposure modeling relied heavily on extrapolations from occupational settings to nonoccupational exposures. What EPA termed "analogous" exposure data would be extremely difficult and costly to validate but easy (as the Fifth Circuit Court demonstrated) to find fault with. The agency used, for example, data on asbestos levels measured at garages where brakes were repaired and at highway tollbooths to estimate exposure levels for the general public. Another complicating factor is that data on workplace asbestos fiber concentrations are generally given as number of fibers longer than 5 μm per unit volume, whereas data on ambient concentrations have been expressed as mass per unit volume, and accurate conversions between mass measurements and fiber concentrations depend on a variety of factors for which data are generally unavailable (NRC 1984).

Although EPA did conduct sensitivity analyses on many of its exposure assumptions to examine if different assumptions would substantially change the projected health benefits, according to an EPA official, in retrospect, the agency "should have had real measurements" from the field to strengthen the nonoccupational exposure assessment. Another EPA official, however, points to the extremely high level of resources the agency devoted to asbestos ($10 million for the cost-benefit analysis alone) and concludes that the agency cannot afford to spend such sums of money on substance-specific analyses given the tens of thousands of chemicals in commerce.[25]

Combining its dose-response and exposure analyses, EPA estimated that thousands of asbestos workers and members of the general population were exposed to risks on the order of 1 in 1,000 (10^{-3}) from asbestos released from products subject to the rule and that millions of people were exposed to risks on the order of 1 in 1 million (10^{-6}). However, as an EPA official suggests, public asbestos exposure resulting from the end use of products was estimated to be "tiny relative to occupational exposure and product fabrication." As the figures suggest, EPA estimated that the relatively low health risk from asbestos exposure to any particular member of the general public aggregated over the entire population roughly corresponded to the total risk of the smaller population exposed to relatively high individual risks (that is, $10^6 \times 10^{-6} = 10^3 \times 10^{-3}$). Altogether, the agency projected that the ban and phaseout rule would avoid approximately 160–200 cancer cases over thirteen years, as well as yield other health benefits for which quantitative estimates were not available (for example, cases of asbestosis) (*Federal Register* 54: 29460).[26]

Note that the final estimated health benefits of the 1989 asbestos ban and phaseout were considerably lower than those included in the proposal. Additional scientific analysis in combination with changes in the market place were responsible for the change. EPA's 1986 proposal estimated approximately 2,560 avoided cancer cases versus 160–200 avoided cases in the final rule. The primary reasons for the difference were (1) removal of asbestos products from the market—some of the products (for example, vinyl-asbestos floor tile) included in the proposal were no longer manufactured or imported in the United States by 1989 (accounting for about 475 cases), and, in addition, the estimated health benefits of the proposed ban were nearly halved as a result of projected asbestos exposures being "preempted" by the market's reaction to asbestos liability and in anticipation of the ban; and (2) modifications to the health effects model—EPA used a lower dose-response constant for mesothelioma resulting from the average of many studies rather than the estimate from one large study)—that resulted in an estimate of health benefits approximately 20% lower for the final rule than for the proposal (accounting for about 200 cases).[27]

Substitutes: Risk-Risk Trade-Offs

Some critics of the proposed asbestos ban suggested that the poorly understood toxicity of substitute fibrous materials might be comparable to or worse than that of asbestos.[28] EPA concluded that, on the basis of available toxicological and epidemiological data, fibrous substitutes are generally less biologically active and pathogenic than asbestos. The agency also concluded that occupational exposures due to synthetic fibrous substitutes would probably be lower than that from naturally occurring substitute fibers because the mining and milling of the latter tend to be "dusty" operations. Another aspect of synthetic substitutes that EPA believed made them less risky was that their diameter could be controlled to reduce the presence of fibers in the respirable range (under a few microns) (*Federal Register* 54: 29460).

EPA did not, however, explicitly estimate safety and some noncancer risks due to the reduced performance or higher costs of products containing asbestos substitutes. According to an EPA official, the first risk substitution issue raised was brake pads. There was concern that banning the use of asbestos in brake linings might increase traffic accidents and fatalities due to longer braking distances.[29] (EPA did convene meetings with the National Highway Traffic Safety Administration [NHTSA] to consider this question, and NHTSA did not object to EPA's final rulemaking [*Federal Register* 54: 29494–20495].) Others raised concerns that the asbestos ban might limit access to safe drinking water in less developed countries because asbestos cement pipe is the cheapest form of building structurally sound water mains.[30]

THE PROCESS WITHIN EPA

Setting the Agenda

The asbestos ban officially got onto EPA's regulatory agenda in 1979, when the agency granted a public petition (under the provisions of TSCA, Section 21) to prohibit the future use of asbestos cement pipe in water systems. In the same year, EPA published its advance notice of proposed rulemaking. As witnessed by its prominence in the earliest regulatory actions of EPA and OSHA, however, asbestos was on the public agenda long before the Carter administration. The scope of the rulemaking expanded in 1984, when the Natural Resources Defense Council petitioned EPA to prohibit further use of asbestos in motor vehicle brakes (*Federal Register* 51: 3740).

Assessing the Science: Asbestos

Three out of five interviewees responded that there was adequate science to inform EPA's decision to ban or phase out asbestos in 1989. Respondents generally agreed that the areas of greatest uncertainty were nonoccupational exposure levels and the risks posed by substitutes for some of the lower-volume asbestos uses. One respondent felt that the scientific information available to assess nonoccupational risks was sparse, poor in quality, and highly uncertain and characterized EPA's treatment of the science as poor. Otherwise, EPA's scientific assessment got good marks.

As has been indicated, EPA's assessment of the health risks posed by asbestos took place over many years and was a large enterprise, so large, in fact, that the Office of Pesticides and Toxic Substances (OPTS) employed a complicated matrix management tasking scheme to conduct components and subcomponents of the regulatory analysis and to organize staff for particular working groups and assignments. Although the practice was very much in keeping with the latest business management trend, the result was (perhaps unavoidably) fragmented. According to an EPA official, "Top people didn't necessarily know what was going on, nor did the staff scientists."

The so-called matrix management approach was necessitated by OPTS's need to be flexibly staffed with analysts having adaptable functional skills (for example, exposure assessment) necessary to conduct or oversee components of any substance-specific assessment. As one EPA official laments, EPA does not have nearly enough resources to house experts for every suspected toxic substance. Thus, OPTS could not afford to amass (and then disband) a team of asbestos experts to work closely as an interdisciplinary team in a comprehensive scientific assessment. The agency, in fact, had no in-house asbestos expertise. Instead, it contracted for expertise as needed. In short, then, the agency traded off expertise for flexibility, and the result was a fragmentary analytical process. Fortunately for EPA, riding herd over the asbestos assessment from 1983 to 1989 was Assistant Administrator John Moore, a toxicologist by profession, and a former deputy director of the National Toxicology Program.

EPA's initial assessment of asbestos (*Exposure to Asbestos*) was prepared in 1979.[31] After meeting with asbestos industry representatives to discuss the report's methods and conclusions, EPA informed the Asbestos Information Association that the agency did not plan to use the report as a reference document in support of future rulemaking (Gaheen 1995). In 1980–1982, EPA issued a series of three draft documents on asbestos health effects and the magnitude of indoor air exposures in support of rulemaking on asbestos in school buildings. Hughes and Weill (1989) suggest that the changing risk estimates from each subsequent draft added to public confusion over the risks posed by exposure to asbestos in schools.[32] With no real in-house asbestos expertise, EPA turned to external scientists and blue-ribbon panels.

In 1982, during the Gorsuch administration, EPA requested that the National Academy of Sciences undertake a study to evaluate nonoccupational exposure to asbestos and to determine whether differences among fiber types should be incorporated into the agency's risk assessment. The committee convened in 1983 and issued its final report in 1984 (NRC 1984). (The National Research Council is the Academy's principal operating arm.) As discussed above, this report served as the basis for EPA's decision to treat different types of asbestos fibers as toxicologically equivalent. In addition, the committee estimated lung cancer and mesothelioma risks resulting from given ambient asbestos levels for smokers and nonsmokers, males and females. Of thirteen members, however, the NRC panel included only one expert on quantitative risk assessment methods. Several calculation errors were discovered in the committee's original report, which was withdrawn, and other errors remained undetected in the revised version.[33] As a result, using the committee's own underlying assumptions, the report was later found to have underestimated risks for mesothelioma by a factor of 17.4 and for lung cancer by a factor of 4.5 (Breslow et al. 1986). As a former senior EPA official allows, "The report was not an example of the academy at its finest." Needless to say, the report did not provide EPA with a recognized authoritative basis for its quantitative conclusions.[34]

In 1984, the EPA's Office of Research and Development (ORD) Environmental Criteria and Assessment Office contracted out the *Airborne Asbestos Health Assessment Update* to William Nicholson, an epidemiologist and colleague of Irving Selikoff at Mt. Sinai Hospital.[35] EPA's critics suggested that Nicholson's assessment was unbalanced and overstated the health risks from asbestos, largely due to the lack of distinction made between types of asbestos.[36] An academic mineralogist insinuated that the choice of Nicholson to conduct the asbestos health assessment was due to then ORD Assistant Administrator Bernard Goldstein being a like-minded "Selikoff guy." Nicholson's report was, however, released in draft in 1984 and 1985 for public comment and was peer-reviewed by the Environmental Health Committee of the EPA Science Advisory Board in 1984 prior to revisions and release in 1986.[37] According to an EPA official, the agency moderated and significantly changed the early draft of the asbestos health assessment "in response to external reviewers, primarily from industry, who objected wildly."

All of these externally prepared reports, however, fell short of generating the type of scientific information and analysis EPA needed under TSCA—an exposure assessment. TSCA requires EPA to balance regulatory costs and benefits. Therefore, it was not sufficient to demonstrate conclusively, as one EPA official pointed out "that the stuff [asbestos] kills people," or to provide, as did Nicholson and the NRC panel, disease risk estimates for *given* low levels of ambient asbestos. The agency needed a projected "body count" to compare with the costs of regulatory compliance, and an asbestos exposure assessment estimating the *actual* levels of asbestos in the environment was the missing piece to link available health effects information to regulatory benefits (that is, avoided cases of disease). As an EPA official explains, the agency treated the hazard the same across products but conducted a separate exposure analysis for each product. "It was a tremendous amount of work to identify each product category, and a lot of work was done to figure out the exposure for each product category. The economic model needed exposure work from all parts of the product life cycle." EPA thus confronted the challenge of estimating the occupational and nonoccupational asbestos exposures resulting from the entire life cycle, from the mine to the landfill and every point in between, of hundreds of asbestos products.

Much of the original occupational exposure data EPA used during formulation of the 1986 proposal came as a result of a 1982 reporting rule issued under TSCA (Section 8). These data were supplemented by inspection reports from OSHA and the Mine Safety and Health Administration and studies by the National Institute of Occupational Safety and Health. Predictably, the asbestos industry faulted the quality of the data for regulatory decisionmaking and EPA's use of the data to extrapolate to nonoccupational exposures. In preparation for the final rulemaking in 1989, EPA conducted an asbestos exposure survey in 1986–1987. OPTS's final asbestos exposure assessment was like a garment stitched together from three bolts of cloth: an analysis of occupational exposure to asbestos and asbestos releases from manufacturing and commercial operations; a modeling study estimating ambient exposure levels resulting from the releases from industrial and commercial sources; and an analysis of consumer and ambient exposures resulting from asbestos products in use.[38]

Assessing the Science: Asbestos Substitutes

According to a former senior EPA official, EPA spent "a fair amount of time trying to develop reasonable estimates" of exposure to and health effects from asbestos substitutes and was "very sensitive to the need to look at friction product substitute risks." This source states that the agency's major focus in this area was on friction products; however, for other uses of asbestos, "particularly where the quantity of asbestos used compared to friction products dropped off by orders of magnitude, there was lower scrutiny and less intellectual input" into the risk analysis of substitutes. OPTS's 1988 *Health Hazard Assessment of Non-Asbestiform Fibers* was produced by its Health and Environmental Review Division. Its evaluation was based, in part, on a prior review of epidemiological studies on populations exposed to nonasbestos fibers prepared by the OPTS Economics and Exposure Division. The examples of safety and noncancer risks posed by asbestos substitutes discussed above (that is, from brake pads and cement pipe) suggest the great difficulties inherent in comparing qualitatively different health risks (for example, cancer versus trauma) and underscore the dilemma of establishing the appropriate scope of analysis for substitute risks.

Scientific Review Procedures

TSCA requires that EPA conduct quasi-judicial hearings as part of its rulemaking process. These administrative proceedings included scientific testimony from agency staff and permitted cross-examination of witnesses and, according to an EPA official, forced EPA to pay more attention to particular analytical issues. However, EPA's scientific review process for the asbestos ban and phaseout seems to have revolved principally around its internal working group process. An EPA official described the working group review process as a series of staff briefings for EPA managers in which staff not directly responsible for the analysis would participate and in which staff recommendations would either "be ratified or shot down." As suggested above, EPA also engaged in an informal external scientific review process by widely circulating a series of draft asbestos health assessments and reacting to subsequent feedback (Hughes and Weill 1989). The fact that much of the scientific review of staff analysis was internal and informal, however, does not necessarily mean that it was not rigorous. Regarding the asbestos ban and phaseout rule, an EPA official remarked that Assistant Administrator John Moore insisted on peer review and held staff to a high standard.

The appeals court found, however, that EPA was deficient in one part of the formal scientific review process, public notice, and comment. EPA argued that it gave "constructive" notice by notifying the public of the available exposure information that could be manipulated to estimate exposures. According to the court, however, this did not constitute adequate public notice of the agency's intended use of exposure data. In addition, according to an EPA official, the EPA Science Advisory Board (SAB) was troubled that it was not more involved in the asbestos ban and phaseout decisionmaking process. Since the SAB's Environmental Health Committee was involved in reviewing the agency's Asbestos Health Assessment Update (U.S. EPA 1986), one can infer that the board's concern in this case also related to the asbestos exposure analysis and its lack of opportunity to review and comment on it.

Communicating the Science to Decisionmakers

According to a former senior EPA official, the proposed ban and phaseout rule eventually went forward in 1986, largely on the basis of a review of the epidemiological information provided by Assistant Administrators John Moore (OPTS) and Bernard Goldstein (ORD) to Administrator Lee Thomas. Linda Fisher became Assistant Administrator for OPTS under the Reilly administration in 1989 prior to the final rulemaking, and, according to an EPA official, Reilly administration officials were briefed on the science and requested additional analysis. Moore, however, was a key decisionmaker. Of the asbestos ban and phaseout, an EPA official simply states, "Jack Moore ran that show." As an experienced scientist and science administrator, Moore was effective in creating demand for rigorous scientific analysis by having an expert's grasp on the substance and by appealing to the pride and professionalism of his staff. "If the staff briefed Moore on lousy science, they were embarrassed, and this permeated" throughout the pesticides and toxic substances program, recalls an EPA official.

SCIENCE IN THE FINAL DECISION

Three out of five interviewees believed that the scientific information had a high impact on the agency's final decision to ban or phase out nearly all uses of asbestos, and there appear to have been very few, if any, impediments to the consideration of the science by decisionmakers.[39] In fact, one

EPA official suggests that the role of science in the decision was excessive. "People were too concerned with the science," while too little attention was paid to the "balancing of costs required under the law [TSCA]. Asbestos was so clearly a danger to humans that it short-circuited thinking about other things people should think about in making regulatory decisions, [such as] cost-benefit analysis and substitutes." In the judgment of a former senior EPA official, the science supporting the asbestos ban and phaseout was, on the one hand, "very strong" because of the data on observed health effects in humans, but, on the other hand, it was "not strong in relation to exposure and particular uses." Although OMB had officially signed off on EPA's regulatory analysis when it acceded to the proposal in 1986, the budget office's involvement in the process apparently did not end there. According to an EPA official, "At the end, we were arguing with OMB." Apparently, what was decisive in coming to closure on the rule was that "there was a body count from shipyard workers' exposure." The fact that EPA was going to regulate asbestos on the basis of occupational (human) data rather than bioassays (animal studies) "gave more weight" to the agency's argument to proceed.

CONCLUDING OBSERVATIONS

Regarding the appellate court decision to overturn the asbestos ban, an EPA official speculates, "Once the court thought it was dumb, they were going to find a reason to overturn EPA's decision." A number of commenters have similarly characterized the court's ruling as replacing the administration's policy judgment about what constitutes an unreasonable risk and sufficient evidence with the court's own judgment. An academic defends the court's activism: "EPA's stance was precautionary to the extent of being financially untenable." The agency's analysis was "not founded on the most balanced and rational assessment of what was known, and what was socially responsible." The court's decision, in turn, was a "reasonable response" to EPA's "unacceptable decision." In this view, then, the judicial appeals process is an appropriate, and perhaps necessary, component of political regulatory decisionmaking. However, Davies et al. (1979) argue that "the determination of what constitutes an unreasonable risk is a societal value judgment." Under the constitutional separation of powers, such judgments are reserved for politically accountable decisionmakers.

In terms of a fate and transport analogy, the court can be compared to wildlife that is unintentionally exposed to pesticides in the field—it was not the intended "target" of the scientific information. (In the pesticides field, nonpest wildlife are referred to as nontarget organisms.) The court

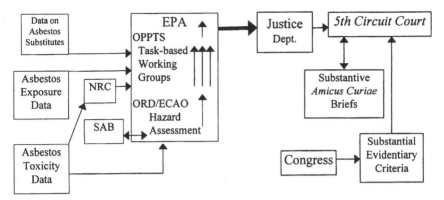

Figure F-1. Fate and transport of science in the asbestos ban and phaseout rule

received science both indirectly as mediated by the Justice Department and EPA and directly through soliciting *amicus curiae* briefs on substantive matters. According to an EPA official, Justice Department lawyers who handled the government's case showed little interest in being briefed by EPA on the scientific substance underlying the asbestos rulemaking. As a result, the Justice Department attenuated the science generated by EPA. Despite the court's limited capability to comprehend and critically analyze, or "absorb and metabolize," the science, it substituted its own science policy judgment for that of politically accountable decisionmakers of the more expert administrative agency. As illustrated in Figure F-1, however, the court was arguably invited to do so by the substantial evidentiary judicial review standard specified under TSCA by Congress.

The Carnegie Commission (1993, p. 57) observes:

> The asbestos decision has provoked considerable debate, and fingers have been pointed in several directions. Regardless of whether the statute, the courts, the agency, or others should be faulted in this case, it is unsettling that EPA could not satisfy TSCA's requirements for promulgating a single rule after a decade's effort. The case raises numerous questions, including whether the executive branch should encourage Congress to revise this legislation, and under what circumstances the agency should devote such a vast amount of time and resources to a single substance at the expense of many other pressing issues in its jurisdiction.

ENDNOTES

[1] Asbestosis was recognized as long ago as 1907 (Hodgson et al. 1988).

[2] Peak U.S. production reached over 136,000 metric tons annually, with imports at approximately 680,000 metric tons per year. By 1993, domestic production had dropped to an estimated 15,000 metric tons, and imports were down to 33,000 metric tons. Historically, asbestos was produced in the United States in Arizona, California, North Carolina, and Vermont. During the 1970s, many companies in these areas discontinued mining operations. By 1991, only two U.S. firms produced asbestos in Vermont and California. Ninety-eight percent of U.S. asbestos imports have come from Canada. Largely as a result of the decline in U.S. asbestos consumption, the Canadian asbestos mining industry has declined considerably (U.S. ATSDR 1995).

[3] According to Cross (1989), until 1973, plaintiffs had little success in recovering damages for occupational asbestos-induced diseases. A modest success in that year, however, marked a key turning point that "opened the floodgates of litigation." In 1982, due to the large number of outstanding injury claims filed against it, Johns-Manville, the major producer of asbestos in the United States was forced into bankruptcy (Bates 1994; Cross 1989). An academic mineralogist suggests that EPA's 1989 asbestos ban and phaseout rule was intended to halt much of the forecasted litigation and compensation claims regarding asbestos-related cancers and disease.

[4] In 1971, the President's Council on Environmental Quality issued a report on toxic substances and recommended comprehensive toxic substances control legislation (TSCA Research Project 1982).

[5] Peck (1989) corroborates that asbestos was salient in TSCA's development. Congress also directly banned the manufacture, processing, and distribution of polychlorinated biphenyls (PCBs) when it wrote TSCA (Shapiro 1990).

[6] Food, food additives, drugs, pesticides, alcohol, and tobacco are exempted from TSCA.

[7] The House report on TSCA specifically states that a quantitative cost-benefit analysis is not required to show unreasonable risk (Shapiro 1990).

[8] In 1979, EPA granted a public petition under TSCA (Section 21 allows individuals to "petition EPA to initiate a proceeding for the issuance, amendment, or repeal of a rule) to prohibit the future use of asbestos cement pipes in water systems (*Federal Register* 44: 60155).

[9] The Asbestos Information Administration in Washington, D.C., was involved in major lobbying efforts against the ban (Peck 1989). Canada also strongly opposed proposals to ban asbestos in the United States. It financed lobbying efforts of the Asbestos Institute (based in Montreal). In 1989, the institute was receiving two-thirds of its $5 million budget from the federal government in Ottawa and the province of Quebec (the center of Canada's asbestos mining industry).

[10] TSCA Section 9 requires EPA to refer chemicals to another agency with regulatory authority, if that agency can manage the associated risks. Most nonoccupational asbestos exposure, however, resulted from the normal use of automobile brake pads and was beyond the purview of OSHA and CPSC. (The 1991 court decision that overturned EPA's 1989 asbestos ban and phaseout rule did find that

it was appropriate for EPA to regulate asbestos under TSCA.) According to press reports, OMB blocked EPA's proposed asbestos ban in 1984, but congressional pressure later overrode OMB in 1985 (*Inside EPA*, February 8, 1985; *Inside EPA*, March 15, 1985).

[11] Long-term databases are often needed to accurately assess chronic health effects that may take decades to manifest.

[12] During the first fifteen years of the TSCA program, production, manufacture, importation, and use of only five chemical substances have been restricted: PCBs (directly by Congress), chlorofluorocarbons in aerosols, nitrates in metalworking fluids, dioxins (disposal requirements), and asbestos (inspection and removal from schools) (Mazurek et al. 1995).

[13] The OSHA standard applies to fibers longer than five micrometers and is averaged over eight hours (gopher://ecosys.drdr.Virginia.EDU:70/00/library/gen/toxics/Asbestos). According to Cross (1989), OSHA's occupational standard was based on technical feasibility, the agency conducted no quantitative risk assessment for asbestos, and the rule was accompanied by very little explanation. Industry successfully contested the 1972 rule in *Industrial Union Department, AFL-CIO v. Hodgson* (D.C. Cir. 1974). OSHA amended the 1972 standard in 1986 (*Federal Register* 51: 22612, cited in U.S. OTA 1995).

[14] In 1973, EPA set a "no visible emissions" standard for asbestos air emissions from five major sources. At the time, EPA declared it impossible to prepare a quantitative risk assessment for airborne asbestos exposures. EPA gave no basis for its use of cost considerations under Section 112, nor did the agency explain how continued emissions of carcinogenic asbestos could be squared with the mandate to provide an ample margin of safety to protect public health (Cross 1989).

[15] In 1981, the U.S. Department of Education promulgated a rule under the Asbestos School Hazard Detection and Control Act, but no funds were appropriated. In 1982, EPA promulgated a rule under TSCA requiring identification and public notification of asbestos in schools. The agency issued its first guidance for controlling asbestos in school buildings (the "Blue Book") the following year. In 1984, Congress enacted the Asbestos School Hazard Abatement Act, establishing grants and interest-free loans for school asbestos projects. The same year, EPA established regulations under the Clean Air Act governing the removal of asbestos from buildings and disposal of wastes generated from removal (*Federal Register* 49: 3658). In 1985, EPA promulgated a worker protection rule for municipal and state workers not covered by OSHA regulations (*Federal Register* 50: 28530). In 1985, EPA issued new guidance (the "Purple Book") for controlling asbestos in school buildings that emphasized that asbestos can often be safely managed in place instead of being removed. This created the perception that EPA was giving mixed signals about the need for regulation. In 1986, prompted by stories of botched asbestos cleanups and the lack of action by many schools, Congress codified the asbestos in schools notification rule and required all asbestos inspection, testing, planning, and response actions be performed by accredited personnel and labs with the Asbestos Hazard Emergency Response Act (AHERA, Title II of TSCA). In response to the 1985 Purple Book, AHERA required EPA to specify measures to reduce

asbestos hazards in schools. TSCA Section 201(a) states that EPA's guidance "is insufficient in detail to ensure adequate responses." EPA promulgated the asbestos-containing materials in schools rule under TSCA Section 203 in 1987 (*Federal Register* 52: 41826). The 1987 rule required local school systems to inspect facilities for asbestos and "select and implement in a timely manner the appropriate response actions." In adopting this regulation, EPA did not rely on a quantitative risk assessment and deflected judgments about asbestos containment and removal to local school districts. EPA's rules on asbestos in schools were upheld on judicial review (Cross 1989). In 1989, Congress approved a hotly debated extension of AHERA deadlines. The following year Congress reauthorized the Asbestos School Hazard Abatement Act (EESI 1996).

[16] According to a former senior EPA official, the asbestos in schools abatement program is a classic example of legalistic decisionmaking that ignores science. In their rush to remove asbestos from the classrooms, policymakers ignored warnings that removal was likely to cause more problems, in part because there were few people who knew how to remove asbestos safely. The response from policymakers, according to this source, was, "No problem. We'll fix it later. If they do it wrong, we'll sue them." An EPA official singled out former U.S. Representative Florio (D-N.J.) for championing the issue of asbestos in schools. An academic suggests that the mishandling of asbestos in schools contributed to the growth in antiregulatory public sentiments.

[17] In some locations, however, water samples may contain 10–300+ million fibers per liter (U.S. ATSDR 1995).

[18] Asbestosis (or pulmonary fibrosis) is a respiratory disorder resulting from inhaling asbestos, characterized by scarring (fibrosis), calcification, and tumors of the lungs. Asbestosis is synergistic with smoking-related cancers (that is, smoking and asbestos are more toxic in combination than individually). Mesothelioma is a tumor, either benign or malignant, arising from the stomach lining or membranous sacs enveloping the heart (pericardium) or lungs (pleura). Pleural mesotheliomas are most common (Hodgson et al. 1988).

[19] The link between fiber type and risk of cancer and disease is said to arise from different rates of fiber clearance from the lung (Bates 1994). Due to greater durability of amphibole fibers in the lung compared with the rapid dissolution of chrysotile (Doll and Peto 1985), it has been argued that amphibole fibers are the sole cause of mesothelioma, due to the presence of traces of tremolite fibers that contaminate commercial chrysotile and cause the small number of mesothelioma cases observed in male workers exposed to chrysotile (Doll 1989). According to Doll, studies strongly suggest that pure chrysotile does not cause mesothelioma. Doll does note, however, that a low occurrence of the disease has been observed in groups of men and women occupationally exposed only to chrysotile (with or without tremolite contamination). With regard to lung cancer, Doll notes that firm evidence to prove a lower risk of lung cancer incidence clearly attributed to chrysotile is lacking. Nevertheless, there is a prominent group of scientists defending an *amphibole hypothesis*, a term coined by Mossman et al. (1990). Mossman and colleagues argue that exposure to chrysotile at the level of current occu-

pational standards does not increase the risk of asbestos-associated diseases. They criticize U.S. federal agencies for not differentiating between different types of asbestos like the European Community, which has more stringent rules for amphiboles. According to Stayner et al. (1996), opponents of the amphibole hypothesis contend that it is not a valid explanation for unexplained variances in effect from asbestos exposure because a mixture of fibers is usually present in commercial asbestos (that is, chrysotile contaminated with tremolite).

[20] In its report on the health effects of asbestos, the Health Effects Institute, which is a public–private sector partnership, also hypothesized that chrysotile asbestos was the cause of most mesothelioma deaths of miners in Quebec (HEI-AR 1991).

[21] Note that EPA is confronting precisely this challenge in its current reassessment of dioxins. The agency has assigned different toxic equivalency factors (TEFs) to different types of dioxins, and commenters have criticized the lack of empirical basis for the different TEF values. Although assigning different uncertainty distributions to the toxicity of various dioxin congeners seems to be a reasonable alternative to TEFs, it is not hard to imagine similar challenges being raised about the lack of empirical basis for the selection of distributions.

[22] HEI is a nonprofit research organization established in 1980 by EPA and the automotive industry. It created HEI-AR in response to the congressionally requested asbestos study to restrict HEI's civil liability and protect it against too close an involvement in the regulatory politics of asbestos (Jasanoff 1990).

[23] This statistical estimation procedure is similar to what EPA used in conducting the quantitative risk assessment for arsenic in drinking water using epidemiological data. This procedure differs somewhat from the standard procedure EPA uses in analyzing animal study data. It is somewhat less conservative than, but generally consistent with, agency cancer risk assessment guidelines (see Appendix B).

[24] EPA has used a similar multimedia approach to evaluating lead exposures, but lead pollution control has been implemented under a variety of statutes, and, at times, there have been inconsistencies among EPA programs in addressing lead pollution. See Appendixes A and H.

[25] Dahl (1995) reported that the TSCA inventory included more than 72,000 chemicals and that EPA receives more than 2,000 premanufacture notices per year. Shapiro (1990) estimates that a modest battery of physical/chemical properties tests would cost $250,000 per substance. The costs of exposure analysis would be additional.

[26] EPA did not quantify avoided asbestosis cases because there was no evidence of "disabling" asbestosis occurring at low occupational exposure levels (*Federal Register* 54: 29470). Asbestosis that is not so severe as to disable, however, might significantly affect quality of life.

[27] The remainder of the difference in estimated health benefits between the proposed and final rule was accounted for by revised final exposure assessments for some asbestos products that were lower than those used in the proposal and a

two-year reduction in the time frame for the regulatory analysis (fifteen years for the proposal versus thirteen years for the final rule) (*Federal Register* 54: 29486).

[28] For example, Interior Department scientist Malcom Ross suggested that replacement substances, such as rock wool and fiberglass, could be more harmful than asbestos and accused EPA's plan to prohibit the asbestos of creating a public "crisis," only because a small number of people had hypothesized nonoccupational exposures to asbestos as harmful (*Inside EPA*, March 22, 1985).

[29] According to a DuPont official cited by (Peck 1989), there were very few uses of asbestos for which there were no substitutes available, with the exception of brakes for heavy machinery. DuPont is a manufacturer of asbestos substitutes. The observation that substitutes were unavailable for heavy-duty brakes tends to support the idea that substitutes were not as effective for lighter vehicles.

[30] One of the legal challenges to the asbestos ban was EPA's failure to consider risks borne outside the country, but the court held that this was a proper construction of TSCA. As a former senior EPA official remarked, "TSCA doesn't regulate Guatemala."

[31] IIT Research Institute (of Chicago) prepared the assessment under contract to EPA.

[32] The 1980 draft estimated 100–8,000 premature deaths attributable to school exposures, while this was reduced to 40–400 deaths in the 1981 draft, without any changes in the underlying exposure assumptions. The 1982 draft contained no quantitative risk estimate.

[33] Hughes and Weill (1989) note that the National Research Council (NRC) report failed to use life-table methods (taking into account age at onset of exposure and disease) for estimating mesothelioma risk or to adjust lung cancer risk to continuous (24 hr/day) rather than workplace exposure (8 hr/day).

[34] In contrast to the approach adopted by EPA and NRC, in 1982 CPSC appointed a panel of seven scientists, including three experts in quantitative methods, from a list of scientists nominated by the National Academy of Sciences to prepare a report on asbestos (CPSC 1983). The estimates of mesothelioma risks provided by the CPSC panel were about an order of magnitude higher than those in NRC 1984 (see U.S. EPA 1986, Table 7-1). Also, unlike EPA, which released a series of widely circulated draft reports during 1980–1982, it was CPSC policy not to publicly release preliminary drafts. Only the final draft was released for public comment (Hughes and Weill 1989). As a result, the broad uncertainty in CPSC's estimates was revealed to the public in one fell swoop, whereas EPA's sequence of draft estimates changed over time.

[35] Although this assessment document was developed nominally for the purposes of updating the scientific criteria for the national emission standard for asbestos as a hazardous air pollutant under the Clean Air Act (asbestos was one of the few substances regulated by EPA as a hazardous air pollutant prior to the 1990 Clean Air Act Amendments), it was a principal supporting document for the asbestos ban and phaseout rule under TSCA. Nicholson also served as the consultant to OSHA for a 1986 asbestos health assessment (Hughes and Weill 1989). See

Stone (1991) for a discussion of the two staunchly opposing camps of asbestos health effects scientists.

[36] The EPA risk estimate (U.S. EPA 1986) for lung cancer is also approximately 50% higher (across all groups) than that of the CPSC (1983). The difference is due primarily to different assumptions concerning the background risk of lung cancer in the absence of asbestos exposure (Hughes and Weill 1989).

[37] Nicholson later criticized the 1991 HEI-AR report for downplaying the health risks of asbestos fibers less than five micrometers in length. An HEI-AR panel member, Nicholson refused to sign the report.

[38] The analysis of occupational exposures was conducted by ICF Incorporated under contract to OPTS. The OPTS Economics, Exposure, and Technology Division would presumably have been responsible for the exposure assessments.

[39] An academic mineralogist suggested that EPA ignored scientific information generated by the asbestos industry that did not support its decision. An EPA official suggests that the principal impediment to considering science in this case was that the agency underestimated the opposition and failed to effectively engage asbestos exporters from Canada and Australia.

REFERENCES

Augustyniak, C. 1997. Asbestos. In *Economic Analyses at EPA: Assessing Regulatory Impact*, edited by R. Morgenstern. Washington, D.C.: Resources for the Future, 171–204.

Bates, D. 1994. *Environmental Health Risks and Public Policy: Decision Making in Free Societies*. Seattle: University of Washington Press.

Breslow, L., S. Brown, and J. Van Ryzin. 1986. Letters. *Science.* 234: 923.

Carnegie Commission. 1993. *Risk and the Environment: Improving Regulatory Decisionmaking*. A Report of the Carnegie Commission on Science, Technology, and Government. New York: Carnegie Commission.

Cross, F. 1989. *Environmentally Induced Cancer and the Law: Risks, Regulation, and Victim Compensation*. Westport, Conn.: Greenwood Press, Inc.

CPSC (Consumer Product Safety Commission). 1983. *Report to the Consumer Product Safety Commission by the Chronic Hazard Advisory Panel on Asbestos*. Washington, D.C.: CPSC.

Dahl, R. 1995. Can You Keep a Secret? *Environmental Health Perspectives* 103(10): 914–916.

Davies, J., S. Gusman, and F. Irwin. 1979. Determining Unreasonable Risk under the Toxic Substance Control Act. An Issue Report. Washington, D.C.: Conservation Foundation.

Doll, R. 1989. Mineral Fibres in the Non-Occupational Environment: Concluding Remarks. In *Non-Occupational Exposure to Mineral Fibres*, edited by J. Bignon, J. Peto, and R. Saracci. IARC Scientific Publication No. 90. Proceedings of an

International Agency for Research on Cancer (IARC) symposium held on September 8–10, 1987, in Lyon, France, 511–518.

Doll, R., and J. Peto. 1985. *Effects on Health of Exposure to Asbestos.* London: Her Majesty's Stationery Office.

EEPC (Energy and Environmental Policy Center). 1989. *Summary of Symposium on Health Aspects of Exposure to Asbestos in Buildings.* Cambridge, Mass.: Harvard University, December 14–16, 1988.

EESI (Environmental and Energy Study Institute). 1996. *Briefing Book on Environmental and Energy Legislation.* Washington, D.C.: EESI.

Gaheen, M. 1995. *Cost-Benefit Analysis and the Regulatory Process: A Case Study of the EPA's Asbestos Ban Regulations, 1979–1991.* Ph.D. dissertation. University of Maryland.

HEI-AR (Health Effects Institute–Asbestos Research). 1991. Asbestos in Public and Commercial Buildings: A Literature Review and Synthesis of Current Knowledge. Cambridge, Mass.: HEI.

Hodgson, E., R. Mailman, and J. Chambers. 1988. *Dictionary of Toxicology.* New York: Van Nostrand Reinhold Company.

Hughes, J., and H. Weill. 1989. Development and Use of Asbestos Risk Estimates In *Non-Occupational Exposure to Mineral Fibres,* edited by J. Bignon, J. Peto, and R. Saracci. IARC Scientific Publication No. 90. Proceedings of an International Agency for Research on Cancer (IARC) symposium held on September 8–10, 1987, in Lyon, France, 471–475.

Jasanoff, S. 1990. *The Fifth Branch: Science Advisors as Policy Makers.* Cambridge, Mass.: Harvard University Press.

Mazurek, J., R. Gottlieb, and J. Roque. 1995. Shifting to Prevention: The Limits of Current Policy, In *Reducing Toxics: A New Approach to Policy and Industrial Decisionmaking,* edited by R. Gottlieb. Island Press. Los Angeles: Pollution Prevention Education and Research Center, UCLA, 58–94.

Mossman, B., J. Bignon, M. Corn, A. Seaton, and J. Gee. 1990. Asbestos: Scientific Developments and Implications for Public Policy. *Science* 24: 291–301.

NRC (National Research Council). 1984. *Asbestiform Fibers: Non-Occupational Health Risks.* Washington, D.C.: National Academy Press.

Peck, L. 1989. EPA Blues. *Amicus Journal* 11(2): 18.

Shapiro, M. 1990. Toxic Substances Policy. In *Public Policies for Environmental Protection,* edited by P. Portney. Washington, D.C.: Resources for the Future, 195–241.

Stayner, L., D. Dankovic, and R. Lemen. 1996. Occupational Exposure to Chrysotile Asbestos and Cancer Risk: A Review of the Amphibole Hypothesis, *American Journal of Public Health* 86: 179–186.

Stone, R. 1991. No Meeting of the Minds on Asbestos. *Science,* 254: 928–931.

TSCA Research Project (Toxic Substances Control Act Policy Research Project, Lyndon B. Johnson School of Public Affairs.) 1982. *The Toxic Substances Control Act:*

Overview and Evaluation. Policy Research Project Report 50. Austin: University of Texas.

U.S. ATSDR (U.S. Agency for Toxic Substances and Disease Registry). 1995. *Toxicological Profile for Asbestos (Update).* Washington, D.C.: U.S. Public Health Service.

U.S. EPA (U.S. Environmental Protection Agency). 1986. Airborne Asbestos Health Assessment Update. EPA 600/8-84/1003F. Prepared by the Environmental Criteria and Assessment Office, Research Triangle Park, N.C.

U.S. OTA (U.S. Office of Technology Assessment). 1995. *Gauging Control Technology and Regulatory Impacts in Occupational Safety and Health: An Appraisal of OSHA's Analytic Approach.* OTA-ENV-635. Washington, D.C.: U.S. OTA.

Appendix G

Control of Dioxins and Other Organochlorines from the Pulp and Paper Industry under the Clean Water Act

BACKGROUND

The term *dioxin* encompasses a family of organic chemical compounds known as dibenzo-*p*-dioxins. The dioxins of greatest environmental and public health concern are halogenated dioxins.[1] Because they are the most common, most attention is focused on the group of seventy-five chlorinated dioxins. Dioxins are not deliberately manufactured, but are a by-product of combustion, some chemical manufacturing, some bleaching of pulp and paper, and other industrial processes involving chlorine and other halogens. In the United States, municipal and medical waste incineration are the dominant known sources of dioxin (U.S. EPA 1994a), but the total releases of dioxin from all sources (including natural sources, such as forest fires) are highly uncertain. Dioxin became notorious in the 1970s when it was identified in the United States as "the most potent animal carcinogen ever tested." As one observer phrased it, dioxin earned the reputation as the "Darth Vader" of chemicals (Roberts 1991). In the 1980s, however, Canada and European countries set dioxin limits less stringent than EPA's by two or three orders of magnitude. Officials in these countries concluded that a different cancer model applied to dioxin. More recently, attention has focused on the environmental and non-cancer effects of dioxin and dioxin-like substances that may mimic hormones and act as "endocrine disrupters."

Dioxin discharges into surface waters from pulp and paper mills arose unexpectedly as a regulatory issue more than a decade ago. In 1982, EPA promulgated Clean Water Act (CWA) effluent limitations and technology-based standards ("effluent guidelines") for most of the pulp, paper, and paperboard industry.[2] A year later, as part of EPA's overall "dioxin strategy," the agency initiated a national survey of environmental dioxin levels. In the process of testing what were believed to be "reference streams" to determine background dioxin concentrations in fish in relatively uncontaminated waters, the agency detected surprisingly high levels of dioxin.[3] According to an EPA official, the reference streams where fish had elevated dioxin concentrations had one feature in common, "When you looked upstream, they all had chlorine bleaching pulp and paper plants."

Over time, the list of toxic water pollutants of concern related to chlorine pulp bleaching was broadened to include a variety of more abundant chlorinated organic compounds (organochlorines). These include polychlorinated phenolic compounds, which are considered representative of various polychlorinated organic materials that may accumulate in food chains, and chloroform, a volatile organic compound. Indicative of the relative magnitude of their production by the U.S. pulp and paper industry, discharges of dioxins and a group of dioxin-like chemicals called furans are measured in terms of grams per year, while the discharges of other organochlorines are expressed in units of metric tons per year. Despite this disparity in the quantity of environmental releases, dioxins and furans have dominated the debate over regulatory controls of the effluents from pulp and paper plants that use chlorine bleaching because chlorinated phenols and volatile organochlorines are estimated to be very much less toxic. Some individuals and groups remained concerned, however, about the heterogeneous soup of organochlorines discharged in bulk from pulp and paper mills because most of these compounds have not been toxicologically analyzed and because the chemical transformations organochlorines undergo in the environment are not fully understood. Staking out a precautionary position in the face of scientific uncertainty, some parties argue that all organochlorines should be considered "guilty until proven innocent."

The regulatory control of dioxins, furans, and other organochlorines discharged from pulp and paper mills into surface waters traces its origins to October 1984, when the Environmental Defense Fund (EDF) and the National Wildlife Federation (NWF) filed a citizen's petition under the Toxic Substances Control Act (TSCA, Section 21). The petition requested that EPA regulate dioxins and furans from all known sources.[4] (At the time, despite the questions raised by the detection of dioxin in streams below pulp and paper mills, the bleaching plants were not yet recognized as a source of dioxins and furans.) EPA denied the petition, prompting a 1985 lawsuit by EDF and NWF (*EDF v. Thomas*, DC Dist. Court, Civ. No. 85-0973). Following a series of news reports about EPA's cooperation with industry to investigate the formation and release of dioxins at pulp and paper plants and a 1987 front-page story in the *New York Times* regarding the detection of dioxin in household paper products, EPA signed a consent decree with the plaintiffs in 1988. The agreement required EPA to perform a comprehensive risk assessment of dioxins and furans considering sludges, water effluent, and products made from pulp produced at 104 bleaching pulp mills. The agreement also required the agency to propose regulations under TSCA (Section 6) to control pulp sludge disposal and under the Clean Water Act to address discharges of dioxins and furans into surface waters from the mills by October 31, 1993 (as amended in 1992). The agency's 1993 proposal to control dioxin and furan releases into surface waters is the primary focus of this case study. The proposal was submitted as a combined set of water effluent limitations and standards and national emission standards for hazardous air pollutants for the pulp, paper, and paperboard industrial sector (also called the proposed "pulp and paper cluster rule" [*Federal Register* 58: 66078–66216]). The pulp and paper cluster rule had not been finalized as of press time. But it appears that the crucial subplot for the effluent limits involves an arcane debate over a Swedish water quality test measure called the adsorbable organic halide indicator.

Regulation of Toxic Water Pollutants

The goal of CWA (also known as the 1972 Federal Water Pollution Control Act [FWPCA] Amendments) is to eliminate entirely discharges of pollutants from point sources (that is, individual discharging facilities) into surface waters. Although eliminating pollutant discharges may be achievable under some circumstances through process changes that prevent pollutant formation or recycle wastes, the goal is largely rhetorical. The statutory goal of eliminating discharges influences the use of science because achieving the goal does not require point sources to eliminate all discharges into surface waters. Attainment depends, to some extent, on which substances are classified as pollutants subject to regulation under the statute. Consequently, the discharge elimination goal has the potential to encourage or distort the use of science in specifying regulated pollutants. Pursuing discharge elimination from one point source may result in offsetting releases of pollutants. For example, on-site waste recovery to prevent surface water discharges may require extra energy inputs, resulting in additional releases of contaminants to the atmosphere.

CWA contains both "water quality–based" regulatory controls, which vary according to the designated use (for example, drinking water source, fishable, swimmable) and attributes (for example, volume and rate of flow) of the receiving water body, and "technology-based" effluent standards that are achievable using available pollution control technology. Legally, the environmental quality standards dominate the technology-based standards in the sense that additional regulatory action may be required if the technology-based limits do not achieve the ambient quality standard in a specific location. Since 1972, the technology-based standards have been emphasized, though ambient standards are becoming increasingly important (see discussion below). EPA's traditional reliance on technology-based standards is due, in part, to the practical difficulties experienced prior to 1972 with state attempts to control surface water pollution. FWPCA relied on water quality standards that required state regulatory authorities to demonstrate that a given level of pollution was "unreasonable" or "unacceptable" under local environmental and socio-economic conditions. Under CWA, Congress has emphasized the approach of the technology-based effluent standards that "do not quibble with judgments of reasonableness" (Fogarty 1991). The emphasis on technology-based standards also avoids the potentially greater time and cost associated with developing, administering, and complying with myriad geographically specific pollutant discharge limits that must be tailored to meet ambient water quality standards.

Under the 1972 provisions, EPA was to develop a list of national standards for toxic water pollutants that would be applied without regard to industrial source. Implementation of this chemical-by-chemical approach was more difficult than Congress expected, and dissatisfaction with the progress led to litigation and, eventually, a 1976 consent decree between the Natural Resources Defense Council (NRDC) and EPA.[5] The approach laid out in this settlement was ratified in the 1977 CWA Amendments. Section 307 of CWA now requires best available technology (BAT) economically achievable by industrial sector to limit toxic pollutant effluents from point sources into surface waters. The settlement originally identified a list of sixty-five "toxic" chemicals and classes of chemicals, which were later subdivided into 129 individual substances or "priority" pollutants (U.S. CRS 1993).[6] Dioxin (TCDD—2,3,7,8-tetrachlorodibenzo-p-dioxin) was originally placed on both lists of toxic pollutants.[7]

CWA directs EPA to develop BAT for toxic water pollutants "that will result in reasonable further progress toward the national goal of eliminating discharges" (Section 301[b][2]). Factors to be considered in developing BAT for toxic water pollutants include the affordability of achieving effluent reductions ("economic achievability"), engineering criteria, nonwater quality environmental impacts, and "such other factors as the Administrator deems appropriate" (Section 304[b][2]). The BAT basis for regulating toxic pollutants is in contrast to the control of "conventional" pollutants (for example, suspended solids and fecal coliform). Under Section 304 of CWA, conventional pollutant limits are achieved by best conventional pollutant control technology (BCT). Determination of BCT depends on the relationship between costs and benefits (essentially a BAT standard moderated by a test of economic reasonableness) (Fogarty 1991). Thus, BAT control of toxic pollutants is intended to be less sensitive to cost considerations than BCT, but it acknowledges that alternative technologies can be compared in terms of environmental benefits. (That is, for one technology to be the "best" it must achieve environmental benefits superior to another technology.) Section 307(a) also allows EPA to impose more stringent toxic effluent standards if the BAT standard is inadequate to protect human health with an "ample margin of safety." For some toxic pollutants, however, the only means of providing any margin of safety (ample or otherwise) may be to prohibit discharges altogether because there may be no discernible threshold level of incremental exposure below which no adverse effects will occur.[8]

Through its 1993 proposed pulp and paper effluent regulations, EPA sought to limit the precursors to the formation of dioxins, furans, and other organochlorines in the pulp and paper manufacturing process. The technology-based approach proposed by the agency involves (1) substituting elemental chlorine with chlorine dioxide or other bleaching agents (for example, peroxide or ozone) and (2) reducing the extent of chlorine bleaching required to achieve a given quality of product through alternative means of pulp delignification (that is, extended cooking or oxygen delignification prior to chlorine bleaching). The agency estimates that its proposed effluent limits for the pulp and paper industry would reduce, but not eliminate, exceedances of health-based water standards for dioxins and furans (see Table G-4 below). However, EPA's ambient water quality criteria are not necessarily the last word. Under the Clean Water Act, EPA and the states share responsibility and authority for setting risk-based ambient water quality standards, identifying specific segments of water bodies where technology-based pollutant controls may be inadequate to achieve uses designated by the states, and developing strategies for achieving ambient water quality standards in these impaired waters.

Section 303 of the 1987 Clean Water Act Amendments required states to adopt binding numeric criteria for all priority pollutants in cases in which discharges could reasonably be expected to interfere with the designated use of water bodies. Congress also authorized EPA to set the criteria if states failed to do so by February 1990 or to develop replacement standards if the agency believes a state's standards do not meet minimum requirements (Copeland 1993; Fogarty 1991). In practice, EPA has permitted the states some discretion in developing their criteria. Under EPA's 1983 revisions to water quality regulations, states retain the right to modify EPA criteria to reflect site-specific conditions or adopt numerical values based on "other scientifically defensible methods" (Executive Enterprises 1984, citing 40 CFR 13.11[b][1]).

In 1990, for example, the State of Maryland proposed a water quality standard for dioxin tenfold higher than EPA's numeric criteria based on an allowable 1-in-100,000 (10^{-5}) cancer risk. By EPA's reckoning,

Maryland's proposed standard suggested a cancer risk (10^{-4}) of potential concern. Acknowledging that there are a variety of equally defensible scientific assumptions that can be made, however, the agency approved Maryland's standard. In the state's proposal, many of the scientific assumptions were the same as those of EPA; where they differed (for example, the estimated carcinogenic potency of dioxin), Maryland used alternative assumptions employed by the Food and Drug Administration (Moore et al. 1993; Thompson and Graham 1997). Thus, the Clean Water Act is unusual among federal environmental statutes in the extent to which EPA and the states share authority to set risk-based public health standards.

Section 304 of the 1987 Clean Water Act Amendments directed states to develop lists of their impaired waters by 1989. Impaired waters are those bodies that do not meet or are not expected to meet ambient water quality standards, even after implementation of technology-based controls implemented by point sources. The states were also required to identify point sources causing the water quality impairments and develop individual control strategies to control those sources further (Copeland 1993). Under CWA, developing these controls is to be done by setting the total maximum daily load (TMDL), the maximum quantity of a pollutant a water body can receive daily without violating ambient water quality standards under local conditions. The TMDL is then to be allocated among the various sources contributing to the problem. Finally, the National Pollution Discharge Elimination System permits for regulated point sources are to be revised, as warranted.[9] If the states failed to identify a list of impaired waters and develop TMDLs, the 1987 CWA Amendments required EPA to develop a priority list for the state and make its own TMDL determination. In response, EPA mandated that the list of impaired waters include those receiving discharges from pulp and paper mills and called for specific limits on dioxin discharges by 1992 (Thompson and Graham 1997).

In 1990, at the request of Idaho, Oregon, and Washington, EPA established its first TMDL for eight pulp and paper mills discharging into the Columbia River basin (which includes the Snake and Willamette Rivers). Each of the states had adopted the same ambient water quality standard for dioxin (0.013 parts per quadrillion [ppq]).[10] Based on considerations of regional hydrology, other sources of dioxin, and so forth, EPA set a TMDL for dioxin (of 5.97 milligrams per day) and allocated 35% of the load to U.S. pulp and paper mills operating in the river basin. Environmental groups sued EPA for not setting a more stringent TMDL, and the pulp and paper mills sued the agency for setting the TMDL before finalizing new effluent guidelines for the entire industry (Thompson and Graham 1997). In 1995, the U.S. Ninth Circuit Court of Appeals upheld EPA's

TMDL for dioxin in the Columbia River basin (*Environment Reporter*, June 30, 1995, 493).

In general, the impaired waters listing process/TMDL program has labored under the Clean Water Act's system of shared EPA-state responsibility. With CWA requiring EPA to serve as a backstop, state environmental agencies may have little incentive to allocate limited resources to the program and take the heat for controversial decisions. The TMDL program has come under increased fire from environmental groups, tribes, industry, and local communities. A series of recent court decisions citing EPA's failure to complete the tasks after states failed to do so within the statutory time limits could force the agency to make an incredible number of geographically specific determinations under demanding time, data, and resource constraints. (In the state of Idaho alone, for example, a federal district court has required EPA to set TMDLs for over 900 water segments in a five-year time period [*Inside EPA*, October 4, 1996, 4].) For EPA and state environmental agencies, the analytically and politically daunting task of setting and allocating innumerable TMDLs makes it all the more appealing to formulate national technology-based effluent guidelines so as to limit the number of water bodies expected to exceed ambient water quality standards. Environmentalists seek to avoid the cost and delay involved in case-by-case regulation and are wary that states may be reluctant to impose additional controls on firms within their borders. Individual firms or plants also have an interest in ensuring that geographically specific pollution controls do not put them at a competitive disadvantage. Thus, the CWA provisions requiring EPA and the states to consider geographically specific conditions may influence the use of scientific information in national rulemaking.[11]

Mirroring the shared regulatory authority between EPA and the states, CWA is implemented in an environment in which Congress and the executive branch continuously wrestle for control over regulatory policy. Spurred by a series of executive orders dating back to the Nixon administration requiring some form of economic analysis for proposed regulations (but in particular, the 1981 Reagan administration Executive Order 12291 requiring Office of Management and Budget [OMB] review of new regulations), EPA has deemed it "appropriate" to *consider* cost-effectiveness comparisons when proposing BAT for toxic water pollutants (see discussion of Section 304[b][2] above). According to sources in the EPA Office of Water, however, the program generally regards cost-effectiveness analysis as an imprecise tool that only permits a rough screening of regulatory options, and the agency has not explicitly made any BAT decisions on the basis of cost-effectiveness.

Thus, while CWA prods EPA to do what is "doable" to reduce toxic water pollution, OMB pulls the agency toward what it thinks is "reasonable." As discussed in greater detail below, the projected benefits of the nominally "technology-based" regulations to limit dioxins and other organochlorines from pulp and paper mills are estimated using the tools of environmental science and risk assessment. Disagreements about the agency's regulatory proposals are often conducted in the language of science and technology and are, in part, over how to properly assess the environmental benefits of regulation. The subtext, however, is whether those benefits are reasonably associated with compliance costs.

In addition to its prominent role in the proposed pulp and paper cluster rule, dioxin has a long and highly publicized history. As Finkel (1988) noted, our national preoccupation with dioxin stems largely from the notoriety of TCDD as the most potent animal carcinogen ever tested, and its ubiquity as a contaminant of pesticides, incinerator smoke and ash, and bleached paper consumer products, such as diapers and coffee filters. More recently, the dioxin story has segued into the broader debate over "endocrine disrupters," a class of hormonelike chemicals suspected of having a variety of reproductive and other noncancer effects. Endocrine disrupters are the subject of the much-discussed popular science book *Our Stolen Future*, which argues that background levels of chlorinated organics and other industrial chemicals may play a role in development of breast cancer, falling sperm counts and other male reproductive disorders, and developmental effects in wildlife and humans (Colborn et al. 1996).[12]

Forty years ago, a European researcher identified the impurity TCDD as causing the skin disease chloracne in chemical workers involved in the production of the herbicide 2,4,5,-T (Moore et al. 1993). But, dioxin first came to public light in the early 1970s as a result of concerns about the exposure of Vietnam veterans and South Vietnamese children to the defoliant Agent Orange (which included 2,4,5,-T).[13] EPA promulgated a partial ban on the herbicide in 1971. The animal studies that resulted in dioxin (TCDD) being labeled as the most potent carcinogen were conducted in 1978. One year later, EPA issued a controversial reanalysis of an epidemiological study conducted in Alsea, Oregon (Alsea II), which associated miscarriages with herbicide spraying, leading to accusations that EPA had "cooked" the data to inflate the risks (Whelan 1985), and the agency suspended essentially all remaining uses of 2,4,5-T.[14] The problem of dioxin emissions from municipal waste incinerators was identified in 1979 and gained public notoriety as the main plank of Barry Commoner's 1980 presidential campaign platform. (Later, debates over the location and siting of incinerators gave impetus to the environmental justice movement.)

In 1974, the federal Centers for Disease Control (CDC) identified dioxin as a toxic substance in Missouri waste oil. In 1982, EPA detected high dioxin levels from TCDD-contaminated oil sprayed on streets in

Times Beach, Missouri, and dioxin was implicated in illnesses in horses and possibly children. Flooding in December raised concerns about contamination spreading to other sites (though it did not) (U.S. OTA 1991). In 1983, CDC and the Missouri Division of Health recommended that the town be evacuated, and EPA and the Missouri Department of Natural Resources paid $36 million to buy all 801 homes in Times Beach and relocate its residents because of the unavailability of demonstrated treatment technologies and the uncertainty about when the cleanup would be completed. A $200 million cleanup of the town's 400 deserted acres was later initiated. In the summer of 1990, Vernon Houk, head of the Centers for Disease Control's Center for Environmental Health and Injury Control, told a congressional committee that new evidence suggested the risk of dioxin historically was vastly overstated.[15]

EPA's first health assessment of dioxin was conducted in 1981 and was revised in 1985. Animal studies by Dow Chemical Co. researchers (Kociba et al. 1978) and the National Toxicology Program (NTP 1982) were important sources of scientific information for the agency's assessment. The 1985 assessment is the current *official* basis of dioxin cancer risk estimates used by EPA for all regulatory decisionmaking, including the 1993 proposed pulp and paper cluster rule. However, EPA has been in the process of reassessing the risks of dioxin for several years. During the 1980s, some researchers postulated that dioxin might "promote" rather than "initiate" cancer and that, as a result, EPA may have overestimated the cancer risks from dioxin. In 1986, with the backing of Assistant Administrator for Pesticides and Toxics John Moore, an ad hoc EPA committee recommended moderating the dioxin cancer risk estimate. At about the same time, OMB highlighted the large scientific uncertainty of dioxin cancer risk estimates in its annual report on federal regulatory programs (Moore et al. 1993; Roberts 1991). The following year, EPA issued a draft reassessment suggesting that the risk of cancer from dioxin was seventeen times less than the agency had assumed. According to Finkel (1988), however, the agency developed its revised estimate not on the basis of any new data, but by essentially splitting the difference between two "fundamentally irreconcilable theories about the carcinogenicity of dioxin." Regardless of whether the decision was "right for the wrong reasons," as some felt, the agency's approach could not withstand scrutiny. In reviewing the agency's draft, the EPA Science Advisory Board (SAB) criticized EPA's current cancer risk assessment methodology but found no new data to support changing the dioxin cancer risk estimate (Moore et al. 1993).

In 1990, Robert Scheuplein of FDA, toxicologist Michael Gallo of the Robert Wood Johnson Medical School in New Jersey, and Dutch scientists organized the "Banbury Conference" (held at New York's Cold Spring

Harbor Laboratory), which formally marked a new scientific consensus about a series of biological steps occurring at the molecular level that precede most, if not all, of the observed effects of dioxin and other similar chemicals (Roberts 1991).[16] Some scientists interpreted this to mean that the very low levels of dioxin in the environment would result in negligible cancer risks. In 1991, an epidemiological study conducted by National Institute of Occupational Safety and Health (NIOSH) researchers (Fingerhut 1991) reported a statistically significant increased cancer risk in U.S. chemical workers exposed to high levels of dioxin but detected no increase in workers exposed to low levels. As a result of the Banbury Conference and the NIOSH study, external pressures mounted for EPA to move beyond research and initiate a formal reassessment of dioxin. According to press reports, the paper industry was a leading voice in persuading the agency to revisit dioxin (*Rachel's Environment & Health Weekly*, August 31, 1995, 1). In April 1991, EPA Administrator William Reilly announced that the agency would comprehensively reassess the cancer and noncancer risks of exposure to TCDD and related compounds.

While EPA slightly moderated the cancer risk estimate for dioxin and similar compounds in its *draft* reassessment released in 1994, it also concluded that there was potential for a variety of adverse noncancer effects in the range of current background exposures to dioxin and similar compounds (U.S. EPA 1994b). In reviewing the draft reassessment, a majority of SAB members concluded that the agency tends to overstate the possibility for danger at near ambient levels, but several SAB members regard the agency's characterization of the risks as appropriately conservative within the context of public health protection (U.S. EPA/SAB 1995). An environmentalist now says, "Reilly's decision to conduct the dioxin reassessment did not turn out the way he dreamed it would. Industry and he thought they would have a slam dunk on dioxin's carcinogenicity."

Although it appeared likely to many in 1988 when EPA began to formulate the new pulp and paper effluent limits that our "national preoccupation" with dioxin would wane, the agency's subsequent dioxin reassessment has highlighted the noncancer effects of dioxins and helped launch the issue of endocrine disrupters onto the environmental regulatory agenda. Some environmentalist groups (notably Greenpeace) have responded by calling for a ban on chlorine. This proposal was afforded a measure of mainstream legitimacy in February 1994 when the International Joint Commission, the Canadian-American bilateral organization established to monitor the Great Lakes Water Quality Agreement, recommended phasing out the use of chlorine and chlorine-containing compounds as industrial feedstocks. It is in this context that EPA will try to finalize the effluent regulations for the pulp and paper industry. Table

G-1 provides a summary background of dioxin science and policy. Table G-2 summarizes the development of the pulp and paper cluster rule.

SCIENTIFIC ISSUES

Dioxin and Related Compounds

The major scientific controversy over dioxin and its chemical cousins is not whether high levels of exposure can cause cancer in humans but rather the risks posed by background levels and incremental releases of all dioxin-like compounds. Although dioxins and other organochlorines have been associated with a variety of noncancer effects, the conventional focus of scientific investigation has been on cancer. According to Lucier et al. (1993), several long-term bioassays have been conducted on TCDD in several species. All studies have produced positive results. It is clear that TCDD is a multisite carcinogen in both sexes of rats and mice. It is also a carcinogen in the hamster, which is considered the most resistant species to the acute toxic effects of TCDD. TCDD is also found to increase cancer incidence in animals at doses well below the maximum tolerated dose. While TCDD appears to be a chemical that strongly promotes cancer development once initiated, it seems to have weak or no potential to initiate cancer itself. The general consensus is that TCDD is an example of a carcinogen whose action is mediated by a specific receptor within cells, suggesting that there may be a threshold dose below which dioxin is not carcinogenic. It may be possible, however, that noncancer health effects result at levels below the threshold dose for cancer.

A considerable body of studies of people exposed to dioxin provides suggestive evidence of its human carcinogenicity, but, according to an EPA official, the epidemiological evidence is inconclusive due to a number of factors. First, scientists cannot be certain of the amount of dioxin and other chemicals to which the subjects were exposed. Second, in most studies, the numbers of people exposed through accidents or in the workplace have been too small to allow scientists to detect substantial changes in cancer rates. Third, those individuals who were exposed to dioxin (mostly healthy adult males) may not have been the most sensitive group. Finally, not enough time may have elapsed between exposures and study completion for most cancers to develop (many cancers only develop 15–30 years after exposure). The first dioxin epidemiological study sufficiently large enough to detect a substantial increase in cancer doses, according to this EPA official, was Fingerhut et al. (1991). This NIOSH study, which took nearly thirteen years to complete and

Table G-1. Summary Background of Dioxin Science and Policy

1949	U.S. Department of Agriculture registers 2,4,5-T as a pesticide.
1957	TCDD is identified as causing chloracne.
1966	USDA and the Food and Drug Administration establish residue tolerances for 2,4,5-T in food.
1969	Initial laboratory studies link 2,4,5-T and TCDD with birth defects.
1970	United States halts use of Agent Orange in Vietnam.
1971	EPA restricts domestic use of 2,4,5-T.
1972	Controversy over EPA's 2,4,5-T decisionmaking process crystallizes congressional support for Federal Advisory Commission Act.
1974	Centers for Disease Control and Prevention identifies dioxin as toxic substance in Missouri waste oil.
1976	Industrial accident releases large quantities of dioxin in Sveso, Italy.
1977	Clean Air Act Amendments list dioxins and furans as hazardous air pollutants.
1978	First EPA study regarding linkage between miscarriages and herbicides in Alsea, Oregon.
	Dow Chemical Co. researchers report that TCDD is a carcinogen in laboratory studies.
1979	Alsea II reevaluates miscarriage–herbicide data. EPA accused of inflating risks.
	EPA suspends use of 2,4,5,-T. Vietnam veterans start class action suit.
	Dioxin and furans identified in emissions from municipal waste combustion plants.
1980	Barry Commoner's presidential campaign elevates concerns about dioxin releases from waste incinerators.
1981	Sveso five-year report finds no dioxin effects other than chloracne.
	EPA's Cancer Assessment Group estimates that dioxin is one of the most potent carcinogens known.
1982	National Toxicology Program reports results of dioxin animal cancer study.

examined 5,172 male U.S. chemical workers exposed to dioxin on the job from 1942 to 1984, presented what many consider the strongest evidence that dioxin is a human carcinogen—but, perhaps, only at very high doses (Roberts 1991). The EPA Science Advisory Board has agreed that although human data are limited, dioxin is a probable human carcinogen under some exposure conditions (U.S. EPA/SAB 1995). In February 1997, an International Agency for Research on Cancer working group also concluded that TCDD should be considered carcinogenic to humans (http://www.iarc.fr/preleases/115e.htm).

Table G-1. Summary Background of Dioxin Science and Policy *(Continued)*

1983	Times Beach, Missouri, buyout.
	EPA issues congressionally mandated national strategy to investigate, identify, and remediate dioxin-contaminated areas.
1984	EPA cancels 2,4,5-T registration.
	Hazardous Solid Waste Act requires EPA to evaluate risks posed by dioxin emissions from municipal waste combustion facilities.
1985	EPA revises its dioxin health assessment, lowering the cancer risk estimate by more than a factor of two but retains the agency's default linear cancer model.
1986	Ad hoc expert committee advises EPA that linear cancer model is inappropriate for dioxin.
1987	EPA scientific group recommends moderating cancer risk estimate.
	EPA develops toxic equivalency factors (TEFs) for dioxin and dioxin-like chemicals.
1989	EPA Science Advisory Board finds no new data to support change in cancer risk estimate; critical of current cancer model; accepts TEFs as an interim approach.
1990	Banbury Conference supports receptor-mediated event for dioxin activity.
	EPA promulgates new source performance standards for municipal waste combustion facilities requiring best management practices to limit total dioxins and furans to 30 ng/m^3.
1991	National Institute for Occupational Safety and Health epidemiological study suggests that dioxin is a human carcinogen, but perhaps only at high levels of exposure.
	EPA initiates dioxin reassessment.
1994	EPA draft dioxin reassessment reports potential for adverse noncancer health effects within the range of current background levels.
	Chlorine ban proposed by Henry Waxman (D-Calif.), Barry Commoner, and others (*Environment Reporter*, September 30, 1994, 1133).

Extrapolating from rodent studies using a linear model of cancer risk, EPA's Cancer Assessment Group derived an extraordinarily high cancer potency factor $(4.25 \times 10^{-5}$ [mg/kg/day]$^{-1})$ for dioxin in 1981. An important basis of this estimate was a reanalysis of the pathological evidence from the Dow Chemical researchers' rat study (Kociba et al. 1978) performed by Robert Squires of Johns Hopkins University Medical School.[17] Squires's reinterpretation of the tissue samples resulted in a cancer potency factor approximately two times higher than the one derived using the original diagnoses. In 1985, EPA revised its dioxin cancer potency estimate downward by more than a factor of two (to 1.56×10^{-5}

Table G-2. Development of the Pulp and Paper Cluster Rule

1983	EPA initiates national dioxin survey, detects elevated dioxins downstream from pulp and paper mills.
1984	Environmental Defense Fund (EDF) and National Wildlife Federation (NWF) file TSCA petition requesting EPA to regulate dioxins and furans from all known sources. EPA denies petition.
	EPA issues ambient water quality criteria report for dioxin.
1985	EDF and NWF file lawsuit.
1986	June: EPA, the National Council of the Paper Industry for Air and Stream Improvement, and American Paper Institute (API) agree to undertake the "Five Mills Study," detect TCDD and TCDF in effluents, pulp, and sludges of pulp and paper mills.
	December: Information on the agreement between EPA and the pulp and paper industry reported.
	Greenpeace initiates Freedom of Information Act (FOIA) request seeking all available information on the pulp mill dioxin problem.
1987	Clean Water Act (CWA) Amendments establish deadlines for EPA and states to address toxic pollutants.
	January: Letter from EPA to API leaked to environmentalists indicates EPA officials had agreed to notify the industry "immediately" of receipt of any requests under FOIA and that, barring such requests or results indicating a potential threat to human health, the agency did not intend to release any results until publication of the final report on the study.
	August: Greenpeace USA releases report alleging an EPA cover-up.
	September: *New York Times* front page story reports traces of dioxin detected in household paper products. Report based on the "Five Mills Study" and analyses of dioxin in paper products.
1988	EDF, NWF, and EPA sign consent decree requiring agency to perform a comprehensive risk assessment of dioxins and furans considering sludges, water effluent, and products made from pulp produced at 104 bleaching pulp mills and (as amended in 1992) to propose regulations addressing discharges of dioxins and furans into surface waters from the mills by October 31, 1993.
	EPA issues "interim strategy" to address dioxin emissions from pulp mills, which included requiring pulp mills to monitor for dioxins and adopt short-term control measures (Hanmer 1988; U.S. EPA-V 1988).
	EPA and industry begin the "104 Mills Study."
	Swedish studies generate adsorbable organic halide indicator (AOX) used in EPA's 1993 proposed effluent limits.
1989	EPA initiates interagency, interoffice assessment of pulp and paper sludges, effluents, and consumer products.

Table G-2. Development of the Pulp and Paper Cluster Rule *(Continued)*

1989 *(cont.)*	Office of Technology Assessment report discusses Swedish pulp mills' compliance with more stringent regulatory standards for organochlorine emissions (U.S. OTA 1989).
	March and June: First results of 104 Mills Study released.
1990	EPA issues *Assessment of Risks from Exposure of Humans, Terrestrial and Avian Wildlife, and Aquatic Life to Dioxins and Furans from Disposal and Use of Sludge from Bleached Kraft and Sulfite Pulp and Paper Mills.* Based on the 104 Mills Study, assessment estimates that preventing adverse wildlife effects would require TCDD soil concentrations 4–400 times lower than levels needed to prevent unacceptable human health risks.
1991	May: Under court consent decree, EPA proposes pulp and paper mill sludge rule under TSCA Section 6. Proposal would set a 10-ppt maximum allowable dioxins/furans concentration for land application (resulting in an estimated human health risk of less than 10^{-4}) and includes provisions for mills to submit annual reports and maintain records on land, application, and laboratory analysis.
	July: Office of Management and Budget objects to proposal's information collection request (*Environment Reporter*, August 16, 1991, 1058).
1992	EPA announces it will seek a voluntary agreement with industry on the pulp and paper mill sludge rule (*Environment Reporter*, December 24, 1993, 1545–1546).
1993	September: Natural Resources Defense Council and fifty-five other environmental groups petition under CWA Section 307(a) for EPA to ban dioxin discharges by the pulp and paper industry by prohibiting the use of chlorine rather than managing dioxin through best-available-technology (BAT) standards under pulp and paper cluster rule (*Environment Reporter*, September 17, 1993, 889–890).
	December: EPA proposes pulp and paper cluster rule based on BAT standards.
1994	February: At hearing on proposed cluster rule, industry representatives claim that EPA's environmental benefits analysis does not employ sound science and overstates benefits. Future EPA Assistant Administrator for Research and Development Robert Huggett reports that substitution of chlorine dioxide for elemental chlorine reduces chemicals that accumulate in fatty tissues to the limits of detectability (*Environment Reporter*, February 18, 1994, 1783–1784).
	April: EPA and pulp and paper industry announce voluntary agreement regarding land disposal of dioxin-tainted sludge, formalizing best management practices. No restrictions on use of sludges if concentration of dioxin and furan is less than 10 ppt. For pasture lands, the concentration limit is 1 ppt (that is, background levels). At 50 ppt, sludge cannot be land applied.

Continued on next page

Table G-2. Development of the Pulp and Paper Cluster Rule *(Continued)*

1996	On the basis of new data regarding the environmental performance of pulp and paper mills that have completely substituted chlorine dioxide for elemental chlorine, EPA announces that it is considering two BAT options for the major pulp and paper subcategory (bleached paper-grade kraft and soda).
1997	November: EPA selects the less stringent of the two BAT options, which calls for the complete substitution of chlorine dioxide for elemental chlorine but does not require the additional and more expensive process of oxygen delignification.

[mg/kg/day]$^{-1}$) by adjusting for the early mortality of study animals observed in Kociba et al. (1978) and by essentially splitting the difference (taking the geometric mean) between the original pathology assessment and Squires's reanalysis (Thompson and Graham 1997). The agency had moderated its dioxin hazard assessment somewhat, but it still estimated that a 1-in-1,000,000 (10^{-6}) cancer risk was associated with exposure to the infinitesimally small quantity of 0.006 pg/kg/day (picograms [10^{-12} g] per kilogram body weight per day).

Although EPA indicated in 1985 that there was inconclusive evidence that dioxin was a mutagen (able to initiate carcinogenesis), the agency determined that the available data on dioxin's biological activity (carcinogenic mechanism and pharmacokinetics) were insufficient to support deviation from the default linear dose-response model for cancer. Canada and European countries, however, rejected the linear cancer model as inappropriate for dioxin because it is not considered genotoxic (that is, dioxin does not directly initiate cancer by causing mutation or DNA damage) and set their limits at 1–10 pg/kg/day. There were also differences in dioxin cancer potency estimates within the U.S. government among EPA, CDC, and FDA. FDA's cancer potency estimate is almost an order of magnitude smaller than EPA's 1985 estimate, and CDC's is intermediate between the two. The inconsistent estimates resulted from the agencies' applying the same linear cancer model but making a variety of different scientific assumptions and data treatments.[18]

In the 1970s, Alan Poland of the University of Wisconsin initiated the first studies on dioxin's biological mechanisms (Thompson and Graham 1997). At the 1990 Banbury Conference, scientists agreed that the biological activity of dioxin and dioxin-like compounds was mediated by first binding to a specific molecular receptor in cells, the aryl hydrocarbon (Ah) receptor (an intracellular protein).[19] Theoretically, dioxin molecules may have to occupy many Ah-receptor sites before any biological response is seen, and, even once activity begins, the cell's internal regulation system has some capacity to adapt to changing hormonal levels and maintain the mix within the range of tolerance. In the view of some scientists, this theoretical argument suggests a threshold below which dioxin cannot cause cancer and implies that EPA's linear cancer model is invalid for dioxin. "If we can't do it [depart from the linear default model] for dioxin, for which we have so much information, then we probably can't do it for anything," said Banbury Conference organizer Michael Gallo (quoted in Roberts 1991).[20]

However, there may be considerable variability among individuals in the threshold level at which carcinogenesis begins. In addition, a continuum of biological activity occurs beginning at relatively low levels of Ah-receptor occupancy. There is, however, considerable controversy regarding the health *significance* of the activities initiated at lower levels of occupancy. (In practical terms, this means that setting dioxin limits low enough to prevent cancer may be insufficient to prevent other biological effects, but the "so what?" question has yet to be resolved by scientific consensus.) In response to the 1991 decision to conduct a dioxin reassessment, scientists at EPA's Office of Research and Development and the National Institute of Environmental Health Sciences began research to characterize a threshold for dioxin in humans. The results, reported in 1992–1993, suggested that enzyme induction occurs at existing background levels of dioxin-like compounds (Thompson and Graham 1997). Instead of cancer being initiated at the lowest dose levels, it is now hypothesized that reproductive, developmental, and immune-system impairments may be the most sensitive health effects of dioxin. For these noncancer effects, says an environmentalist, the old toxicological adage that "the dose makes the poison" may not apply.[21] Instead, the timing— not the quantity—of exposure may be the critical factor. This source is concerned, for example, that exposure to a trace quantity of dioxin that might be irrelevant in terms of cancer risk could result in a substantial developmental risk if maternal exposure occurs at a critical period of fetal development.

Further, the biological system responds to the cumulative exposure of dioxin and similar chemicals that bind to the Ah receptor rather than to the exposure to any single dioxin-like compound. As a result, much disagreement now centers on just how close existing background levels of all dioxin-like compounds occurring in the environment and stored in human tissues are to the levels required to cause adverse health effects.[22] As Thompson and Graham (1997) suggest, the significance of this dispute is that the concept of a threshold level of Ah-receptor occupancy may be irrelevant to decisions about additional releases of dioxin-like compounds if typical body burdens already exceed the threshold.

Of the group of seventy-five chlorinated dioxins, only TCDD has been subjected to long-term animal carcinogen experiments. To account for the cumulative exposure to compounds that, like dioxin, would bind to the Ah receptor, in 1987, the EPA Risk Assessment Forum developed toxic equivalency factors (TEFs). These TEFs derive from a relative ranking scheme based on assigning a TEF of 1.0 to TCDD, since it shows the greatest affinity for binding to the Ah receptor. Other dioxin-like compounds are assigned a fractional weight proportional to their binding affinity relative to that of TCDD. The TEFs are intended to be additive weighting factors. The TEF for TCDF, for example, is 0.1—its affinity for binding to the Ah receptor is 1/10th that of TCDD (U.S. EPA 1989). (Thus, 5 g of TCDD plus 5 g of TCDF yields the estimated equivalent of 5.5 g of TCDD.) There is not a perfect correlation, however, between Ah-receptor binding affinity and the potency for various toxic effects. Consequently, there is considerable uncertainty about how accurately TEF equivalent weights reflect cumulative effective exposures.[23]

As indicated earlier, dioxins are produced in very small quantities. EPA estimates annual emissions from known sources for the entire United States at 3,300–26,000 *grams*, with the total possibly being as high as 50,000 g/yr (U.S. EPA 1994a).[24] However, dioxins are extremely insoluble in water, are environmentally and biologically stable, persist in the environment for long periods, and tend to accumulate in animal tissues. Thus, the predominant route of human exposure is probably through the food chain rather than inhalation or drinking water. Relative to other foods, measurements of background levels of dioxin are particularly high in fish. Currently, bleaching pulp and paper mills are the only significant known source of dioxins released into surface waters (U.S. EPA 1994a). According to an EPA official, the agency estimated a very wide range of risks resulting from dioxins and furans released from pulp and paper mills, much of which was explained by the size of the receiving water body into which plant effluent was being discharged.

Formation of Dioxin and Other Organochlorines from Bleaching Pulp

Lignin is a natural polymer that binds and supports cellulose fibers of woody plants, but it discolors and weakens paper products. Chemical pulping dissolves a large fraction of lignin using nonoxidizing chemicals (for example, alkalis or sulfites), while preserving a large fraction of the desired cellulose fibers. Various forms of chlorine and other bleaching agents are used to further remove lignin from pulp to produce durable white paper products (like this page). For many decades, elemental chlorine (Cl_2) has been the bleaching agent of choice for much of the U.S. pulp and paper industry due to its relatively low cost. Chlorine dioxide

(ClO_2) is more selective for lignin and, thus, can achieve the same level of pulp bleaching with a substantially lower input or "charge" of chlorine, but it costs more than elemental chlorine. Using a process called oxygen delignification (OD), oxygen may also be used as an initial bleaching agent to reduce the chlorine charge required to achieve a given level of pulp brightness. However, OD is a capital-intensive technology.[25]

When dioxin was first detected in streams below pulp and paper mills, the first culprits identified were oily defoamers and wood chips treated with polychlorophenols. Addressing these sources, however, did not eliminate dioxin formation from bleaching pulp and paper mills. This suggests that some dioxin and furan precursors might occur in trees naturally (Berry et al. 1991). It now appears that the only way to entirely prevent formation of dioxins, furans, and other organochlorines by the pulp and paper industry is to eliminate the use of chlorine as a bleaching agent. By substituting the more lignin-selective ClO_2 for elemental chlorine, however, the formation of organochlorines—and, particularly, the persistent, bioaccumulable polychlorinated organics of greatest concern—can be dramatically reduced.

According to Berry et al. (1991), of the chlorine used in pulp bleaching, about 90% ends up as common salt (for example, calcium chloride) and about 10% binds to organic material removed from the pulp. About 80% of this organically bound chlorine occurs in high–molecular weight material that does not permeate cell walls and is relatively water soluble.[26] Most of the organically bound chlorine that occurs in low–molecular weight compounds that can permeate cell walls is relatively water soluble and is readily hydrolyzed or metabolized. A small fraction (about 1%) of the total organically bound chlorine is relatively fat soluble and potentially bioaccumulable and toxic. A component of particular concern in this fraction is the polychlorinated organic material, which includes dioxin, furan, and polychlorinated phenolic compounds. The polychlorinated phenolic compounds, however, are considered much less toxic than dioxin. For example, EPA estimates the cancer potency of 2,4,6-trichlorophenol to be seven orders of magnitude lower than that of TCDD (U.S. EPA 1993a, Table 3-1).

Because chlorine atoms are added to organic precursors in a largely sequential process (with the dichlorinated organics most likely to be formed before trichlorinated organics, trichlorinated organics most likely to be formed before tetrachlorinated organics, and so forth), Berry et al. (1991) concluded that a threshold level of chlorine charge would be required for any TCDD and TCDF formation to occur. They further suggested that 100% substitution of ClO_2 for Cl_2 (called "complete substitution") could prevent such formation. However, more recent data from mills employing complete substitution show detectable levels of TCDD and TCDF in bleach plant effluents (ERG 1996). Given the huge number of randomly interacting molecules present in commercial-scale pulp bleaching, one would expect some trace (perhaps undetectable) amounts of trichlorinated organics (such as trichlorophenol) and tetrachlorinated organics (such as TCDD and TCDF) to be formed at even the lowest chlorine charges, particularly if the pulp and chlorine are not uniformly mixed.[27] Thus, complete substitution of ClO_2 for elemental chlorine would not entirely eliminate dioxin and furan formation. Complete substitution does appear, however, to reduce dioxin and furan formation to the flat portion of the curve, well beyond the point of diminishing returns. (See the data presented in Berry et al. 1991.)

Berry et al. (1991) also observed that the formation of dioxin and furan is little affected by the lignin content of unbleached pulp. This conclusion has been reinforced by the more recent environmental performance data. TCDD and TCDF were not detected in any industry-supplied sample results from bleached paper-grade kraft mills employing complete substitution (of Cl_2 with ClO_2). But, TCDD and TCDF were detected in EPA-collected samples at several mills using both complete substitution *and* oxygen delignification (ERG 1996). Therefore, while complete substitution may not entirely preclude dioxin formation, initiating the bleaching process with OD (and thereby further reducing the required chlorine charge) does not appear to prevent it either. In contrast, the lower lignin content of pulp prior to bleaching plays a decisive role in the reduced formation of the less toxic but more abundantly formed chlorinated phenolics (Berry et al. 1991).

Releases and Detection of Dioxin and Other Organochlorines from Bleaching Pulp

For contaminants like dioxin that are toxic at trace concentrations, damages may occur at environmental levels resulting from the cumulative releases of multiple sources that, when considered individually, may discharge undetectably low concentrations of the pollutant. EPA's 1993 proposed BAT for pulp and paper effluents was expected to yield nondetectable concentrations of TCDD for two subsectors of the industry and of pentachlorophenol for three subsectors of the industry (Table G-3). However, assuming human consumption of both water and organisms, the agency estimated that the proposed effluent limits would reduce, but not eliminate, exceedances of the most stringent federal health-based ambient water quality criteria (AWQCs) for dioxin, furan, and other chlorinated organic priority pollutants (Table G-4).[28] Comparing modeled dioxin fish tissue concentrations with the various advisory action levels adopted by states, EPA estimates that the BAT proposed in 1993 would substantially

Table G-3. Effluent Limits (Maximum for Any 1 Day) for Existing Plants Using Proposed BAT Process

Subsector	TCDD (ng/kkg)	TCDF (ng/kkg)	Pentachlorophenol	AOX (kg/kkg)
Bleached paper-grade kraft and soda	ND	359	ND	0.267
Dissolving sulfite	ND	1,870	ND	3.13
Dissolving kraft	300	415	ND	0.65
Paper-grade sulfite	N/A	N/A	N/A	0.1

Notes: ng/kkg is nanograms per metric ton (1 ng = 10^{-9} g; 1 metric ton = 10^{6} g); kg/kkg is kilograms per metric ton (1 kg = 10^{3} g, 1 metric ton = 1,000 kg, or about 2,200 lbs); ND is no detection limits of the analytical methods for TCDD and TCDF are 10 pg/l (pg = 10^{-12} g), or 10 ppq;[a] AOX is adsorbable organic halides.

[a] A 70-kg (154-lb) person drinking 2 liters of water per day containing 10 pg/l would receive a dose of 0.29 pg/kg/day. This figure is more than an order of magnitude (over fortyfold) higher than EPA's 1985 1-in-1,000,000 cancer risk–specific dose of 0.006 pg/kg/day.

Source: Federal Register 58: 66078–66216.

Table G-4. Estimated AWQC Exceedances: Number of Streams Below Plants Exceeding AWQC by Industry Subsector

Subsector	Priority Pollutant	Baseline	Proposed BAT
Bleached paper-grade kraft and soda	TCDD	80	71
	TCDF	59	15
	Chloroform	24	0
	Pentachlorophenol	19	2
	2,4,6-Trichlorophenol	5	0
Dissolving sulfite	TCDD	5	5
	TCDF	4	3
	Chloroform	3	2
	Pentachlorophenol	3	1
Dissolving kraft	TCDD	3	2
	TCDF	1	1
	Chloroform	1	0
	Pentachlorophenol	1	1
	2,4,6-Trichlorophenol	1	0
Paper-grade sulfite	TCDD	9	0
	TCDF	5	0
	Chloroform	1	0
	Pentachlorophenol	1	0

Notes: Baseline is estimated current AWQC exceedances. Proposed BAT is estimated AWQC exceedances after implementation of proposed BAT. In addition to TCDD and TCDF, of the organochlorines listed, chloroform, pentachlorophenol, and 2,4,6-trichlorophenol are regulated as priority pollutants.

Sources: Federal Register 58: 66078–66216; U.S. EPA 1993a (Attachment A-12).

reduce (by 70%–95%), but not eliminate the number of state dioxin-related fish advisories in place (U.S. EPA 1993a).

Because dioxin and other organochlorines may be toxic in trace amounts, EPA proposed to establish effluent limitations for these pollutants measured at the bleach plant within the mill rather than at the end of the pipe. This permits greater detection of these pollutants before they are diluted in downstream milling processes or wastewater treatment. Given a large number of samples in which toxic pollutants are not detected, their estimated concentration in the pulp mill bleach plant effluent will be sensitive to how the "no-detect" measurements are treated. (The no-detects signify that the actual concentration lies somewhere between zero and the analytical detection limits.) Consistent with the agency's standard procedures, EPA analyzed TCDD and TCDF sample data from bleach plant effluents assuming one-half detection limit values for those contaminants not detected in the effluent. The agency noted that a "significant portion of [the estimated] risk is associated with the use of one-half the EPA designated detection limit [5 pg/l] for these" pollutants (U.S. EPA 1993a). Any particular value (or point estimate) one could apply to the nondetect samples could be regarded as arbitrary. A probabilistic approach would employ a distribution of values ranging from zero to the detection limit. This approach is conceptually preferable to the simpler point-estimate approach, but it remains unclear how to ascertain the precise form of the distribution of pollutant concentrations below the analytical detection limits.

Alternatively, all chlorinated organic compounds in an effluent sample can be measured collectively at the end of the pipe using an indicator, such as the concentration of adsorbable organic halides (AOX). AOX is a test measure used by Swedish researchers in studies conducted between 1977 and 1985 to evaluate some dramatic effects on fish populations located near bleached kraft paper mills in that country (for example, fish kills). The Swedish Environmental Protection Agency (SEPA) began regulatory action in 1986 to reduce the total organochlorine discharges from pulp and paper mills. In its regulation, Sweden relied on AOX because it is a relatively inexpensive and reliable measurement technique and because essentially all of the halides emitted from pulp and paper mills are chlorinated compounds (Thompson and Graham 1997). SEPA established an AOX discharge limit of 1.5–2.0 kg/air-dried ton for Swedish mills (Berry et al. 1991).

According to EPA, however, the distribution of observed effects in fish populations does not appear to correlate well with AOX measurements, and there is no statistically significant relationship between the level of AOX and specific chlorinated organic compounds, such as TCDD and TCDF (U.S. EPA 1993a). Berry et al. (1991) concluded that AOX is essentially linearly related to the amount of chlorine used in bleaching, while the polychlorinated organic materials (dioxin, furan, and polychlorinated phenols) show a greater rate of reduction than AOX as the chlorine charge is reduced. Furthermore, according to an industry official, unlike pulp and paper mills in the United States, Swedish mills do not employ secondary wastewater treatment that can biologically degrade some organochlorines. As a result, a Swedish mill would discharge a larger amount of organics (which deplete dissolved oxygen in water that fish require for respiration) and a different mix of pollutants.[29] On the basis of the Five Mills Study and the integrated risk assessment of dioxin from pulp and paper mills required under the consent decree, EPA concludes that "although AOX concentrations can be used to determine the removal of chlorinated organics to assess loading reductions, they do not provide information on the potential toxicity of the effluent" (U.S. EPA 1993a).

An environmentalist argues that because discharges of dioxin and pentachlorophenol below the analytical detection limits will not meet stringent, federal health-based ambient water quality criteria, it is justifiable to use reductions achieved in AOX loadings as a surrogate for comparing the efficacy of alternative pollution control technologies. The preference for AOX is also motivated by concerns that only a few of the numerous organochlorines discharged by bleaching mills have been identified or toxicologically tested and that many of the uncharacterized compounds could be environmentally hazardous. An industry official, on the other hand, focuses on the lack of correlation between AOX and effluent toxicity and argues that if alternative pollution control technologies yield similar concentrations of dioxin and furan, it is invalid to compare them on the basis of reductions achieved in AOX. The reductions in AOX, says this industry representative, do not achieve any measurable or monetizable environmental benefits. In addition, the estimated AWQC exceedances for dioxin are based on nonbinding federal criteria that are substantially more stringent than the binding ambient criteria established by some states and approved by EPA. According to an academic, industry has demonstrated that it can meet the dioxin AWQC of states like Maryland without adopting all of the measures proposed by EPA in 1993.

This seemingly arcane disagreement over the use of AOX will perhaps be the pivotal issue in finalizing EPA's pulp and paper effluent regulations. As would be required under EPA's regulatory proposals, the pulp and paper industry has already begun to convert many of its bleaching plants from elemental chlorine to chlorine dioxide. As discussed above, in some plants where complete substitution of ClO_2 for elemental chlorine is currently being used, the result has been TCDD and TCDF concentrations in effluents that are below the limits of detectability. Dioxin and furan have been detected in other plants employing both complete sub-stitution and oxygen delignification (AET 1994; ERG 1996). This informa-tion was unavailable at the time of the proposal because no pulp bleach-ing plants were operating using 100% chlorine dioxide prior to the proposal. (See discussion below.)

As originally proposed in 1993, the regulations would require large segments of the industry to employ OD (or extended delignification). Based on the new environmental performance data for mills employing complete substitution, the agency announced in 1996 that for the bleached paper-grade kraft mills, two BAT options were being consid-ered: complete substitution with and without additional delignification (*Federal Register* 61: 36835–36858). In terms of organochlorines, the main *measurable* difference between the alternative technologies is a substantial reduction in AOX achieved by adding delignification.[30] Because there is no discernible threshold chlorine charge for dioxin formation in the pulp bleaching process, some imperceptibly small reduction in the formation of dioxin-like compounds may be associated with a reduction in AOX. How much reduction there would be in concentrations below the detec-tion limit is speculative.

THE PROCESS WITHIN EPA

Setting the Agenda

According to an environmentalist, the inadvertent detection of high lev-els of dioxin in fish downstream of pulp and paper mills in 1983 was pri-marily responsible for getting the effluent regulations on the agency's agenda. An EPA official observes, "Even in the absence of [detecting] dioxin, there was an internal schedule within the water program that would have had them [the water office] review the effluent guidelines for pulp and paper mills. But the dioxin issue changed the pollutant of con-cern and added new impetus to that exercise."[31] Harrison and Hoberg (1991) conclude, however, that while the preliminary findings in 1983 were sufficient to get pulp mill dioxins quickly onto EPA's *research* agenda, when exposure of the findings to a wider audience occurred as a result of the 1987 *New York Times* article by Philip Shabecoff, pulp mill dioxins were elevated to the *regulatory* agenda.[32] Prominent press reports in 1989–1990 concerning dioxin in milk from bleached paper cartons and dioxin-related fish advisories below pulp and paper mills kept up the pressure (Thomp-son and Graham 1997).

Sources from inside and outside EPA indicate that the Reilly adminis-tration made the decision to combine the hazardous air pollutant and clean water rulemaking for the pulp and paper industrial "cluster" for two reasons. First, there were political and administrative advantages to dealing with the industry in one broad regulatory package. Second, the Reilly administration sought to take an integrated look at reducing environmental risks across environmental media from the sector as a whole.

Dioxin Formation Disclosed: The Five Mills Study

Initially, EPA suspected that the source of dioxins detected in the 1983 national survey reference streams was the use of dioxin-contaminated chlorophenols as "slimicides" on pulp mill machinery, rather than formation of dioxins during the process (Harrison and Hoberg 1991). In 1985, the Environmental Defense Fund and the National Wildlife Federation filed suit against EPA for denying the environmentalists' petition to regulate dioxins and furans under TSCA. Meanwhile, EPA tested wastewater treatment sludge from pulp and paper mills and found that dioxin levels were highest in the sludges of bleached kraft pulp mills. This suggested that dioxin was probably being formed as a by-product during the bleaching of wood pulp with chlorine. (*Federal Register* 58: 66092). In 1986, EPA, the American Paper Institute (API), and NCASI (National Council of the Paper Industry for Air and Stream Improvement, the industry's research arm) agreed to undertake the "Five Mills Study." The study results detected TCDD and TCDF in effluents of four of five mills, pulps of all five mills, and wastewater treatment plant sludges of all five mills.

Environmentalists learned of the agreement between EPA and the paper industry to conduct the Five Mills Study, and, in December 1986, Greenpeace initiated a Freedom of Information Act (FOIA) request seeking all available information on the pulp mill dioxin problem. In January 1987, a letter from an EPA official to API was leaked that suggested that EPA had agreed to notify the industry immediately of receipt of any FOIA requests and that, barring such requests or results indicating a potential threat to human health, the agency did not intend to release any results until publication of the final Five Mills Study report. The following August, Greenpeace released a report alleging an EPA cover-up (Harrison and Hoberg 1991). In September, based on the Five Mills Study and analyses of dioxin in paper products that were later added to the study, the *New York Times* ran a front-page story reporting that traces of dioxin had been detected in household paper products (Shabecoff 1987). Until this point, dioxin was associated in the public mind primarily with pesticides and combustion processes. The prominent disclosure that dioxin was formed by pulp bleaching and was present in common household products undoubtedly provided EPA with leverage to gain industry's cooperation to support additional research.

Integrated Assessment, the Consent Agreement, and the 104 Mills Study

According to an EPA official, when the agency's water program first realized that dioxin was being formed by pulp and paper bleaching, it immediately recognized that the problem went beyond its jurisdiction and approached the toxic substances program saying, "We have a problem to share with you." After a quick review, the agency decided that the issue spilled over into the risk management jurisdictions of other agencies. An interagency working group—including EPA, FDA, the Consumer Products Safety Commission (CPSC), and the Occupational Safety and Health Administration—was formed to identify the data needs to determine the extent of the problem. The interagency group met with API to lay out its plans for a multimedia, multipathway assessment of effluents, pulp, sludge, occupational, and consumer risks and to ask industry to bear the burden of the costs. An industry official says that EPA proposed using its TSCA (Section 4) authority to require the pulp and paper sector to provide the data, and noted that the authority supplied the agency "with an arrow in their quiver" during the negotiations. The American Forest and Paper Association (AF&PA, a newly consolidated industry trade association) "hit the roof" because of the TSCA threat, says this source. "It was a question of trust." EPA did not exercise its TSCA authority but retained it as a negotiating point. Thus began the "104 Mills Study." The assessment officially began in 1989, and an EPA official noted that the forest products industry spent $3–$4 million on the study.

According to an EPA official, to get a handle on the extent and magnitude of the problem, the interagency group insisted on representative samples from all 104 mills that used chlorine in bleaching. EPA assumed the coordinating role and concentrated on the risks from effluents, sludge, and occupational exposures. The EPA coordinator was Dwain Winters, an analyst in the Office of Toxic Substances. FDA focused on food contact papers and medical devices. CPSC looked at writing papers, diapers, and other consumer products. Overall, the group identified more than 153 separate dioxin exposure pathways to be analyzed. The interagency group developed a quality assurance/quality control protocol for conducting samples, and NCASI conducted the study.

In securing broad coverage to provide a strong analysis for the purposes of scoping and identifying the dioxin problem from bleaching mills, the interagency group traded off in-depth analysis at individual plants that would have been more useful in formulating a technological remedy. An academic observes that this investigative strategy also permitted the agency to avoid having to generalize between different mills using similar processes. Industry, on the other hand, wanted to do an intensive study of particular types of bleaching mills in order to better evaluate processes that caused dioxin formation and to identify the key steps that were responsible. Late in 1988, the industry designed and conducted an intensive study of twenty-two bleaching plants independent of the 104 Mills Study (Thompson and Graham 1997).

While the negotiations with industry were ongoing, the agency began separate negotiations with EDF and NWF regarding their suit by describing the planned study. In addition to writing the 104 Mills Study into the consent decree, EPA agreed to determine whether regulatory actions were required for the pulp, sludge, and effluent and, if so, to identify the information needed for regulatory decisionmaking. These decisions were subject to judicial review. An academic notes that by involving EDF and NWF in negotiating the plan for the 104 Mills Study, the scope of the study was broadened and the process was somewhat delayed.

According to an EPA official, as a result of the 104 Mills Study, dioxin in pulp and paper wastewater was identified as the route of major health concern, with land disposal of sludge being a secondary health concern. Agency analysts, however, flagged land disposal of sludge as the primary route of ecological concern. The agency did not view dioxin in paper products as a major risk, but it was "on the borderline of concern," so the agency referred food contact papers to FDA under TSCA (Section 9). FDA accepted the referral and pursued voluntary reductions, according to this source. To arrive at these conclusions, the agency scientists combined the results of the 104 Mills Study with information regarding the health hazards of dioxin and analyses of various exposure pathways. Some of this information was taken as given (for example, the estimated carcinogenic potency of dioxin supplied by the 1985 EPA hazard assessment), some of it represented departures from EPA's normal assessment procedures for water quality criteria (for example, assumptions made regarding fish consumption), and some of it resulted from original research (for example, a study was conducted on uptake of dioxin through the skin using cadaver tissue samples).

Hazard Assessment

EPA had established its official position on the environmental and health effects of dioxin before the integrated assessment got under way. The agency issued its ambient water quality criteria document for dioxin in 1984.[33] According to the hazard assessment issued by EPA in 1985, dioxin was to be regarded by all agency programs as a probable human carcinogen on the basis of adequate animal data and limited human data. The dose associated with an increased cancer risk of up to 10^{-6} was officially 0.006 pg/kg body weight/day. The agency's 1983 statement permitting states to use "other scientifically defensible methods" to modify EPA

ambient water quality criteria, as well as the agency's initial stab at a
dioxin reassessment in 1988, invited the states to exercise some scientific
discretion in assessing the hazards of dioxin. For the EPA regulatory program offices, however, using the results of the 1985 assessment remains
nondiscretionary. According to an EPA water official, it is clear that the
toxicological information on dioxin available through the agency's Integrated Risk Information System (IRIS) was to be used in the program's
analyses: "We use whatever ORD tells us to use."

The latest dioxin reassessment was under way by 1991, and the water
program has acknowledged in its regulatory proposals that new information might become available, but there seems little chance that the agency
will reach closure on the review of dioxin's toxicity before it finalizes the
pulp and paper effluent guidelines. According to an EPA official, however, the minor adjustment made to the dioxin risk-specific factor in the
1994 draft reassessment (from 0.006 pg/kg/day to 0.01 pg/kg/day for a cancer risk of 10^{-6}) would not have significantly affected decisions about the
pulp and paper sector. The existing reference dose (RfD) on the IRIS for
dioxin noncancer effects is at about the level where cancer risks approach
10^{-4} (1 in 10,000). Because the agency was concerned with individual risk
levels below this (10^{-5}–10^{-6}), says this source, estimated cancer effects
drove the determination that the pulp and paper effluent was the prime
human health concern. An EPA water official comments, "At this stage, on
the effects of dioxin, as a user of science, I feel somewhat more certain.
But for every question we've answered, we've raised new ones."

The scope of EPA's hazard assessment was limited by the complex
and variable chemical composition of pulp and paper effluents and by
the lack of toxicity data for those substances that were identified. Based
on an evaluation of pulp and paper effluent sampling data collected by
EPA (both independently and in cooperation with industry), the agency
identified twenty-six organic chemicals (including dioxin and furan) as
contaminants of concern. Of these twenty-six contaminants, twenty-four
are organochlorines and six are priority pollutants. Only eleven have
RfDs and six have cancer potency factors available using EPA's primary
toxicological databases, IRIS and the Health Effects Assessment Summary
Tables.[34] Due to a lack of data on human health toxicity, only thirteen of
the twenty-six contaminants could be evaluated for their potential
human health impacts (U.S. EPA 1993a).

Exposure Assessment

The dioxin exposure assessment was conducted by a consultant (Tetra
Tech) under contract to the EPA Office of Water/Office of Science and
Technology/Standards and Applied Science Division. To conduct the

assessment required estimates of several factors. The factors and the sources of the estimates are provided in Table G-5. An examination of the table reveals a grab bag of types and sources of scientific information: peer-reviewed literature, gray literature, default assumptions, ad hoc assumptions, and professional judgments produced by all levels of government, industry, and academia.

The exposure assessment for dioxin in pulp and paper effluents marked a departure from EPA's standard operating procedures for developing water quality criteria. Normally, the water office would assume fish consumption levels based on national averages. In this case, however, the agency focused on exposure scenarios it expected would occur near pulp and paper mills, including highly exposed subpopulations—sports anglers and subsistence anglers. According to an EPA official, for the fish consumption rates of highly exposed subpopulations, risk analysts can provide only some reasonable upper-bound estimate that is subject to a high degree of uncertainty due to the lack of data. In contrast, there are good data available on average fish consumption rates by the general population. This source allows that "the way we chose to calculate the risk for fish consumption became complicated. The normal procedure under water quality criteria assumed a certain level of average consumption, but we had to make assumptions that went beyond that to look at specific subpopulations like subsistence anglers. Nobody believed that the average consumption rates would be representative of real exposures, so the solution was to provide a number of consumption rates."

EPA also evaluated the exposure scenarios under two alternative models for estimating how dioxin concentrations would be diluted in receiving water bodies. The first was a simple dilution model, which assumed that all carcinogenic pollutants discharged into a receiving stream are available for uptake by fish. The other—called the DRE (dioxin reassessment evaluation) model—is a more complex dilution model under development by EPA. It assumes that the uptake of dioxins and furans depends on the levels of suspended solids and the partitioning of the pollutants between fish tissue and sediment. As a result, the partitioning model produces lower exposure estimates.

In EPA's 1996 notice that it was considering two BAT options for the major pulp and paper subsector, the agency also stated that it was considering using only the DRE model for estimating dioxin and furan concentrations in fish for the final rule. However, the agency stated that it would modify the DRE model to reflect ongoing contamination. The modification entailed replacing the biota to suspended solids accumulation factor (BSSAF) of 0.09 with a BSSAF of 0.2 (*Federal Register* 61: 36846). This might seem trivial, but the BSSAF is a critical model parameter in predicting the accumulation of dioxin in fish. The BSSAF value of 0.09 represents the

Table G-5. Sources of the Estimates Used in Dioxin Exposure Assessment

Exposure Factor	Source of Estimate
Dioxin discharges from plants	104 Mills Study
In-stream dioxin concentrations	Site-specific flow data for sixty-eight streams, dilution factors for seventeen mills discharging to open waters (for instance, oceans, estuaries, lakes), and two dilution models. Dilution factors provided by Office of Water publication and regional EPA personnel. One of the dilution models was developed by EPA/Office of Research and Development/Exposure Assessment Group and was still under EPA review at the time.
Dioxin concentrations in fish	Bioconcentration factor (BCF) for TCDD in trout from a laboratory study included in 1991 Banbury report edited by Gallo et al.
	BCF for TCDF derived from 1988 article in *Environmental Toxicology and Chemistry*. Biota to suspended solids accumulation factor (BSSAF) of 0.09 derived from EPA Lake Ontario study. BSSAF of 0.02 derived from NCASI study.
Fish consumption rates	For recreational anglers, estimates based on a 1988 New York State survey; a 1981 Tacoma, Washington, County Health Department report; and a 1989 University of Michigan technical report.
	For subsistence anglers, 145 g/day assumed and said to be "consistent" with a 1982 U.S. Department of Agriculture report.
Change in fish consumption due to state fishing advisories	EPA assumed a 20% decrease for recreational anglers and no effect on subsistence anglers based on a 1990 University of Michigan Master's thesis and a 1990 paper presented at an American Fisheries Society meeting.
Size of fish-consuming populations	Number of fishing licenses in a county multiplied by average family size (based on Census Bureau data) of 2.63 to yield fish-consuming population in each county. EPA assumed that 95% of fishing licenses were issued to recreational anglers and 5% to subsistence anglers.
Exposure duration, body weight, and so on	EPA Office of Solid Waste and Emergency Response standard default exposure factors.

agency's default for dioxin based on Lake Ontario data that are primarily from historical sources. To support its conclusion that a BSSAF of 0.2 was more appropriate for ecosystems subject to ongoing contamination, the agency cited U.S. EPA 1994c, one of the volumes of the agency's draft dioxin reassessment. Although U.S. EPA 1994c indicates that BSSAFs for aquatic systems in which contamination is ongoing might be greater than for systems in which the contamination is primarily historical, it stops short of endorsing any particular value for all water bodies where contamination is ongoing. U.S. EPA 1994c (Section 7.2.3.6) describes an analysis of the impact of pulp and paper mill effluents on fish tissue concentrations conducted by NCASI using data from the 104 Mills Study. This study found that *for the thirty-eight mills discharging into smaller water bodies*, using a BSSAF of 0.2 (up from 0.09) improved the predictive performance of the model. According to U.S. EPA 1994c (Table 4-1), BSSAFs for dioxin observed in the literature range from 0.009 to 2.94.

Therefore, EPA's assumption that a BSSAF of 0.2 is appropriate for *all* water bodies downstream of discharging mills, regardless of the size of the receiving water body and other factors, appears to have a somewhat tenuous basis. More importantly, however, this analytic anecdote illustrates the problems associated with attempting to conduct an environmental risk assessment for a national rulemaking while taking into account site-specific conditions. On the basis of at least some relevant empirical data, EPA is adjusting its default model to take into consideration differences between historical and ongoing dioxin contamination. Making a finer distinction could potentially require a large number of field studies to determine the BSSAF for every water body downstream of a bleaching pulp and paper mill. Furthermore, despite the importance of the BSSAF in predicting the dioxin accumulation in fish tissue, it is only one of many exposure factors to be evaluated.

The Regulatory Impact Assessment: The Numbers Behind the Proposal

The Water Office's regulatory impact assessment (RIA) estimated that the proposed BAT for the pulp and paper industry would result in a national annual reduction of five to thirty-five cancers in recreational and subsistence anglers (U.S. EPA 1993b). The lower value in the range is only for TCDD and TCDF and was estimated using the partitioning dilution (DRE) model. The upper value is calculated using the simpler, conservative dilution model and is still dominated (99%) by the effects of TCDD and TCDF. Using a range of $2 to $10 million for the value of a life, EPA estimated the benefits of the proposed effluent regulations to be $10–$350 million per year (U.S. EPA 1993b).[35] Using the simple dilution approach, EPA estimated that greater than 99% of the noncancer hazard (as indicated by the mills exceeding RfDs for recreational and subsistence angler populations) can be attributed to dioxin and furan (U.S. EPA 1993a).

Senior EPA sources have suggested that the industry has framed the issue narrowly by focusing on cancer effects and failing to consider non-cancer effects of dioxins. An EPA water official, however, notes that the program was unable to monetize the benefits of noncancer health effects in the RIA due to the lack of quantitative dose-response data for non-cancer effects. (This situation is generic to noncancer effects and not unique to dioxins.) According to this official, "We did a very good job with what science we had." Not surprisingly, an industry official charac-terizes the RIA as a "poor job." An environmentalist complains that the Office of Water feels compelled to conduct a detailed risk and economic assessment for a "technology-based" rule. However, EPA is required by executive order to conduct a regulatory impact analysis for all proposed regulations with an annual economic impact of more than $100 million.

Communicating the Science to Decisionmakers

Because Assistant Administrator for Water Robert Perciasepe was confirmed only days before the 1993 proposed effluent regulations were issued in com-pliance with the consent decree, he was unable to engage in the decision-making. According to an EPA water official, the key decisionmakers included Acting Assistant Administrator Martha Prothro and Tudor Davies, Director of the Water Office of Science and Technology (then acting Deputy Assistant Administrator). Reportedly, EPA Administrator Carol Browner was also briefed. The issue was of sufficient importance that it was not com-pletely delegated to an Acting Assistant Administrator (AA). "The cluster rule was high stakes, high visibility," says this source, adding "this involved the 'D-word' [dioxin]. It had everybody's attention." There were presenta-tions for Prothro from NCASI, AF&PA, and academics. "The AA attended mill [plant site] visits—that's unusual." Due to the high level of decision-maker interest, according to this source, it was "not difficult to get their attention." In addition, because dioxin is regarded as an important multime-dia pollutant, the decisionmakers "knew a lot about it; they were sensitized and could mesh things together about dioxin other than what they just heard about in the context of pulp and paper."

THE PROPOSAL, INDUSTRY'S RESPONSE, AND A NEW FRAMEWORK

In announcing the 1993 proposal, Assistant Administrator Perciasepe esti-mated that dioxin discharges from the pulp and paper sector into surface

Table G-6. 1993 Proposed BAT and Estimated Impact for Pulp and Paper Subsectors

Subsector	Principal Products	Proposed BAT	Size/Est. Cost
Bleached paper-grade kraft and soda	Paper-grade kraft market pulp, paperboard, coarse papers, tissue papers, and fine papers for business, writing, and printing	OD* and 100% ClO_2	78 mills $260 annualized 1–3 plant closures 500–4,000 jobs
Dissolving sulfite	Pulps used for rayon, cellophane, and cellulose products	OD and 100% ClO_2	5 mills $5 annualized 1 plant closure
Dissolving kraft	Pulps used for rayon, acetate, and other cellulose products	OD and 70% ClO_2	3 mills $11.9 annualized No plant closures
Paper-grade sulfite	Tissue paper, fine papers, newsprint	TCF	10 mills $25 annualized 2 plant closures

Notes: OD* is oxygen delignification or extended cooking of pulp prior to bleaching. OD is oxygen delignification prior to pulp bleaching. 100% ClO_2 is complete substitution of elemental chlorine (Cl_2) for chlorine dioxide (ClO_2) in the bleaching process. 70% ClO_2 is 70% of Cl_2 for ClO_2. TCF is totally chlorine-free bleaching using peroxide (H_2O_2) or ozone (O_3).

Costs in millions of 1991 dollars for all the mills in the subsector, as estimated by EPA. According to an industry official, EPA underestimated the capital costs of the proposed BAT by more than 40% because OD, in combination with some of the best management practices EPA has proposed (particularly spill controls), may require plants to install larger recovery furnaces (costing approximately $100 million per plant).

waters would be reduced from over 300 grams to less than 30 grams per year (*Environment Reporter*, November 5, 1993, 1227–1228). The BAT effluent regulations proposed in 1993 for four pulp and paper subsectors organized by production process are summarized in Table G-6.

In 1993, NCASI reported data on the pulp and paper industry's progress in reducing the dioxin content of effluents, pulps, and wastewater treatment sludges (NCASI 1993). In 1994, a study prepared for a supplier of chlorine dioxide technology, the Alliance for Environmental Technology, reported that substitution of chlorine dioxide for elemental chlorine reduces the formation of chemicals like dioxin that accumulate in fatty tissues to the limits of detectability (AET 1994). Among the study's authors was Robert Huggett, a marine biologist at the College of William and Mary, and soon-to-be EPA Assistant Administrator for ORD.[36]

The new data on the performance of alternative technologies made it clear to EPA sources that *human health* risks resulting from land disposal of pulp and paper sludge were moot. Agency officials felt confident that whatever the final outcome of the cluster rule, it would simultaneously control human health risks associated with sludge land disposal. EPA estimated that TCDD soil concentrations of 0.12 ppt and 12.0 ppt would produce human cancer risks of 10^{-6} and 10^{-4}, respectively (U.S. EPA 1990). The new performance data suggested that alternative technologies achieved overall decreases in the formation of dioxins. This meant that the dioxins were not simply being shifted from the wastewater to the sludge.

Some EPA official, however, remained concerned about possible *ecological* effects of sludge land disposal. EPA estimated that if sludge were to be land applied, dioxin concentrations as low as 0.03 ppt would be necessary to protect the "most sensitive" terrestrial species, identified as the American woodcock (*Scolopax minor*) (U.S. EPA 1990). The bird would be highly exposed to dioxin in soils because the contaminant accumulates in the woodcock's primary food source, worms.[37] Some EPA staff felt that the agency's decision to retract the proposed regulation of sludge land disposal did not give due consideration to ecological risks. They also felt that environmental advocates who had been engaged in the issue lost interest once human health risks appeared under control. According to an industry official, however, AF&PA insisted on an SAB review of the ecological risk assessment. SAB determined that while the agency had identified a hazard to the woodcock, it had not substantiated that population-level effects would result from land disposal of dioxin-tainted pulp and paper sludge.[38] In 1994, EPA negotiated a voluntary agreement with the pulp and paper industry (Gilman and U.S. EPA 1994) that places no land disposal restrictions on sludges containing less than 10 ppt dioxin and serves primarily to formalize best management practices.

In 1995, EPA announced the availability of sampling data characterizing the performance of bleached paper-grade kraft and paper-grade sulfite mills employing complete substitution of elemental chlorine with ClO_2 with and without oxygen delignification (*Federal Register* 60: 34938–24940). In July 1996, EPA announced that its review of new data on the performance of chlorine dioxide bleaching indicated that dioxins and furans in wastewater discharges from bleached paper-grade kraft and soda mills could be reduced by 95% and 99%, respectively (U.S. EPA 1996). The agency also announced a "new framework" for the pulp and paper effluent guidelines. The primary elements of this new regulatory framework included (1) consideration of two BAT options for two pulp and paper subsectors, bleached paper-grade kraft and paper-grade sulfite; (2) deferral of a final decision for the dissolving sulfite and dissolving kraft subsectors; and (3) a voluntary incentives program to reward mills that exceeded regulatory requirements in reducing discharges (*Federal Register* 61: 36835–36858).

For the bleached paper-grade kraft subsector, EPA announced that it was considering complete (100%) substitution of ClO_2 for elemental chlorine (Option A) and complete substitution with OD (or extended delignification) (Option B). EPA stated that both options appear to reduce dioxins and furans in wastewaters to concentrations at or below the current analytical detection limits. The incremental environmental benefits that EPA attributed to the use of OD (or extended delignification) included reduced chronic toxicity to some aquatic species. EPA stated that the reduced chronic toxicity is probably attributable to a reduction in chemical oxygen demand. (Chemical oxygen demand is a conventional water pollutant, the technological control of which is subject to a test of economic reasonableness, as discussed above.) The agency also stated that the reduced chronic toxicity may reflect an incremental reduction in the potential formation of dioxin and furan below the analytical detection limits, as well as a reduction in all chlorinated compounds loadings, as measured at the end of the pipe by AOX. (Recall that dioxin and furan are priority pollutants, the technological control of which is subject to the less demanding economically achievable standard.) The notice further states:

> Although statistically significant relationships between AOX and a broad range of specific chlorinated organic compounds have not been established, trends in concentrations changes have, however [*sic*], been observed between AOX and specific pollutants, including dioxin, furan, and chlorinated phenolic compounds. Even though dioxin and furan are no longer measurable at the end-of-pipe at many mills, the potential for formation of these pollutants continues to exist at pulp and paper mills as long as any chlorine-containing compounds (including chlorine dioxide) are used in the bleaching process. . . . EPA expects that [reductions in] AOX discharges . . . will in turn further reduce the likelihood of the formation and discharge of these chlorinated organic pollutants.

Thus, EPA appears to reject the claim that there is a threshold level of chlorine below which no dioxin formation occurs in bleaching pulp. The agency also implicitly suggests that it is considering requiring reductions in priority pollutants beyond the point of diminishing returns because such reductions are achievable. (Recall that the determination of BAT involves consideration of both technological and economic achievability.)

Table G-7 summarizes the proposed effluent limits under the 1996 Options A and B for the bleached paper-grade kraft and soda subsector.

Table G-7. Bleached Paper-Grade Kraft and Soda Plant (Daily Maximum) Limitations

Pollutant	Option B 1993 OD* and 100% ClO$_2$	Option A 1996 100% ClO$_2$	Option B 1996 OD* and 100% ClO$_2$
TCDD	ND	ND	ND
TCDF[a]	359 (ng/kkg)	24.1 (pg/l)	24.1 (pg/l)
Chlorinated phenolics[b]	ND	ND	ND
Chloroform (g/kkg)	5.06	5.33	5.33
AOX (kg/kkg)[c]	0.267	0.769	0.236

[a] The difference in units for TCDF between the 1993 proposal and the 1996 options reflects a change from a production-normalized bleach plant limitation to an effluent concentration-based limitation.

[b] ND for two pollutants (trichlorosyringol and 2,4,6-trichlorophenol) (mg/kkg).

[c] Whereas the other pollutant limits are measured at the bleach plant, that for AOX is an end-of-pipe measurement.

Notes: OD* is oxygen delignification or extended cooking of pulp prior to bleaching. ND is no detection.

Table G-8 summarizes EPA's estimates of the economic impacts under Options A and B (*Federal Register* 61: 36840–36841).

CONCLUDING OBSERVATIONS

Sources interviewed for this case study suggest that the treatment of AOX is the principal outstanding science policy issue regarding the final BAT determination for the pulp and paper effluent guidelines. In terms of a fate and transport analogy, the information provided by AOX was first transported to EPA from Sweden in the early 1980s via international scientific journals. In 1986, the use of AOX as a regulatory parameter achieved a measure of international legitimacy when the Swedish EPA adopted it. Australia, Austria, Belgium, Finland, and Germany have followed, and AOX was thoroughly assimilated by EPA into the 1993 proposal. However, the fate of the information provided by AOX varied considerably across countries. The AOX limits proposed by EPA (0.448 kg/ton or 0.162 kg/ton based on a monthly average limitation) were slightly to three times lower than Sweden's (0.5 kg/ton), currently the strictest national-level standard for softwood pulp mills (EKA 1996; Thompson and Graham 1997).[39] The final AOX standard set by EPA for continuous discharging bleached paper-grade kraft mills was a daily maximum of 0.951 kg/ton (http://www.epa.gov/OST/pulppaper, Section 430.44), which is closer to the standards of Australia, Austria, and Germany (1 kg/ton).

Table G-8. Estimated Impacts for the Bleached Paper-Grade Kraft and Soda Subsector under 1996 Options A and B[a]

Option B 1993 OD* and 100% ClO$_2$	Option A 1996 100% ClO$_2$	Option B 1996 OD* and 100% ClO$_2$
78 mills	85 mills	85 mills
$223.2 annualized	$140 annualized	$155 annualized
1–3 plant closures	1 plant closure	3 plant closures
500–4,000 jobs	500 jobs	4,100 jobs

[a] Differences in estimated impacts between the 1993 proposal and Option B in 1996 in Table G-6 reflect a revised economic impact analysis. The difference between Tables G-6 and G-8 in the annualized cost estimate for the 1993 proposal is due to expressing the costs in different base-year dollars (1991 or 1995). As indicated above, industry sources believe that EPA has substantially underestimated the incremental capital costs of OD.

Notes: OD* is oxygen delignification or extended cooking of pulp prior to bleaching. Costs in millions of 1995 dollars, as estimated by EPA.

The fate and transport of AOX in international pulp and paper effluent limits is illustrated in Figure G-1.

Although the formation of dioxin and other organochlorines from pulp bleaching seems plausible at even minimal levels of chlorine input, the effect that AOX has on EPA's comparison of dioxin control alternatives is impeded by the agency's inability to transform AOX into toxicological equivalency units. According to one EPA official, "On AOX, we have a lot more information, and some specific questions have been addressed, but I don't think we've moved too far in being more certain about its environmental significance. While the uncertainty has gone down some, the level of controversy may have heightened. As the analysis now stands, we would continue to use AOX as a regulatory parameter." This is a case, adds this source, in which perfect information might make the decision for the decisionmaker.

An environmentalist suggests that the impact of science has been very high—but negative—in this case. The Clean Water Act "is a technology-based statute, and EPA has digressed from the technology basis." This source points out that requiring OD in addition to complete substitution is superior in preventing organochlorine discharges into the environment and complains that the "health effects analysis has tied [EPA] up in knots." This source blames industry and OMB for pressuring EPA to emphasize cost-effectiveness criteria and to stray too far from the congressional intent to rely on technology-based standards to achieve environmental results in the face of scientific uncertainty. This source also points out that the Natural Resources Defense Council still has a 1993 petition pending with EPA to ban all discharges of dioxin from the pulp and paper industry under CWA Section 307(a). NRDC's position is that

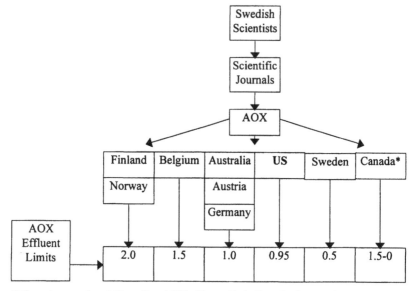

*There is no national Canadian AOX effluent limit. It varies by province.

Figure G-1. Fate and transport of AOX in international pulp and paper effluent limits (kg/ton for softwood pulp mills)

due to the technological inability to detect dioxins at concentrations believed to be toxic, the ban can be achieved only by prohibiting the use of chlorine in pulp bleaching.

An EPA official appreciates that when one evaluates a technology-based standard in terms of its adequacy to protect health and safety, it may be seen as falling short of the mark. "For some people, *any* exposure to dioxin is unacceptable." The pulp and paper industry, once widely perceived as environmentally recalcitrant, has taken some voluntary and negotiated steps to demonstrate how it could dramatically reduce dioxin discharges. These steps were taken in an attempt to get out ahead of the regulatory curve and in response to consumer demand for environmentally friendly products and potential tort liability concerns.[40] The industry may have also sensed that the ongoing dioxin reassessment would not produce substantial regulatory relief.

An April 1997 article appearing in the *Washington Post* suggested that the "Pulp Friction" saga had spilled over into the halls of Congress (Skrzycki 1997). In light of the performance of newly adopted pulp-bleaching technologies, EPA was faced with the decision of whether the substantial and *measurable* reductions in discharges of dioxins and other organochlorines were sufficient to scale back from its 1993 proposal to impose more stringent and costly pollution control technologies. To a considerable extent, the decision hinged on the marginal "unmonetizable" benefits of technologically imperceptible reductions in the formation of dioxin.

In November 1997, EPA Assistant Administrator for Water Robert Perciasepe announced that the agency would not require bleached paper-grade kraft mills to employ oxygen delignification in addition to complete substitution of chlorine dioxide for elemental chlorine. The rule retained mandatory monitoring requirements and effluent limits for AOX but also provided mills with incentives (additional time to comply with their pollutant discharge permit limitations, reduced monitoring requirements, reduced inspections, and reduced penalties) for voluntarily adopting advanced technologies that would result in AOX discharges below the discharge limits set in the final rule (http://www.epa.gov/OST/pulppaper). Although the agency noted that it was not required under the Clean Water Act to balance the costs and benefits of its technology-based regulations, in announcing the rulemaking, Assistant Administrator Perciasepe said that the additional cost of oxygen delignification was difficult to justify because it would result in only one gram of reduction of dioxin releases while costing industry $1 billion (*Risk Policy Report*, November 21, 1997, 7).

Some observers have hailed the pulp and paper cluster rule as a model of flexible, negotiated rulemaking that goes "beyond compliance" to achieve environmental benefits at reduced costs. However, not all interests have been satisfied. Feeling that EPA should prevent dioxin discharges to the maximum extent possible, the National Wildlife Federation and other environmental organizations quickly announced their intent to mount a legal challenge to the pulp and paper cluster rule (*Inside EPA*, November 21, 1997, 7–8). In the eyes of some, dioxin remains the Darth Vader of chemicals.

ENDNOTES

[1] Halogens include chlorine, bromine, iodine, and so forth.

[2] Effluents are wastewater discharges into surface waters.

[3] For example, dioxin concentrations in fish in a Wisconsin reservoir were more than 50 ppt (parts per trillion), leading the state to close a commercial fishery. Samples in Maine and Minnesota found dioxin concentrations in fish of up to 85 ppt (Harrison and Hoberg 1991). By comparison, measured background levels of dioxin in fish are 0–2 ppt (U.S. EPA 1994a).

[4] The term *furans* refers to chlorinated dibenzofurans.

[5] *NRDC et al. v. Train*, 8 ERC 2120 (D.D.C. 1976). Later modified as 12 ERC 1833 (D.D.C. 1979).

[6] See 40 CFR 401.15 and *Federal Register* 57: 60911–60915, respectively, for complete lists. EPA has since reduced the number of priority pollutants to 126. Priority

pollutants are carcinogens, suspected carcinogens, or pollutants known to be seriously toxic at low levels. The priority pollutant list originated from a 1975 EPA water toxics regulatory strategy developed in response to the NRDC lawsuit (U.S. CRS 1993).

[7] Although furan (TCDF—2,3,7,8-tetrachlorodibenzofuran) is not explicitly listed as a priority pollutant, EPA treats it as a dioxin-like compound. The International Joint Commission has identified TCDD and TCDF as two of eleven critical pollutants for the Great Lakes (AET 1995).

[8] Toxic effluent standards are pollutant specific, nationally uniform, and applicable across all categories of industry and all dischargers. By 1976, EPA had promulgated such standards for aldrin/dieldrin, DDT, endrin, toxaphene, benzidine, and polychlorinated biphenyls, but stringent procedural and scientific requirements have prevented more extensive development of toxic effluent standards under Section 307(a)(2) (Fogarty 1991).

[9] Although nonpoint sources, such as runoff from most farms and roadways, and atmospheric deposition may contribute to exceedances of ambient water quality criteria, enforceable limits can be placed only on point sources. Nonpoint and mobile sources may contribute to background levels of dioxins and other organochlorines that end up in surface waters, sediments, and aquatic organisms.

[10] 1 ppq is 1×10^{-15}.

[11] In 1996, EPA began drafting a strategy to administratively reform the TMDL program and convened a Federal Advisory Committee Act group to develop recommendations. (See *Inside EPA*, November 22, 1996, 4–6 for a summary of the agency's draft strategy.)

[12] See Hirshfield et al. (1996) for a thoughtful review of *Our Stolen Future* and comparison to Rachel Carson's 1962 *Silent Spring*, which publicized the environmental effects of pesticides and is associated with the birth of environmentalism as a mass movement.

[13] A 1969 National Cancer Institute study found a link between TCDD and birth defects. According to Smith (1992), other studies by U.S. scientists critical of the Vietnam War also reported teratogenic effects of TCDD. Restrictions on domestic uses of 2,4,5,-T were first announced in 1970 by the Secretary of Agriculture. In promulgating a partial ban on the herbicide in 1971, EPA Administrator William Ruckelshaus rejected the advice of an ad hoc scientific panel chaired by Emil Mrak, Chancellor of the University of California, Davis, and accepted the counsel of a group of Food and Drug Administration (FDA) scientists who had conducted earlier animal tests on 2,4,5,-T. Critics of the Mrak panel had received leaked copies of the report prior to its release. Both advisory groups were informally convened prior to the advent of the 1972 Federal Advisory Committee Act (FACA), and the episode crystallized support for FACA (Smith 1992, 24–25).

[14] See Jasanoff 1990 (24–26) for a more balanced discussion of the 2,4,5-T controversy.

[15] Two sources interviewed for the overall study of science in environmental regulation volunteered the Times Beach buyout as an EPA decision in which science played little or no role.

[16] Consensus broke down, however, on just what such a biologically based model would predict in terms of dioxin's cancer risks (Roberts 1991). See the arsenic in drinking water case study (Appendix B) for a discussion of Gallo's role in promoting EPA's use of biologically based risk assessment models. According to a former senior EPA official, industry—notably the Chlorine Institute—played a role in initiating the Banbury Conference.

[17] Pathology includes laboratory analysis of animal tissue slides to characterize and enumerate abnormalities, such as tumors. It is traditionally descriptive and can be fairly imprecise, but standardized protocols and quantitative and chemical techniques have been developed to promote consistency and precision.

[18] The agencies' estimation procedures differed in how to extrapolate from rat to man (body weight or surface area); which pathology results were used (Kociba and colleagues', Squires's, or both); whether early mortality was taken into account; the assumed average human body weight (80 kg or 70 kg); and how the dose was measured (concentration in the tissue or administered dose) (Thompson and Graham 1997). Using surface area to scale the administered dose between animals and humans leads to a higher potency estimate than does using body weight as a scaling factor. Currently, EPA uses a scaling factor of body weight raised to the two-third power. According to an academic, there is a proposal for all federal agencies to adopt a scaling factor of body weight raised to the three-fourth power, but FDA continues to scale on the basis of body weight.

[19] In 1995, the EPA Science Advisory Board reported that it was also possible that dioxin may produce toxic responses that are not mediated through the Ah receptor (Thompson and Graham 1997).

[20] Similar statements have been made regarding departure from the linear model for ingested arsenic. See Appendix B.

[21] This means that too much of anything—even something essential to life in normal doses—can be harmful.

[22] Background levels would include accumulations of both natural and anthropogenic sources.

[23] According to a former Science Advisory Board member, the Environmental Defense Fund encouraged EPA to develop the TEF scheme. When the board reviewed the scheme in the late 1980s, says this source, "The SAB said, 'We'll accept that as an interim procedure, but more research is needed to substitute for TEFs.' Now, the TEFs are getting locked in, and the research wasn't done. People get used to using the old numbers, and they take on a life of their own. There's a 'check the box' mentality, a resistance to revisiting old decisions. Risk assessment needs to be an iterative process."

[24] These figures are for all dioxin-like compounds weighted by toxic equivalency factors, but they are dominated by TCDD (about 90% of the total).

[25] Because oxygen is relatively unselective for lignin, OD results in more dissolved organic material. Extended cooking has a similar effect. Consequently, pulp and paper mills using these delignification technologies require more recovery boiler capacity than those mills that do not.

[26] Berry et al. (1991) surmise that it is highly improbable that the high–molecular weight chlorinated lignin material would be broken down and transformed in the environment into problematic, polychlorinated compounds because the potentially troublesome aromatic (6-carbon ring) structure of the residual lignin would be largely destroyed by oxidation in the bleach plant. Berry et al. (1991) add, however, that further investigation of the environmental fate of this fraction of the organochlorines is needed to confirm that neither it nor its decomposition products are harmful.

[27] Berry et al. (1991) noted that thorough mixing and good process controls would be essential to ensure that no portions of the pulp are exposed to higher than the minimum chlorine charge.

[28] The 1993 water quality assessment of the proposed pulp and paper effluent guidelines cites the AWQC for TCDD for consumption of water and fish as 1.3×10^{-9} µg/l (equivalent to 0.0013 ppq (U.S. EPA 1993a, Table 3-1). In its 1984 water quality criteria report for dioxin, EPA recommended ambient levels of dioxin in the 10^{-5}–10^{-7} cancer risk range, with 1.3×10^{-9} µg/l corresponding to EPA's estimated risk level of 1×10^{-7} (U.S. EPA 1984, xi). Thus, the agency's 1993 assessment based its estimates of AWQC exceedances for the proposed pulp and paper effluent guidelines on nonbinding federal ambient criteria at the lowest end of the recommended range. As noted above, however, EPA has approved binding, numeric state ambient water quality criteria for dioxin that are even less stringent than the range recommended in 1984. EPA also cites an AWQC for TCDF for consumption of water and fish as 8.10×10^{-8} µg/l (0.081 ppq) (U.S. 1993a, Table 3-1). According to an EPA water program official, however, the agency has no official ambient water quality criterion for TCDF. It appears that the AWQC for TCDF has been inferred from the AWQC for TCDD on the basis of a TEF (that is, 0.1) and a different estimated bioconcentration factor (BCF). (See U.S. EPA 1993a, 20. The BCF is used to estimate the concentration of a substance in fish tissue based on its concentration in water.) EPA also estimates that the proposed BAT would result in no remaining exceedances of the AWQCs for pollutants other than TCDD or TCDF if human consumption is assumed to be limited to fish and not to include drinking water (U.S. EPA 1993a, Attachment A-12).

[29] Secondary biological treatment can have a strongly reduced polychlorinated phenol discharge levels. However, the efficiency of chlorophenol removal by secondary treatment varies greatly. Secondary treatment has a weak effect on dioxin and furan levels (Berry et al. 1991).

[30] EPA has estimated that bleached paper-grade kraft mills using complete substitution and OD can achieve undetectable levels of TCDD and AOX concentrations of approximately 0.25 kg/kkg. In its 1996 notice, EPA proposed to set the daily maximum limitation for bleached paper-grade kraft mills using complete substitution at 0.769 kg/kkg (*Federal Register* 61: 36842). According to an industry official, however, using complete substitution alone yields AOX concentrations of approximately 0.5 kg/kkg. Therefore, the incremental reduction in AOX appears to be on the order of 50%–70%. The reductions in AOX are also associated with reductions in conventional pollutants, such as chemical and biological oxygen

demand. NRDC (1996) reports that EPA has estimated that requiring OD in addition to complete substitution would reduce the cumulative loading of chlorinated phenolic compounds by 2,000 kg per year. In EPA's 1996 notice, however, the agency indicated that both options were expected to achieve undetectable daily maximum bleach plant limits for specified chlorinated phenolics (*Federal Register* 61: 36841).

[31] CWA Section 307(a)(3) requires effluent standards to be reviewed every three years.

[32] Shabecoff and his successor at the *New York Times*, Keith Schneider, have been prominent figures in the recent debate over the appropriate role of environmental journalists. Many viewed Shabecoff and other environmental journalists as being too sympathetic to environmental groups, and Schneider, who wrote a highly publicized series of articles on EPA's overestimation of the cancer risks of dioxin, went on to become the leading voice of a revisionist camp of environmental journalists.

[33] The 1984 criteria document was prepared jointly by the EPA Offices of Water and Research and Development (ORD). The process was managed by the ORD Environmental Criteria and Assessment Office located in Cincinnati, Ohio. The health effects chapter acknowledged forty-four contributors from EPA and other federal agencies, international institutions, academia, industry, and environmental groups.

[34] The priority pollutants are chloroform, methylene chloride, pentachlorophenol, TCDD, TCDF, and 2,4,6-trichlorophenol.

[35] According to Viscusi (1996), on average, workers receive $3–$7 million in compensation per statistical life lost.

[36] Huggett recused himself from involvement in the pulp and paper dioxin rule.

[37] The woodcock is a favored game bird. Adult woodcocks are approximately one foot long and weigh half a pound and can eat their weight in worms daily. In response to the drop in woodcock numbers over the last twenty years in some parts of its range (from Georgia to the Canadian Maritime Provinces and west to Minnesota), the U.S. Fish and Wildlife Service announced the North American Woodcock Management Plan in 1990. Because the woodcock's breeding habitat is successional scrub forest, the principal cause of its decline appears to be the maturation of hardwood forests growing on abandoned farmlands in the Northeast (http://alloutdoors.com; //www.im.nbs.gov).

[38] It should be noted that estimating population-level effects is extremely uncertain and can require lengthy and costly site-specific studies.

[39] Softwoods are evergreens. Hardwoods are deciduous. In the United States, most pulp is produced from softwood. Sweden has set an AOX limit of 0.3 kg/ton for hardwood pulp mills to be achieved in 2000–2005 (EKA 1996).

[40] See Thompson and Graham (1997) for further discussion of the possible role of tort liability risk in the pulp and paper industry's decision to adopt advanced bleaching technologies.

REFERENCES

AET (Alliance for Environmental Technology). 1994. *A Review and Assessment of the Ecological Risks Associated with the Use of Chlorine Dioxide for the Bleaching of Pulp.* Washington, D.C.: AET.

———. 1995. Five Great Reasons Why We Care: The Pulp and Paper Industry's Virtual Elimination Strategy. (http://aet.org/science/5reasons.html), September.

Berry, R., C. Luthe, R. Voss, et al. 1991. The Effects of Recent Changes in Bleached Softwood Kraft Mill Technology on Organochlorine Emissions: An International Perspective. *Pulp and Paper Canada* 92(6): 43–55.

Colborn, T., D. Dumanoski, and J. Peterson Myers. 1996. *Our Stolen Future: Are We Threatening Our Fertility, Intelligence, and Survival?* New York: Dutton.

Copeland, C. 1993. *Toxic Pollutants and the Clean Water Act: Current Issues.* Congressional Research Service Report for Congress. 93-849 ENR. Washington, D.C.: Congressional Research Service.

EKA (EKA Chemicals, Inc.). 1996. *EKA Chemicals Comments on Cluster Rule Proposal.* Marietta, Ga.: EKA Chemicals, Inc., August 12.

ERG (Eastern Research Group, Inc.). 1996. *Pulp and Paper Mill Data Available for BAT Limitations Development.* Prepared under contract for U.S. EPA Office of Water. July 10. Lexington, Mass.: ERG.

Executive Enterprises. 1984. *Clean Water Act Permit Guidance Manual.* Washington, D.C.: Executive Enterprises Publications Co., Inc.

Fingerhut, M., W. Halperin, B. Marlow, et al. 1991. Cancer Mortality in Workers Exposed to 2,3,7,8-Tetrachlorodibenzo-*p*-Dioxin. *New England Journal of Medicine* 342: 212–218.

Finkel, A. 1988. Dioxin: Are We Safer Now than Before? *Risk Analysis*, 8(2): 161–165.

Fogarty, J. 1991. A Short History of Federal Water Pollution Control Law. *Clean Water Deskbook*, 2nd ed. Washington, D.C.: Environmental Law Institute, 5–20.

Gilman and U.S. EPA. 1994. Memorandum of Understanding between Gilman Paper Company and the U.S. Environmental Protection Agency Regarding the Land Application of Pulp and Paper Mill Materials. Washington, D.C.: U.S. EPA.

Hanmer, R. 1988. Memorandum from Rebecca W. Hanmer, Acting Assistant Administrator for Water, to Water Management Division Directors, regarding Interim Strategy for the Regulation of Pulp and Paper Mill Dioxin Discharges to the Waters of the United States, August 9.

Harrison, K., and G. Hoberg. 1991. Setting the Environmental Agenda in Canada and the United States: The Cases of Dioxin and Radon. *Canadian Journal of Political Science* 24: 3–27.

Hirshfield, A., M. Hirshfield, and J. Flaws. 1996. Problems Beyond Pesticides. *Science* 272: 1444–1445.

Jasanoff, S. 1990. *The Fifth Branch: Science Advisors as Policymakers*. Cambridge, Mass.: Harvard University Press.

Kociba, R., D. Keyes, J. Beyer, et al. 1978. Results of a Two-Year Chronic Toxicity and Oncogenicity Study of 2,3,7,8-Tetrachlorodibenzo-*p*-Dioxin in Rats. *Toxicology and Applied Pharmacology* 46: 279–303.

Lucier, G., G. Clark, C. Hiremath, et al. 1993. Carcinogenicity of TCDD in Animals. *Toxicology and Industrial Health* 9: 631–668.

Moore, J., R. Kimbrough, and M. Gough. 1993. The Dioxin TCDD: A Selective Study of Science and Policy Interaction. In *Keeping Pace with Science and Engineering*, edited by M. Uman. Washington, D.C.: National Academy of Engineering, 221–242.

NCASI (National Council of the Paper Industry for Air and Stream Improvement). 1993. *Summary of Data Reflective of Pulp and Paper Industry Progress in Reducing the TCDD/TCDF Content of Effluents, Pulps, and Wastewater Treatment Sludges*. New York: NCASI, June.

NRDC (Natural Resources Defense Council). 1996. *Option B Has Significant Environmental Benefits*. Washington, D.C.: NRDC.

NTP (National Toxicology Program). 1982. *Bioassay of 2,3,7,8-Tetrachlorodibenzo-p-Dioxin for Possible Carcinogenicity (Gavage Study)*. Technical Report Series No. 102. Research Triangle Park, N.C.: NTP.

Roberts, L. 1991. Dioxin Risks Revisited. *Science* 251: 624–626.

Shabecoff, P. 1987. Traces of Dioxin Found in Range of Paper Products. *New York Times*, September 24, A1.

Skrzycki, C. 1997. Pulp Friction: The EPA's Tussle Over Paper Pollutant Rules. *Washington Post*, April 11, G1–G2.

Smith, B. 1992. *The Advisers: Scientists in the Policy Process*. Washington, D.C.: Brookings Institution.

Thompson, K. M., and J. D. Graham. 1997. Producing Paper without Dioxin Pollution. *The Greening of Industry: A Risk Management Approach*. Cambridge, Mass.: Harvard Center for Risk Analysis, 203–268.

U.S. CRS (U.S. Congressional Research Service). 1993. Toxic Pollutants and the Clean Water Act: Current Issues. CRS Report 93-849 ENR. Washington, D.C.: Congressional Research Service.

U.S. EPA (U.S. Environmental Protection Agency). 1989. *Interim Procedures for Estimating Risks Associated with Exposures to Mixtures of Chlorinated Dibenzo-p-Dioxins and Dibenzofurans (CDDs and CDFs) and 1989 Update*. Washington, D.C.: U.S. EPA Risk Assessment Forum.

———. 1990. *Assessment of Risks from Exposure of Humans, Terrestrial and Avian Wildlife, and Aquatic Life to Dioxins and Furans from Disposal and Use of Sludge from Bleached Kraft and Sulfite Pulp and Paper Mills.* Washington, D.C.: U.S. EPA Office of Pesticides and Toxic Substances, July.

————. 1993a. *Water Quality Assessment of Proposed Effluent Guidelines for the Pulp, Paper, and Paperboard Industry.* EPA-821-R-93-022. Washington, D.C.: U.S. EPA Office of Water.

————. 1993b. *Regulatory Impact Assessment of Proposed Effluent Guidelines and NESHAP for the Pulp, Paper, and Paperboard Industry.* Washington, D.C.: U.S. EPA Office of Water.

————. 1994a. *Estimating Exposure to Dioxin-Like Compounds. Volume I: Executive Summary.* External Review Draft. Washington, D.C.: U.S. EPA.

————. 1994b. *Health Assessment Document for 2,3,7,8-Tetrachlorodibenzo-p-Dioxin (TCDD) and Related Compounds.* Volume III of III. External Review Draft. Washington, D.C.: U.S. EPA.

————. 1994c. *Estimating Exposure to Dioxin-Like Compounds. Volume III: Site-Specific Assessment Procedures.* External Review Draft. Washington, D.C.: U.S. EPA.

————. 1996. EPA Releases New Data on Pulp and Paper Discharges; Considers Regulatory and Voluntary Options. Press Release. Washington, D.C.: U.S. EPA, July 3.

U.S. EPA-V (U.S. EPA Region V). 1988. *U.S. EPA Bench Scale Wastewater Treatability Study, Proposed Interim Control Measures, Interim NPDES Permit Strategy.* Chicago, Ill.: U.S. EPA Region V.

U.S. EPA/SAB (U.S. EPA Science Advisory Board). 1995. *A Second Look at Dioxin.* Washington, D.C.: U.S. EPA.

U.S. OTA (U.S. Office of Technology Assessment). 1989. *Technologies for Reducing Dioxin in the Manufacture of Bleached Wood Pulp.* OTA-BP-0-54. Washington, D.C.: U.S. Government Printing Office.

————. 1991. *Dioxin Treatment Technologies.* OTA-BP-0-93. Washington, D.C.: U.S. Government Printing Office.

Viscusi, K. 1996. Alternative Institutional Responses to Asbestos. *Journal of Risk and Uncertainty* 12: 147–170.

Whelan, E. M. 1985. *Toxic Terror.* Ottawa, Ill.: Jameson Books.

Appendix H

Lead in Soil at Superfund Mining Sites

BACKGROUND

This case study deals with an environmental hazard, lead in soil, for which EPA has no universally applicable standard that regional and state officials can take from a "lookup table" and "plug in" to determine the "preliminary remediation goal" (PRG) for contaminated site remediation. Establishing the PRG is the first step that Superfund remediation project managers take in the remedy selection process to begin to address the question of "How clean is clean?" Although technical feasibility, cost, the permanence of the treatment, applicable or relevant and appropriate requirements (ARARs), and other factors are considered by remediation project managers in negotiating the final remedy selection with the potentially responsible parties (PRPs), community, and other agencies, the PRG can serve as an important benchmark in the negotiations. Because there is no discrete number indicating "safe" levels of lead in soil, this case study does not describe the typical process by which EPA uses scientific information in site-specific decisionmaking. Instead, it provides some insight into the challenges and opportunities facing contaminated site programs if they were to more broadly and consistently develop and use site-specific scientific information.

The Comprehensive Environmental Response, Compensation and Liability Act (CERCLA) of 1980 created the EPA Superfund program. CERCLA had two primary objectives: to identify and clean up sites contaminated with hazardous substances throughout the United States and to assign the costs of cleanup directly to those parties—called responsible parties—who had something to do with the sites (Probst et al. 1995). Although Superfund is not limited solely to closed or abandoned sites, the program has focused primarily on these types of sites. The number of old mineral extraction, milling, and processing sites within the United

States is unknown, but large, with estimates ranging between 100,000 and 400,000 sites. As of 1990, the Superfund National Priorities List (NPL) contained about 1,200 sites, of which sixty were mining sites.[1] An additional 220 mining sites are listed by the Comprehensive Environmental Response Compensation and Liability Information System (CERCLIS) (Tilton 1994).[2] Lead in soil is a contaminant of concern at over 400 Superfund sites (Ryan and Zhang 1996).

The Resource Conservation and Recovery Act (RCRA) of 1976 deals broadly with hazardous waste produced at operating industrial sites. Mineral extraction and ore milling generate 2 billion tons of solid waste a year, representing nearly 40% of the country's total solid waste, and dwarfing the 270 million tons of hazardous waste generated outside the mineral sector. Responding to concerns of the mining industry, however, Congress passed the "Bevill amendment" in 1980 (RCRA Section 3001[b][3][A][i-iii]) to exempt "high-volume, low hazard" solid wastes from RCRA Subtitle C hazardous waste federal regulation. Currently, wastes generated from extraction and milling are classified as nonhazardous wastes and subject to state regulation under RCRA Subtitle D, while many (but not all) mineral-processing wastes fall under Subtitle C (Tilton 1994). While much of the ongoing discussion concerning lead in soil regulation occurs in the context of the Superfund program, the potential liability for future RCRA "corrective action" motivates the owners of active industrial operations to engage in the debate.

Many old mining sites are relatively small and isolated, but the contamination at some mining and processing sites (including areas in the "fallout zone" downwind of smelters) is extensive. The sheer magnitude of many mining sites, relative to most contaminated sites, presents EPA with an extraordinary risk management challenge. EPA has identified approximately fifty "large-area lead sites." The Jasper County, Missouri, Superfund site, for example, includes ten different areas covering a total of 7,000 acres.[3] At a Superfund site located in Midvale, Utah, the old tailings pond alone covers 260 acres and contains 14 million tons of material in uncovered piles.[4] In some cases, these large sites overlap residential areas. At the California Gulch Superfund site in Leadville, Colorado, mountainous tailing piles lie in close proximity to residences in the small, remote mining town. According to Murray (1995), current or historical mining areas of the intermountain west often encompass tens or hundreds of square miles.

Although there is a variety of risk management measures that can be taken to limit exposure to lead-contaminated soil for extended periods (for example, capping small sites with asphalt[5]), the cleanup standards provisions of the 1986 Superfund Amendments and Reauthorization Act (SARA, Section 121) explicitly state a congressional preference for permanent treatment. Currently, the only proven, permanent remedial alternative for lead-contaminated soil is excavation and disposal.[6] The cost and extent of the cleanup at large-area lead sites depend on the level or concentration of lead in soil that is considered hazardous (that is, the lower the level of soil lead that EPA considers a health concern, the greater the extent and cost of the cleanup). Given the enormous volume of material involved, excavation and disposal could cost liable parties hundreds of millions of dollars for some large-area lead sites.[7]

Opposition to EPA's proposed remedies for these sites has not come exclusively from the PRPs, however. In some cases, local governments and citizens have objected. The city of Midvale, for example, objected to EPA's proposed solution because it would have precluded future commercial use of the site. Leadville residents voiced concerns about property values and jobs.[8] At the Smuggler Mountain site near Aspen, Colorado,[9] local residents challenged the need for EPA's proposed remedy by questioning the basic scientific model that EPA uses for assessing the health risks from soil lead (Tilton 1994). (The agency's Integrated Exposure Uptake Biokenetic [IEUBK] model for lead in children is discussed in greater detail below.) The Superfund Coalition Against Mismanagement, an umbrella group of citizens representing forty-five communities around the country, filed suit in 1994 against EPA for issuing de facto soil-lead cleanup regulations for CERCLA and RCRA sites as Office of Solid Waste and Emergency Response (OSWER) guidance without seeking public input (*Environment Reporter*, December 16, 1994, 1608).[10]

A principal basis for the OSWER soil-lead cleanup guidance is the IEUBK model. The agency's default assumption imbedded in its IEUBK model about how much lead in soil ingested by children would be absorbed into their bloodstream is commonly regarded as driving the cost of the proposed cleanups and has been hotly contested in these cases. In general, the higher the rate of lead absorption, the lower the soil-lead preliminary remediation goal indicated by the IEUBK model. As a result of the standoff over the assumed rate of lead absorption, any cleanup decision regarding the large-area lead sites throughout the country was deferred for several years (Tilton 1994) while EPA Region 8 (centered in Denver, Colorado) directed a study on the lead absorbed by juvenile swine after digesting soil samples from eight different sites. Based on previous laboratory studies using rodents, the presumption by many was that EPA's model generally overstated the amount of lead that would be absorbed from soil on mining sites, and that it especially overestimated it for mining sites without significant historical milling or smelting activity.

The juvenile swine experiment results, however, indicated considerable variability among the sites tested, with some higher, some lower, and some about the same as the agency's default assumption, according to an EPA scientist. For at least one site, lead absorption from soils downwind of the smelter was unexpectedly *lower* than that from soils contaminated with milling wastes. Consequently, it appears that EPA cannot generalize across sites where similar mining activities occurred or even draw any general distinctions between different types of mining sites.

According to an EPA official, the total cost (excluding EPA staff time) of the lead absorption studies directed by EPA on eight large-area lead sites was $1.4 million, a small sum compared to the potential liability associated with the sites. However, soils from more than forty sites currently identified as large-area lead sites remain untested, and many active or old mining sites could become subject to future regulation. Furthermore, once the site-specific lead absorption rate is no longer contested, a new variable in the complex risk equation (for example, local children's blood-lead levels or activity patterns) can replace it as the disputed factor that drives remediation costs.[11]

Many Superfund critics have characterized the program's approach to risk assessment as following a "cookbook" and have criticized the program for ignoring site-specific risk assessment information. Under Superfund, EPA, state agencies, or environmental consultants often conduct site-specific exposure assessments of varying scope, depth, and complexity, but the program typically adopts the substance-specific cancer risk values (slope factors), maximum contaminant levels (MCLs), and reference doses (RfDs) supplied by EPA's Integrated Risk Information System in a wholesale fashion. Superfund generally uses slope factors to assess whether a site's cancer risks lie in the 10^{-4}–10^{-6} "action" range. In accordance with the ARARs provision of the 1986 SARA, MCLs often determine groundwater remedial objectives. RfDs are used to evaluate potential noncancer health effects from soil ingestion.

However, EPA's current goal for reducing lead health risks is to ensure that children's blood-lead (PbB) levels do not exceed 10 µg/dl (micrograms per deciliter). (Over the past decade, the PbB level commonly associated with impairment has decreased from 25 µg/dl to 10 µg/dl.) There is no numerical MCL for lead in drinking water, nor is there a single numerical RfD for ingested lead.[12] Consequently, the Superfund program has not had the *option* of following its standard operating procedures for evaluating risks and determining preliminary remediation goals for lead-contaminated sites.[13] This Superfund case study, therefore, illuminates the challenges and opportunities posed by developing and using rigorous site-specific scientific information. Table H-1 provides a summary background on large-area lead sites.

Table H-1. Background on Large-Area Lead Sites

1976	Resource Conservation and Recovery Act (RCRA) enacted.
1980	Comprehensive Environmental Response, Compensation and Liability Act (CERCLA) enacted.
	Congress passes Bevill amendment to RCRA.
1985	EPA concludes that some wastes associated with mineral processing meet the hazardous criteria for regulation under RCRA Subtitle C, but that the high volumes of waste from extraction and milling do not.
	U.S. Centers for Disease Control sets screening level for children's blood lead (PbB) at 25 µg/dl; suggests that lead in soil or dust appears to be responsible for PbB levels in children increasing above background levels when the concentration in the soil or dust exceeds 500–1,000 ppm.
1986	Superfund Amendments and Reauthorization Act requires the Department of Health and Human Services' Agency for Toxic Substances and Disease Registry (ATSDR) to prepare a study of lead poisoning in children. After EPA fails to determine which mineral-processing wastes would come under the jurisdiction of RCRA Subtitle C, the Environmental Defense Fund and the Hazardous Waste Treatment Council sue the agency.
1988	ATSDR suggests a potential risk of developmental toxicity from lead exposure at PbB levels of 10–15 µg/dl or lower; identifies paint and contaminated soil as the principal sources of lead for children most at risk (ATSDR 1988).
	D.C. Circuit Court of Appeals orders EPA to use the high-volume, low-hazard criteria to narrow the scope of the RCRA Bevill amendment exemption for mining wastes.
1989	EPA Office of Solid Waste and Emergency Response (OSWER) guidance recommends a soil-lead cleanup level of 500–1,000 ppm at residential Superfund sites.
	EPA issues final rules during 1989 and 1990 making most mineral-processing wastes subject to RCRA Subtitle C; however, slag from lead and zinc processing is among twenty mineral-processing wastes that remain exempt from federal regulation. Extraction and milling wastes remain classified as nonhazardous wastes and thus fall under RCRA Subtitle D.
1990	U.S. Department of Justice files a Superfund suit against Sharon Steel, UV Industries Liquidating Trust, and Atlantic Richfield Company (ARCO) regarding a closed lead-smelting facility in Midvale, Utah. ARCO claims that soil lead on the Midvale site poses no hazard. EPA Region 8 directs study rejecting ARCO's claim. The companies eventually agree to pay the government $63 million.
1990	EPA Clean Air Science Advisory Committee recommends a maximum safe PbB level for children of 10 µg/dl.
	OSWER issues RCRA program guidance on soil-lead cleanup describing three alternative methods for setting cleanup levels: (1) use preliminary results of the Integrated Exposure Uptake Biokenetic (IEUBK) model, (2) use 500–1,000 ppm range, or (3) use "background" levels at the facility.

Continued on next page

Table H-1. Background on Large-Area Lead Sites *(Continued)*

1991	EPA formalizes its multimedia strategy for reducing lead exposures, adopting the 10-µg/dl PbB level goal.
	EPA Science Advisory Board weakly endorses IEUBK model for developing soil-lead cleanup levels at CERCLA and RCRA sites.
1992	Residential Lead-Based Paint Hazard Reduction Act requires EPA to develop standards defining hazardous levels of lead in lead-based paint, household dust, and soil.
	OSWER circulates draft guidance setting 500-ppm lead in soil as the preliminary remediation goal (PRG) for Superfund remediations and media cleanup standards (MCSs) for RCRA corrective actions.
1994	Urban Soil Lead Abatement Demonstration Project (the five-year, $15 million "Three Cities Study") finds a link between soil-lead levels and blood-lead levels in children. However, the study suggests that soil abatement alone (without lead-based paint stabilization and/or household dust abatement) will have little or no effect on reducing exposure to lead unless there is a substantial amount of lead in soil (as in Boston test sites, where soil-lead levels averaged 2,400 ppm) and unless soil lead is the primary source of lead in house dust (*Science*, October 15, 1993, 323; U.S. EPA 1995a).
	OSWER issues "revised interim guidance" under CERCLA and RCRA for residential soil lead setting a "screening level" of 400 ppm. The screening level is based on the IEUBK model using national average inputs and on a goal of limiting exposure to soil-lead levels such that a child would have an estimated risk of no more than 5% of exceeding the 10-µg/dl PbB level. The screening level may be selected as the soil cleanup goal (PRG/MCS), or a site-specific assessment using the IEUBK model can be used to develop soil cleanup goals. The guidance recommends remedial action when the goal is exceeded but adds that soil excavation may not be necessary. Research to determine the bioavailability of soil lead is encouraged for mining sites without significant past milling/smelting activity.
	Citizens' group sues EPA for issuing de facto regulation as guidance without public comment.
1995	House Commerce Committee Chair Thomas Bliley (R-Va.) expresses concern about EPA's soil-lead cleanup guidance and objects to the agency's reliance on the IEUBK model to justify the lead policy (*Environmental Executive Report*, October 30, 1995).
	EPA proposes to tighten controls of many mineral-processing wastes previously exempt from RCRA Subtitle C.
1996	Results of juvenile swine studies directed by EPA Region 8 indicate considerable variability in soil-lead bioavailability among large-area lead sites.

SCIENTIFIC ISSUES

The National Research Council (NRC 1993) reports that the health effects of lead at approximately 10 µg/dl in blood include:

- Impaired cognitive function and behavior in young children;
- Increases in blood pressure in adults, including pregnant women;
- Impaired fetal development; and
- Impaired calcium function and homeostasis[14] in sensitive populations.

NRC (1993) also concludes that somewhat higher PbB concentrations are associated with impaired biosynthesis of heme (a substance required for blood formation, oxygen transport, and energy metabolism) and cautions that some cognitive and behavioral effects may be irreversible.

The primary commercial use for lead is making batteries.[15] In 1990, the United States produced (excluding recycling) approximately 500,000 tons of lead (Young 1992). Environmental releases of lead from the minerals industry results not only from mining lead and lead-zinc ores, but more generally from mining and smelting of nonferrous (noniron) metals. Copper mining, for example, can release lead into the environment. Like all heavy metals, lead accumulates in the environment because it does not degrade. In soils, lead tends to accumulate in surface organic matter (Kabata-Pendias and Pendias 1984). In contrast to EPA's soil-lead screening level of 400 ppm, the lead content of agricultural soils ranges from 1 to 135 ppm with a typical value of 10 ppm (Ryan and Zhang 1996). As a result of industrial and automotive emissions and exterior lead paint, innercity neighborhoods in many of our major cities have elevated accumulations of lead in soil.[16] Because lead is reported to be least mobile in soils among the heavy metals (Kabata-Pendias and Pendias 1984), the leaching of lead into groundwater is generally of less concern than is direct ingestion of soil lead, particularly since children frequently engage in hand-to-mouth activity.[17] According to an EPA scientist, lead and arsenic are generally the principal contaminants of concern to human health at mining sites.

During the mid-1980s, EPA began developing a risk assessment model—the IEUBK model—that strives to take into account simultaneously the various pathways of lead exposure—inhaled automobile and industrial emissions and ingested soil, dust, food, and drinking water.[18] The model was developed by the Office of Air and Radiation Office of Air Quality Planning and Standards (OAQPS), first in the context of reviewing the national ambient air quality standard for lead. The model was later used for the lead/copper drinking water rulemaking (see Appendix A) and is currently managed by the EPA Office of Research and Development's National Center for Environmental Assessment. The model predicts a distribution of children's blood-lead levels as a function of existing blood-lead levels, inputs from the various sources, and different lead "bioavailability" rates. Bioavailability refers to the rate and extent of absorption of a substance. For ingested substances, it is measured by the fraction of the orally administered dose that is absorbed by the body.[19] The default assumption for the bioavailability of ingested lead in soil and dust under the IEUBK model is 30%.[20] According to Murray (1995), site-specific risk estimates for contaminated mining areas are sensitive to variability in assumed bioavailability parameter values. Other things being equal, the higher the rate of lead absorption, the lower the level of soil lead that presents a health concern (and, potentially, the greater the extent and cost of cleanup).

Bioavailability of lead is a function of a number of factors, including the solubility of the chemical form of lead ingested, the soil particle size and surface area-to-weight ratio, the pH of the digestive tract, and nutritional status. The physical and chemical characteristics of soil lead vary from site to site. Based largely on the argument that the soil lead at the Smuggler Mountain Superfund site "is almost certainly less bioavailable" than the "lead present in soils in the vicinity of smelters and in urban areas where most of the studies relating the concentration in soil to body burdens have been conducted," a group of environmental consultants estimated that soil-lead levels of 1,000 ppm would be "safe" (Lagoy et al. 1989).[21] Findings reported by Steele et al. (1990) and Hemphill et al. (1991) suggested that lower-than-default soil-lead bioavailability values may be appropriate generally for mining sites, particularly where the predominant source of lead contamination was not late-stage processing activities, such as smelting and milling, which produce very fine particle sizes.

Lead absorption also varies with age. It is higher for children than for adults because calcium and lead absorption are linked. Children absorb calcium at high rates because their bones are growing, but the human metabolism does not distinguish well between lead and calcium, so lead unfortunately "goes along for the ride." As a result, the most common laboratory animals, mature rodents, are not the best animal model for assessing lead bioavailability if physiological similarity to children is the primary evaluative criterion.[22] Emphasizing this criterion, EPA Region 8 selected juvenile swine as the appropriate animal model for testing soil-lead bioavailability on large-area lead sites.

Swine, however, have some distinct disadvantages as laboratory animals. Notably, they are large (approximately 35 lbs) and expensive. In contrast to the abundance of laboratories certified to conduct tests with small animals (for example, rodents, rabbits, and fish), there are very few facilities across the country that can provide the necessary quality assurance to conduct reliable laboratory studies using large animals.[23] A soils scientist notes that rodent studies cost ten times less, on a per-animal basis, than juvenile swine studies (the cost per soil sample is $5,000 and $50,000, respectively). Consequently, for a given sum of money, researchers using rodents can conduct a larger, more statistically powerful study (that is, one able to detect smaller differences between treatment levels).[24] However, the rodent studies may systematically underestimate soil-lead bioavailability in children. According to an EPA scientist, Superfund PRPs promoted the rodent model, because "conveniently for them, it showed low absorption, but that's expected because [the rodents] are mature."

To a considerable extent, selection of the most appropriate animal model involves tradeoffs between cost, experimental power and control, fidelity to children's physiology, and the value of information for decisionmaking. Determination of the "optimal" animal model may depend on which evaluative criterion is being used. According to a soils scientist, when the U.S. Department of Agriculture (USDA) conducts nutritional research, it begins with rodent studies to contain research costs, and then, if deemed necessary, moves up progressively to "higher," more costly animals, such as swine or primates. Underscoring the subjectivity involved, this source remarks that in the future, using the results of rodent studies calibrated by the available swine data "is good enough for me. With all of the lead and zinc mine wastes [that need to be tested], it will save millions [of dollars]."

Currently, researchers are working to develop cheaper *in vitro* laboratory models to study lead bioavailability.[25] Most parties seem to agree that *in vitro* methods are ultimately needed to get around the need for animal research altogether. But, until a cheap, reliable, and valid *in vitro* model is developed, disputes over selection of the most appropriate animal model will linger. Disagreement among scientists about the *physiological* suitability of the competing animal models has largely subsided, and opinion among toxicologists has apparently converged on the juvenile swine as an appropriate test animal for high-stakes Superfund decisions. But, in the early 1990s, when EPA Region 8 initiated the swine studies, the bulk of previous work on the bioavailability of heavy metals in soils had been conducted with rodents. According to an EPA scientist, "Other investigators in the field were challenged professionally by [EPA's] decision to use a large animal model." Selection of the appropriate animal model was hotly contested, not only by scientists but also by PRPs and their environmental consultants.

THE PROCESS WITHIN EPA

Setting the Agenda

According to an OSWER official, the program was first introduced to the IEUBK model by Jeff Cohen, who was responsible for developing the model as a member of the OAQPS staff and later directed the Office of Drinking Water's Lead Task Force during finalization of the 1991 lead/copper drinking water rule. "Our usual approach," says the OSWER official, "is to use RAGS [the Risk Assessment Guidelines for Superfund], site-specific exposure assessments, and then go to IRIS for cancer slope factors, MCLs, and RfDs. For lead, those were not available to us in IRIS, and we needed some other way to evaluate the sites."

EPA Region 8 first applied the IEUBK model in a Superfund context in the late 1980s on a site in East Helena, Montana, where there was a closed lead smelter.[26] According to an EPA scientist, data collected from the site were used to develop agency assumptions in the soil/dust ingestion component of the IEUBK model. For the purposes of the East Helena remedy selection, the results of the IEUBK model were not used to generate the final numerical soil cleanup level. Instead, "the model was used as a means of technical negotiations with the PRP, ASARCO," says an EPA official. "EPA was quite successful with it under those circumstances," in part because one of the ASARCO technical consultants had worked academically in development of the model.

Phase I

In 1989, OSWER issued guidance (OSWER Directive 9355.4-02) recommending a soil-lead cleanup level of 500–1,000 ppm[27] at Superfund sites where the land was currently in residential use or where remediation project managers believe future residential use is possible.[28] Late in the same year, the federal government was preparing for litigation on the Sharon Steel Superfund site in Midvale, Utah. Historically, the site had been a lead-smelting operation, and there were also mining wastes present on the location. One of the PRPs, the Atlantic Richfield Company (ARCO), argued that the particular form of soil lead at the site was not hazardous because it had *zero* bioavailability. The Department of Justice, which is responsible for litigating Superfund cases in court, approached a toxicologist with EPA Region 8, Chris Weis, to ascertain whether ARCO could be right. Weis recommended a short-term experiment using juvenile swine to test ARCO's argument. The results of the study (now referred to as Phase I) rejected ARCO's assertion that the soil lead had zero bioavailability, but also suggested it was somewhat lower (20%–25%) than EPA's default assumption under the IEUBK model (30%).[29]

Unbeknownst to EPA and the Justice Department, ARCO had already conducted some lead bioavailability tests on soil samples from the Midvale site using rodents, and the PRP felt that it had a rock-ribbed scientific argument for minimizing its liability. According to an EPA scientist, "ARCO was furious that we criticized the studies they had done. . . . We exposed their ace."[30] The PRP may have been surprised by EPA's challenge because although EPA regional offices are generally staffed with a complement of environmental engineers and hydrogeologists, they typically contain relatively few experts in the health aspects of environmental risk assessment, such as Weis.[31] An EPA official observes, "They [ARCO] were, for the first time, faced with a trained toxicologist who could challenge their evidence. Now, they had to contend not only with lawyers and engineers, but also with trained scientists." The PRPs "brought their checkbook to court" and settled on the first day of the trial for $63 million. Prior to the swine study, the estimated cleanup cost was more than $200 million, according to an EPA official.[32]

The resolution of the Sharon Steel case did not, however, clear up the dispute about EPA's use of the IEUBK model for setting numerical soil-lead cleanup goals, either for the national Superfund program or for any particular contaminated site. Instead, the terms of debate shifted to attempts to distinguish among types of large-area lead sites where soil lead was expected to be more or less bioavailable (thus, obviating the need for costly and time-consuming site-specific applied research) and to arguments over the appropriate laboratory animal for use in testing soils for lead bioavailability (that is, rats versus pigs).

Phase II

Although OSWER had issued broad soil-lead guidance for Superfund and RCRA in 1989–1990, pressure was mounting for EPA to provide more definitive guidance for lead-contaminated soil cleanups. In 1992, the Residential Lead-Based Paint Hazard Reduction Act required that EPA develop national standards for hazardous levels of lead in soil. Many EPA remediation project managers and PRPs were hopeful that the Sharon Steel site test results indicating lower-than-IEUBK-default soil-lead bioavailability could be applied to the remainder of the large-area lead sites. Others, including scientists who had conducted previous heavy-metal bioavailability research and some PRPs, continued to defend the application of rodent studies. However, EPA Region 8 officials cautioned that a decision to apply the Sharon Steel site test results (Phase I) to all mining sites could not be supported technically due to the variability in site conditions.[33] According to Murray (1995), the PRPs for the Leadville site submitted the results of rodent studies with lower-than-default bioavailability values. Given the results of Phase I, Region 8 was reluctant to depart from the default on the basis of rodent studies. Subsequently, a second tranche (Phase II) of juvenile swine studies was conducted testing lead bioavailability in soil samples from seven different Superfund sites across the country.[34]

For some sites, the results indicated bioavailability rates substantially lower than the IEUBK default of 30%. For example, the bioavailability was 19% at the Bingham Creek Superfund site near Salt Lake City, Utah (Murray 1995). The studies have not, however, consistently yielded bioavailability rates lower than 30%. According to EPA officials, the Leadville site results were not substantially lower than the default. For the Jasper County site, the bioavailability of lead in soils contaminated by smelting emissions was in the 28%–31% range. Surprisingly, the bioavailability of soil lead from mining wastes was higher—in the 40%–50% range. This finding was contrary to the conventional wisdom that smelter fallout would produce the most highly bioavailable lead form present on mining sites.

The Role of External Scientists in the Process

After holding a 1988 conference on lead in soil, the Society for Environmental Geochemistry and Health (SEGH) formed the Lead in Soil Task Force to examine the relationship between blood lead and environmental lead and to develop guidelines for assessing and managing the health risks associated with lead in soil and dust. (For a discussion of the role of SEGH in the arsenic in drinking water case, see Appendix B.) In 1990, SEGH, with support from EPA and other sponsors, convened a scientific meeting in Chapel Hill, North Carolina, on bioavailability and dietary intake of lead. The SEGH task force report (summarized in Wixson and Davies 1994) questioned the use of a single value for lead in soil for use in all cleanup situations. An EPA scientist says, "I got the feeling that it [SEGH] was somewhat dominated by the regulated industry. But to their credit, nobody else was providing a forum for these discussions." The major impact that SEGH had was in focusing EPA's attention on the need to consider differences among sites in soil-lead bioavailability.[35]

In 1991, a year after OSWER issued RCRA program guidance permitting the use of the IEUBK model for setting soil-lead cleanup levels, the EPA Science Advisory Board (SAB) concluded that the IEUBK model represented an improved methodology for assessing total lead exposure and for developing soil-lead cleanup levels at CERCLA and RCRA sites. However, SAB raised concerns about incorrect application of the model and

selection of inappropriate input values for default and site-specific applications. In response, OSWER developed a guidance manual for the
IEUBK model.

SCIENCE IN THE REMEDY SELECTION

The bioavailability of lead in soil is directly related to the *potential*
extent and cost of large-area lead site cleanups. Other factors being
equal, the higher the bioavailability, the lower the soil-lead level indicated by the IEUBK model for the preliminary remediation goal. As
indicated above, a suite of factors is considered in the remedy selection
process, but a lower soil-lead PRG suggests that a greater volume of
dirt would have to be excavated and disposed of to *permanently* meet
the health-based cleanup *goal* using currently *proven* remedial technologies. The juvenile swine studies estimated soil-lead bioavailability
to provide a site-specific refinement of the IEUBK model. Perhaps not
surprisingly, the results indicated considerable variability among the
sites tested, with some higher, some lower, and some about the same as
the agency's default assumption. In terms of the final remedy selection,
however, it appears that none of the results will increase the liability of
large-area lead site PRPs because EPA deems the cost of removing the
contaminated soil to be excessive.

According to an EPA official, although agency scientists had expected
the swine study for the Midvale site to indicate even lower soil-lead
bioavailability, the difference between the study results (20%–25%) and
the default assumption (30%) "was significant in terms of site liability,"
causing EPA "to raise the proposed action level" (that is, the soil-lead concentration of concern) for the site. For the Bingham Creek site near Salt
Lake City, the experimentally estimated soil-lead bioavailability was 19%.
According to Murray (1995), the reduced bioavailability value halved the
estimated cost of site cleanup from $8 million to $4 million. According to
an EPA official, it is questionable whether the slightly reduced soil-lead
bioavailability estimate for the Leadville site was big enough to make an
impact on the risk management decision.

For the Jasper County site, as a result of the unexpectedly higher
bioavailability of mining wastes relative to smelter fallout, the estimated
geographic distribution of effective exposures to lead via soil was
markedly different from what was initially suspected, according to an
official with the Missouri Department of Natural Resources. Consequently, while the swine study did not affect the soil cleanup level, it
altered the geographic emphasis of cleanup activities. According to an
EPA official, the agency has decided to initiate soil excavation at the site

while studying the feasibility of treating the soil with phosphate to reduce the soil-lead solubility. According to a Missouri state health official, public demand for additional swine studies to determine the bioavailability of soil lead after phosphate treatment has helped overcome some bureaucratic reluctance to allocating the necessary funds. Although the juvenile swine study did not dramatically reduce the scale of remedial activity, the EPA official considers the study "money well spent. It settled a lot of disputes and saved a lot of arguing regarding what the real number was." Thus, an important contribution of the swine studies to the Jasper County site has been their capacity as a dispute resolution tool (perhaps lowering transaction costs). The results have been accepted as a means for determining the remedial action plan and for assessing the performance of an alternative treatment.

Despite the fact that tests have estimated soil-lead bioavailability to be higher than the IEUBK default on some sites, the information has been either beneficial or essentially neutral to large-area lead site PRPs, apparently because EPA feels that the cost of removing the contaminated soil is excessive. Regarding the sites where bioavailability may be higher than the default, an EPA official remarks, "It could potentially lead to higher liability, but I doubt it. All the discussion is about lowering the default." Regarding large-area Superfund sites generally, an EPA official concludes, "We will be managing exposure [to environmental hazards] in perpetuity" due to the prohibitive costs of permanently removing the risk. As a result of EPA's implicit policy, the PRPs can take a chance on dramatically reducing their costs by exposing themselves to a relatively modest increase in site liability (EPA can recover the swine study costs), secure in the knowledge that they are not exposing themselves to greater financial risk, even if the science "goes the wrong way."

CONCLUDING OBSERVATIONS

In terms of a fate and transport analogy, this case study again illustrates the mobilization of science (in particular, the IEUBK model) accumulated in different compartments of EPA (the air and drinking water programs) and its assimilation by another compartment (the CERCLA and RCRA programs). While the soil-lead bioavailability studies for large-area lead sites were being done, most of the attention at EPA headquarters at the time was on the high-profile Urban Soil Lead Abatement Demonstration Project (U.S. EPA 1995a), according to an OSWER official. Thus, EPA Region 8 was also able to exploit the institutional learning that was occurring in the context of the urban soil-lead abatement

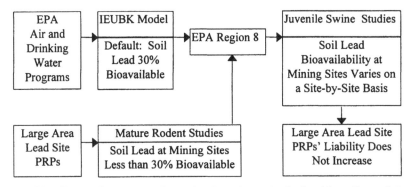

Figure H-1. Fate and transport dynamics for science in the lead in soil at mining sites decisions

study for the purposes of the lead mining sites without the attendant supervision and scrutiny. As in other case studies conducted as part of this project, individual EPA staff (Cohen and Weis) who could bridge media-bound program areas and interface with scientists played key roles in ensuring that the science generated by the agency was put to use in regulatory decisions. Cohen integrated what was known about the effects of lead exposure from various sources (air, drinking water, and soil) that were artificially disaggregated by EPA's organizational structure into three separate programs (the criteria air pollutants, drinking water, and Superfund/RCRA), each with its own parochial concerns. As a trained health scientist on staff at an EPA regional office, Weis was able to design an experiment to *test* the scientific information generated by the regulated community and deposited in the office. These dynamics are illustrated in Figure H-1.

Due to the limited number of experts in the health aspects of environmental risk assessment in EPA regional offices, the current management options often may be limited to either absorbing or rejecting such information produced by potentially responsible parties. Relatively few health scientists may be needed if EPA regional offices (and state agencies with delegated responsibility) are expected to uniformly apply the substance-specific toxicological values provided by IRIS and other central databases. If, on the other hand, environmental agencies are expected to more consistently use site-specific scientific information in the contaminated site remedy selection process, they face the challenge of having adequate scientific capacity available (through an appropriate mix of staff and consultants) to manage the production of site-specific data and to critically evaluate the information produced by the regulated community.

ENDNOTES

[1] As used here, the generic term "mining sites" refers to locations where various stages in the mining process occur. It includes, but is not limited to, sites where mineral-bearing rock called ore is extracted from underground or surface mines. During milling (or beneficiation), ore is processed (or upgraded), usually at or near the site of extraction, into concentrates that are further processed in a smelter or refinery. Final processing sometimes occurs at a considerable distance from the site of extraction.

[2] CERCLIS contains information on over 30,000 potentially hazardous sites (Tilton 1994). According to a report by the Society of Environmental Geochemistry and Health, as of 1993, soil lead had been found at 922 of the 1,300 sites on the NPL (*Environmental Health Perspectives*, November 1994, 912–913).

[3] Jasper County is a historic lead and zinc mining, milling, smelting area located in southwest Missouri. It is located in EPA Region 7, headquartered in Kansas City, Kansas. Mining activities in Jasper County began around 1850 and continued through the 1950s, and millions of tons of mining, milling, and smelting waste materials are dispersed throughout the county, including residential areas. The site was added to the Superfund National Priorities List in 1990 (Baysinger-Daniel 1995).

[4] Tailings are the fine waste particles that are produced during ore milling and typically suspended in water. Under current practices, tailings from surface mines are deposited in a tailing (or settling) pond, while those from underground mines are deposited in the mine itself. Historically, tailings may have simply been piled in waste yards or placed in a settling pond. At abandoned sites, however, uncovered tailing piles often remain after settling ponds have dried up.

The Sharon Steel site is located in Midvale, Utah, a small town twelve miles south of Salt Lake City. Around 1906, the U.S. Smelting, Refining, and Mining Company began processing ores at Midvale, and in the mid-1920s, the company built a mill to process lead-zinc ores and a lead smelter. During World War II, Utah was a major lead-producing region. ASARCO (American Smelting and Refining Company) also operated a lead smelter in Murray, and the International Smelting and Refining Company, a subsidiary of the Anaconda Company, had a lead mill and smelter at Tooele. After the war, lead-zinc mining production in Utah declined. In 1958, U.S. Smelting and International Smelting agreed to share facilities. U.S. Smelting closed its Midvale smelter and shipped its concentrates to Tooele for smelting, and International Smelting closed its mill and shipped its ore to Midvale for milling. In 1971, U.S. Smelting closed its Midvale mill and changed its name to UV Industries; and in 1979, a year before passage of CERCLA, the company went bankrupt. U.S. Smelting sold the Midvale site to Sharon Steel, which planned to use the site for commercial purposes. The mill and smelter had been torn down, but the tailing pond and slag pile from the smelter remained. The tailing pond contained 14 million tons of uncovered piles up to fifty feet deep and covered 260 acres. The fine-grained material blew off site, and 44,000 people lived within two miles of the site. Health effects were first suspected in 1982, and the site was proposed for the NPL in 1984 (Tilton

1994). Site characterization is an important element in assessing what specific wastes (for example, lead carbonate or lead sulfide) and proportions of different wastes (for example, milling wastes versus smelter emissions) may be present on a particular site. Due to the complex history of many mining sites, this can be difficult.

5 According to an EPA official, the agency has decided to initiate soil excavation at the Jasper County Site, with the lead-contaminated soils being deposited in the roadbed of a new highway construction project. Due to their areal extent, capping large-area lead sites with asphalt is not regarded as a feasible alternative.

6 In this context, disposal means deposition in a facility designed to permanently contain the contaminated soil. Federal scientists and academics are researching alternative remediation methods involving the application of phosphate, iron oxide, and sewage sludges intended to permanently render soil lead less soluble. See, for example, Ryan and Zhang (1996). An independent toxicologist comments, however, that while phosphate may temporarily immobilize lead, it can mobilize arsenic, which is commonly associated with lead on sites contaminated by extractive mining.

7 According to an EPA official, the initial remediation cost estimate for the Sharon Steel–Midvale site was in excess of $200 million. All of the sites involved in EPA's Phase II bioavailability study (discussed below) had projected cleanup costs of more than $100 million. Probst et al. (1995) suggest that for the mining industrial sector as a whole, the annual financial burden of Superfund cleanup and transaction costs, estimated at $220.5 million under the current law, is likely to be quite large in relation to the industry's profitability.

8 According to Murray (1995), Leadville homeowners have suffered a substantial drop in property values. An EPA official estimates that 80% of the employment in Leadville remains mining-related and suggests that residents "were parroting whatever the mining company said."

9 The Smuggler Mountain Superfund site covers approximately seventy-five acres and is located about one mile northeast of Aspen, Colorado, in Pitkin County. A mobile home park and two condominium developments are located a few hundred yards from the site (Lagoy et al. 1989).

10 CERCLA does not require EPA to conduct national rulemaking. As a result, EPA relies on administrative guidance to implement Superfund. However, the 1992 Residential Lead-Based Paint Hazard Reduction Act (Title X of the Housing and Community Development Act) required EPA to develop standards defining hazardous levels of lead in lead-based paint, household dust, and soil (adding Title IV to the Toxic Substances Control Act) by 1994. In July 1994, EPA issued guidance for CERCLA and RCRA (OSWER Directive 9355.4-12), as well as interim guidance under the Toxic Substances Control Act (TSCA). The agency has not yet promulgated a formal standard for lead in soil under TSCA. According to press reports, EPA issued TSCA guidance in lieu of a formal rule because existing scientific data could not justify the standards the agency was considering and suitable data may not be available for five years (*Inside EPA*, December 16, 1994, 14).

[11] For example, in the absence of mining activity, background levels of children's blood lead would tend to be higher than normal in areas with naturally high lead content in the rock and soils. Therefore, once the site-specific lead absorption rate has been determined, the debate could shift to ascertaining the relative contribution of the contaminated site to children's measured blood-lead levels. Alternatively, in a probabilistic analysis of the IEUBK model funded by ASARCO, Lee et al. (1994) concluded that the driving variables include bioavailability of lead and dietary intake of lead. Therefore, once the site-specific lead absorption rate has been determined, the debate could shift to estimating the actual dietary intake of lead.

[12] See Appendix B for discussion of the development of EPA's lead policy and regarding EPA's unusual RfD for arsenic. EPA Region 8 is also directing juvenile swine studies to determine the bioavailability of arsenic-contaminated soils.

[13] In principle, Superfund PRGs are developed without consideration of technical feasibility, cost, and public acceptance.

[14] This refers to maintaining calcium at appropriate levels in the body.

[15] Lead is also used for radiation shielding, cable covering, ammunition, chemical reaction equipment, fusible alloys, type metal, vibration damping in heavy construction, foil, and bearing alloys (Hawley 1981). Leaded gasoline, new lead plumbing, and the pesticide lead arsenate have been phased out.

[16] Ryan and Zhang (1996) report that many innercity neighborhoods have mean or median soil-lead concentrations in excess of 1,000 ppm, with values as high as 50,000 ppm being reported. However, an OSWER official questions whether these values are overestimates. By comparison, the Urban Soil Lead Abatement Demonstration Project was performed in areas with median soil-lead concentrations ranging from 237 ppm to 2,396 ppm (U.S. EPA 1995a). Suffice to say that levels of lead in soil in many urban areas, particularly in older neighborhoods, are considerably higher than background levels due to accumulated anthropogenic releases of lead into the urban environment.

[17] However, lead does become more soluble in acid soils (Kabata-Pendias and Pendias 1984), and sulfides (which can produce sulfuric acid) make up more than a third of many nonferrous mineral ores (Young 1992).

[18] According to EPA officials, the IEUBK model builds on research conducted in the late 1970s at New York University by Harley and Kneip, who developed a pharmacokinetic model for lead (describing the absorption, movement, storage, and excretion of lead in various compartments of the body) by conducting studies with juvenile baboons and by analyzing human cadavers. The IEUBK added exposure components (levels of lead in air, food, water, soil, and dust) to the pharmacokinetic model.

[19] The absorption rate can be a function of administered dose. For example, according to an independent toxicologist, the relationship between blood lead and dose becomes curvilinear at higher doses due to either attenuated uptake or more efficient excretion.

[20] By contrast, the IEUBK default bioavailability for the generally more soluble forms of lead present in food and water is 50%.

[21] Lagoy et al. (1989) based their determination partially on 1985 Centers for Disease Control (CDC) guidance that suggested that increased blood-lead levels were associated with soil-lead levels of 500–1,000 ppm. Lagoy et al. (1989) also indicate that various approaches to establishing a "safe" level of lead in the soil yield widely varying results ranging from about 100 ppm to 8,000 ppm.

[22] According to a federal scientist, juvenile swine have been found to be better animals to model human uptake of nutrients and carbohydrates. An EPA scientist notes that rodents mature at six to ten weeks, whereas swine mature more slowly. Other problems with the rodent model include a low "residence time" for food (ingested substances pass through the animals quickly, leaving little time for lead absorption) and the fact that rats eat their feces, resulting in a confusing "redosing" of the animals with excreted lead.

[23] The juvenile swine studies were conducted at the University of Missouri under contract with EPA.

[24] An EPA scientist suggests that the precision of the swine studies has been greater than expected.

[25] Scientists at the University of Colorado Geochemistry Department and the EPA Environmental Criteria and Assessment Office laboratory in Cincinnati, Ohio, are working to develop a "fake GI [gastrointestinal] tract," says an EPA scientist. The bench-top model consists of a series of chambers with flow-through mechanisms, and preliminary results have been "pretty consistent" with those from the juvenile swine model.

[26] In 1986, CDC and EPA issued a report, *East Helena, Montana, Child Lead Study, Summer 1983.*

[27] This range is identical to that in the 1985 CDC guidance to which Lagoy et al. (1989) referred.

[28] The role of land use in Superfund cleanups has been a source of considerable controversy. OSWER directs personnel to "assume future residential land use if it seems possible based on the evaluation of the available information" (U.S. EPA/OSWER 1989, 6–7). The presumption of residential land use affects risk assessment exposure assumptions regarding levels of on-site soil ingestion, and it may also affect assumptions regarding the quality of drinking water drawn from groundwater. According to U.S. EPA 1994, residential land use is the fourth most common current land use at Superfund sites and the second most common expected future land use. According to U.S. EPA 1995b, 15% of Superfund sites have people living on site, 80% have residences adjacent to them, and about 20% of the time, when a commercial/industrial site has residences nearby, EPA will assume a future residential land use. See Hersh et al. (1997) for a discussion of the role of land use in Superfund cleanups.

[29] OSWER headquarters and ECAO each provided $100,000 for Phase I.

[30] According to an independent toxicologist, external scientists—such as Herbert Needleman of the University of Pittsburgh, John Drexler of the University of Colorado, and Paul Mushak of PB Associates—were also prepared to testify as expert witnesses on behalf of the government in the Sharon Steel case. Mushak had submitted a critique of ARCO's arguments that laboratory solubility tests

approximated the behavior of lead tailings in humans. Drexler had performed laboratory analyses of Midvale soil samples indicating that lead was present in very fine particles and, therefore, more likely to be bioavailable, and also consisted of toxic chemical forms. See Appendix A for further discussion of this case and Needleman's role.

[31] Weis has a doctorate in environmental toxicology, is a board-certified toxicologist, and conducted postdoctoral research in physiology and biophysics at the University of Virginia Medical School prior to joining EPA Region 8. According to an OSWER official, Regions 1 and 8 have a comparatively large number of health scientists.

[32] Tilton (1994) estimates that the transaction costs of doing the technical work and negotiating and litigating the case accounted for between a fifth and a third of the total funds devoted to the Midvale site by the PRPs and EPA.

[33] Phase I was also designed simply to quickly *test* the hypothesis that the soil-lead bioavailability at the Midvale site was zero. A larger, more elaborate test was required to *estimate* the actual bioavailability with much precision.

[34] Phase II studies were supported by regional Superfund program budgets (with cost recovery provisions applicable).

[35] SEGH task force member Rufus Chaney, a USDA researcher, was a strong proponent of using rodents for testing lead bioavailability. Chaney's bioavailability studies with rodents were supported by DuPont. According to an EPA official, DuPont's interest stems from the company's interest in bioavailability of residuals and alternative remediation technologies. EPA was represented on the task force by Richard Cothern (see Appendixes A and B for discussion of Cothern's role in the scientific assessment of lead and arsenic in drinking water).

REFERENCES

Baysinger-Daniel, C. 1995. Cooperation and Coordination among Several Agencies and Private Groups at the Jasper County Superfund Site. Poster Presentation. St. Louis, Mo.: State and Tribal Forum on Risk-Based Decision Making, November 12–15.

Hawley, G. 1981. *The Condensed Chemical Dictionary,* 10th ed. New York: Van Nostrand Reinhold Company.

Hemphill, C., M. Ruby, B. Beck, A. Davis, and P. Bergstrom. 1991. The Bioavailability of Lead in Mining Wastes: Physical/Chemical Considerations. *Chemical Speciation and Bioavailability* 3: 135–148.

Hersh, R., K. Probst, K. Wernstedt, and J. Mazurek. 1997. *Linking Land Use and Superfund Cleanups: Uncharted Territory.* Washington, D.C.: Resources for the Future.

Kabata-Pendias, A., and H. Pendias. 1984. *Trace Elements in Soils and Plants.* Boca Raton, Fla.: CRC Press, Inc.

Lagoy, P., I. Nisbet, and C. Schulz. 1989. The Endangerment Assessment for the Smuggler Mountain Site, Pitkin County, Colorado: A Case Study. In *The Risk*

Assessment of Environmental and Human Health Hazards: A Textbook of Case Studies, edited by D. Paustenbach. New York: John Wiley & Sons, 505–525.

Lee, R., W. Wright, W. Haerer, and J. Fricke. 1994. Development of a Stochastic Blood Lead Prediction Model. Abstract from 1994 Society for Risk Analysis Annual Meeting. http://riskworld.com/abstract/ab4me001.htm.

Murray, B. 1995. Bioavailability of Lead in a Juvenile Swine Model to Support Superfund Mine Site Reclamation. Presentation at the State and Tribal Forum on Risk-Based Decision Making, St. Louis, Mo., November 12–15.

NRC (National Research Council). 1993. Measuring Lead Exposure in Infants, Children, and Other Sensitive Populations. Washington, D.C.: National Academy Press.

Probst, K., D. Fullerton, R. Litan, and P. Portney. 1995. *Footing the Bill for Superfund Cleanups: Who Pays and How?* Washington, D.C.: Brookings Institution and Resources for the Future.

Ryan, A., and P. Zhang. 1996. *Soil Lead Remediation: Is Removal the Only Option?* Cincinnati, Ohio: U.S. EPA Risk Reduction Engineering Laboratory. http://128.6.70.23/html_docs/rrel/ryan.html.

Steele, M., B. Beck, B. Murphy, and H. Strauss. 1990. Assessing the Contribution from Lead in Mining Wastes to Blood Lead. *Regulatory Toxicology and Pharmacology* 11: 158–190.

Tilton, J. 1994. Mining Waste and the Polluter-Pays Principle in the United States. In *Mining and the Environment,* edited by R. Eggert. Washington, D.C.: Resources for the Future, 57–84.

U.S. ATSDR (Agency for Toxic Substances and Disease Registry). 1988. *Nature and Extent of Childhood Lead Poisoning in Children in the U.S.: A Report to Congress.* Washington, D.C.: U.S. Department of Health and Human Services, July.

U.S. EPA (U.S. Environmental Protection Agency). 1994. *Swift/Dingell Q&A Document.* Question 10. Washington, D.C.: U.S. EPA, January 26. Mimeographed.

———. 1995a. *Urban Soil Lead Abatement Demonstration Project.* External Review Draft. EPA/600/R-95/139. Washington, D.C.: U.S. EPA.

———. 1995b. *Superfund Administrative Reforms Fact Sheet.* Washington, D.C.: U.S. EPA, May 25.

U.S. EPA/OSWER (U.S. EPA Office of Solid Waste and Emergency Response). 1989. *Risk Assessment Guidance for Superfund.* Vol. 1 of *Human Health Evaluation Manual.* EPA/540/1-89/002. Washington, D.C.: U.S. EPA.

Wixson, B., and B. Davies. 1994. Guidelines for Lead in Soil: Proposal of the Society for Environmental Geochemistry and Health. *Environmental Science and Technology* 28(1): 26A–31A.

Young, J. 1992. Mining the Earth. In *State of the World 1992.* Washington, D.C.: Worldwatch Institute, 100–118.

Appendix I

List of Interviewees and Commenters

Charles Abernathy, EPA/OW
George Alapas, EPA/ORD
Alvin Alm, SAIC, McLean, Va.
Christine Augustiniak, EPA/OPPTS
John Bachmann, EPA/OAR
Donald Barnes, EPA/SAB
Sherry Baysinger-Daniel, Missouri Department of Health
Jay Benforado, EPA/ORD
David Bennett, EPA/OSWER
Clyde Bishop, EPA/ORD
David Blockstein, Committee for the National Institute for the Environment, Washington, D.C.
Craig Boreiko, International Lead Zinc Research Organization, Research Triangle Park, N.C.
John Bowser, EPA/OPPTS
Jeanne Briskin, EPA/OAR
Ken Brown, Kenneth G. Brown Inc., Chapel Hill, N.C.
Gary Carlson, Missouri Department of Health
Rufus Chaney, U.S. Department of Agriculture
Wendy Cleland-Hamnet, EPA/OPPE
Jeff Cohen, EPA/OW
John Cohrssen, Senate Committee on Commerce
Doug Costle, Former EPA Administrator
Richard Cothern, EPA/OPPE
Joseph Cotruvo, EPA/OPPTS
Jim Detjen, Michigan State University
Bill Diamond, EPA/OW
Mark Doolan, EPA/Region 7
Bob Drew, American Petroleum Institute, Washington, D.C.

Elizabeth Drye, Senate Committee on Environment and Public Works
Claire Earnhart, Case Western Reserve University
Jim Elder, EPA/OW (retired)
Gerald Emison, EPA/OAR
Mike Firestone, EPA/OPPTS
Sheila Frace, EPA/OW
Herman Gibb, EPA/ORD
Bernard Goldstein, Rutgers University
John Graham, Harvard School of Public Health
Lester Grant, EPA/ORD
Hank Habicht, Safety-Kleen Corp., Elgin, Ill.
John Haines, EPA/OAR
Fred Hansen, EPA
Fred Hauchman, EPA/ORD
John Hausoul, EPA/OPPTS
Robert Huggett, EPA/ORD
Sheila Jasanoff, Cornell University
Richard Johnson, EPA/Office of Administration and Resources Management
Bruce Jordan, EPA/OAR
Victor Kimm, EPA/OPPTS
Alan Krupnick, Resources for the Future, Washington, D.C.
Bob Lackey, EPA/ORD
Jessica Landman, Natural Resources Defense Council
Marc Landy, Brandeis University
Arthur Langer, Brooklyn College–CUNY
Lester Lave, Carnegie-Mellon University
Ronnie Levin, EPA/Region 1
Morton Lippman, New York University Medical Center
Sylvia Lowrance, EPA/AO
Edmond Maes, Centers for Disease Control, Atlanta, Ga.
Clarence Mahan, EPA/Office of Administration and Resources Management (retired)
Allan Marcus, EPA/ORD
Catherine Marshall, American Forest and Paper Association, Washington, D.C.
Karl Mazza, Office of Senator Moynihan
John McCarthy, American Crop Protection Association, Washington, D.C.
John Melone, EPA/OPPTS
Robert Menzer, EPA/ORD
Fred Miller, Chemical Industry Institute for Toxicology, Research Triangle Park, N.C.
Jack Moore, Institute for Evaluating Health Risks, Washington, D.C.

Karen Morehouse, EPA/ORD
Dick Morgenstern, EPA/OPPE
David Mosby, Missouri Department of Natural Resources
Lawrie Mott, Natural Resources Defense Council
Marcia Mulkey, EPA/Region 3
Paul Mushak, PB Associates, Durham, N.C.
Rashmi Nair, Monsanto Company
Herbert Needleman, University of Pittsburgh Medical Center
Debra Nicoll, EPA/OW
Warner North, Decision Focus Inc., Mountain View, Calif.
Eric Olson, Natural Resources Defense Council
David Ozonoff, Boston University
Dorothy Patton, EPA/ORD
Dalton Paxman, U.S. Congress Office of Technology Assessment
Craig Potter, Vedder, Price, Kaufman, Kammholz and Day, Washington, D.C.
Peter Preuss, EPA/ORD
William Reilly, World Wildlife Fund
Joe Reinert, EPA/OPPE
Harvey Richmond, EPA/OAR
Barbara Rosewicz, Wall Street Journal
William Ruckelshaus, Browning Ferris Industries, Inc., Seattle, Wash.
Amy Schaffer, American Forest and Paper Association, Washington, D.C.
David Schnare, EPA/Office of Enforcement and Compliance Assurance
Joel Schwartz, Harvard School of Public Health
Kristin Shrader-Frechette, University of South Florida
Ellen Silbergeld, University of Maryland and Environmental Defense Fund
Allan Smith, University of California, Berkeley
Robert Sussman, Latham & Watkins, Washington, D.C.
Lee Thomas, Georgia-Pacific Corp., Atlanta, Ga.
William Thomas, EPA/OAR
Russell Train, Former EPA Administrator
Arthur Upton, The University of Medicine and Dentistry of New Jersey–Robert Wood Johnson Medical School, Piscataway, N.J.
Linda Vlier Moos, EPA/OPPTS
Vanessa Vu, EPA/OPPTS
Thomas Wallsten, University of North Carolina
Peter Washburn, Natural Resource Council of Maine
Chris Weis, EPA/Region 8
Paul White, EPA/ORD
Ron White, American Lung Association, Washington, D.C.
LaJuana Wilcher, Winston & Strawn, Washington, D.C.

Dwain Winters, EPA/OPPTS
Paul Wise, Illinois Environmental Protection Agency
Gerald Yamada, Fried, Frank, Harris, Schreiber, and Jacobson, Washington, D.C.
Terry Yosie, E. Bruce Harrison Co., Washington, D.C.
Maurice Zeeman, EPA/OPPTS

COMMENTERS WHO WERE NOT INTERVIEWED

Barbara Beck, Gradient Corporation, Cambridge, Mass.
Maggie Dean, EKA Chemicals, Inc., Washington, D.C.
Robert Fegley, EPA/ORD
Robert Fensterheim, RegNet Environmental Services, Washington, D.C.
Douglas Pryke, Alliance for Environmental Technology, Washington, D.C.
Alan Roberson, American Water Works Association, Washington, D.C.
Alan Rubin, EPA/OW
Michael Schock, EPA/ORD
Tracy Slayton, Gradient Corporation, Cambridge, Mass.
Kimberly Thompson, Harvard Center for Risk Analysis, Cambridge, Mass.
Jim Wilson, Resources for the Future, Washington, D.C.

Appendix J
Science at EPA
Interview Guides

CASE-SPECIFIC INTERVIEW GUIDE

I. Identification of Interviewee

A. Name, organization:

B. Date, location, time of interview:

C. Briefly describe the interviewee:

(Note: In what program areas has the interviewee been most involved [that is, air, water, toxic substances/pesticides]? What is their affiliation [that is, disciplinary and organizational])?

INTRODUCTION: As you know, we are conducting eight case studies of EPA's use of scientific information:

- 1987 Revision of the National Ambient Air Quality Standard for Particulates (NAAQS)
- 1993 Decision Not to Revise the NAAQS for Ozone
- 1991 Lead/Copper Rule under the Safe Drinking Water Act (SDWA)
- 1995 Decision to Pursue Additional Research Prior to Revising the Arsenic Standard under SDWA
- 1989 Asbestos Ban and Phaseout Rule under the Toxic Substances Control Act
- 1983–1984 Suspensions of Ethylene Dibromide under the Federal Insecticide, Fungicide, and Rodenticide Act
- Control of Dioxins from Pulp and Paper Sludge Land Application under the Clean Water Act (as part of the combined air/water "cluster rule" proposed in 1993)
- Lead in Soil at Mining Sites

We will be discussing ___(the specific case study[ies])___ with you today. After we finish with the questions about the case(s), I'll give you the opportunity to make any comments you'd like about the case or about the role of science in EPA environmental regulatory decisionmaking in general.

Would you prefer your comments to be on the record or off the record? If you want the interview to be off the record, but there are specific statements on which you would like to be quoted, just let me know.

(Note: If the interview is off the record, let the interviewee know that the names of the respondents will be listed in an appendix to the report, but their comments would be attributed nonspecifically to "an EPA official noted that. . . .")

Before proceeding, I want to be clear about what we mean by "science." Throughout the interview, by "scientific information," we are referring specifically to the information that could be used in assessing environmental risks (this doesn't include economic or engineering information). This information could take many forms, from contractors' studies, public comments, or staff briefings to peer-reviewed literature or advisory committee reports.

II. Questions on the Case(s)

1.   What got (this case) on the agency's agenda (that is, lawsuit, statutory deadline, well-publicized event, key individual[s], and so forth)?

2.   Please very briefly list the important steps in the process from assessing the available science to developing recommendations for agency decisionmakers in (this case).

   • Who was responsible for the risk assessment?
   • How was the risk assessment conducted?
   • Who communicated the risk analysis to agency decisionmakers?
   • How was the risk communication conducted?
   • Who were the key agency decisionmakers and scientists?

3.   In terms of quantity, quality, and uncertainty, was there adequate scientific information available at the time to inform a decision to (promulgate a rule, ban a substance, revise a standard)? Why?

4.   The quantity of scientific information available at the time for use in (this case) was

   a) Very abundant
   b) Abundant
   c) Moderate
   d) Sparse
   e) Very sparse
   f) Don't know

(Note: By quantity, we mean the number of studies, the size of the database.)

(Note: Reassure the interviewee that we are not treating the questionnaire as a statistical survey instrument and that we are not going to be conducting statistical analysis of their responses. This coding provides just a systematic form of collecting the information.)

5. The quality of the scientific information available at the time for use in (this case) was

    a) Very good
    b) Good
    c) Fair
    d) Poor
    e) Very poor
    f) Don't know

6. The level of scientific uncertainty at the time of (this case) was

    a) Very large
    b) Large
    c) Moderate
    d) Small
    e) Very small
    f) Don't know

7. Was there clear guidance available about what default options or standard assessment procedures to use in the analysis or to fill in the important data gaps in the risk assessment process? If yes, what was the source of this guidance?

8. Was there sufficient scientific evidence justifying departure from the default options or standard assessment procedures? If yes, did the agency do so?

9. For (this case), how would you rate the quality of EPA's treatment of the scientific information available at the time (for example, the risk assessment or the criteria document and risk assessment portions of the staff paper)?

    a) Very good
    b) Good
    c) Fair
    d) Poor
    e) Very poor
    f) Don't know
       Why?

10. What were the principal means by which the key decisionmakers became engaged in the scientific issues of (this case) (for example, decision memos, summaries of scientific assessments, perusal of advisory committee reports, staff briefings, discussions with nonagency scientists)?

11. How would you rate the communication of the scientific information to (the key agency decisionmaker[s]) in (this case)?

   a) Very good
   b) Good
   c) Fair
   d) Poor
   e) Very poor
   f) Don't know
      Why?

12. What were the most important factors impeding or facilitating a thorough consideration of the scientific information in (this case)?

   a) Impeding factors
   b) Facilitating factors

INTRODUCTION: Now I'm going to ask a couple of questions dealing separately with, first, the level of consideration given to the science and, second, the impact the science had on the ultimate decision.

13. (The key agency decisionmaker[s]') consideration of the relevant scientific information in (this case) was

   a) Very thorough
   b) Thorough
   c) Less than thorough
   d) None/no consideration
   e) Don't know
      Why?

   (Note: How much time and energy do you think the decisionmaker devoted to the science?)

14. How would you rate the impact that the scientific information had on the ultimate decision in (this case)?

   a) High
   b) Medium
   c) Low
   d) None

e) Don't know
    Why?

15. How significant was the role of nonagency scientists (including official advisory board members and other scientists from other agencies, academia, industry, or environmental organizations) in influencing the decisionmaking process in (this case)?

   a) Very significant
   b) Significant
   c) Insignificant
   d) Don't know

   (For example, did nonagency scientists force an independent review of EPA's assessment or did they call for additional research, and so forth, or were they in the back room cutting a deal?)

16. How significant was the role of nonagency scientists in legitimizing the decisionmaking in (this case)?

   a) Very significant
   b) Significant
   c) Insignificant
   d) Don't know

17. In ___(this case)___ , what were the principal issues (if any) you regarded as being on the science-policy interface (for example, What constitutes an "adverse effect"? What is an appropriate safety factor? What is the appropriate sensitive population?).

Follow-Up Questions

In your opinion, should the issue(s) have been resolved
   a) Primarily on scientific grounds or
   b) Primarily on policy grounds

Was the issue, in your view, resolved?
   a) Yes. How? Was it resolved satisfactorily (Y/N)?
   b) No. Why? What was the outcome?

Mapping Exercise
INTRODUCTION: Before we move on to discuss the general role of science in EPA decisionmaking, we turn to mapping the flow of scientific information from its source to its ultimate user to better understand the process of how science is actually used in the agency decisionmaking process.

18. Regardless of whether they were considered or not, what, in your opinion, were the (1–3) most important pieces or sources of scientific information in <u>(this case)</u>?

    (Note: If the respondent suggests a class of information, for example, animal bioassay results, ask him or her to discuss a representative of the class.)

- Source: (What was the source(s)? How did the information originate? Who paid?)
- Outside agency: (Describe the external transport and transformation process)
- Exposure: (Discuss how the agency was exposed [for example, What was the "route of exposure"? Who were the "receptors"? Were there barriers to exposure?])
- What happened to it inside the agency: (Describe the internal transport and transformation process)
- How was it delivered and did it impact at the decisionmaker: (Discuss the "delivered dose," was it diluted, and effects)
- Additional comments regarding the case
- Additional comments regarding the general role of science at EPA

III. Supplementary Questions Regarding General Role of Science at EPA

INTRODUCTION: Finally, we would like to discuss the general role of science in EPA's regulatory decisionmaking process. Conclude by posing Questions 1–7, and 10 from the General Interview Guide.

GENERAL INTERVIEW GUIDE

I.   Identification of Interviewee
A. Name, organization:
B. Date, location, time of interview:
C. Briefly describe the interviewee:

    (Note: In what program areas has the interviewee been most involved [that is, air, water, toxic substances/pesticides]? What is their affiliation [that is, disciplinary and organizational]?)

INTRODUCTION: As you know, we are conducting an assessment of the process by which EPA uses science, and we will be discussing the general role of science in environmental regulatory decisionmaking with you today. In addition, we will have questions about some particular cases that we are studying in which you may have been involved.

Would you prefer your comments to be on the record or off the record? If you want the interview to be off the record, but there are specific statements on which you would like to be quoted, just let me know.

(Note: If the interview is off the record, let the interviewee know that the names of the respondents will be listed in an appendix to the report, but their comments would be attributed nonspecifically to "an EPA official noted that. . . .")

Before proceeding, I want to be clear about what we mean by "science." Throughout the interview, by "scientific information," we are referring specifically to the information that could be used in assessing environmental risks. This information could take many forms, from contractors' studies, public comments, or staff briefings to peer-reviewed literature or advisory committee reports.

II. Questions Regarding General Role of Science

1. How would you rate EPA's current overall utilization of scientific information for regulatory decisionmaking?

   a) Very good
   b) Good
   c) Fair
   d) Poor
   e) Very poor
   f) Don't know

2. What rating would you have assigned the agency ten years ago?

   a) Very good
   b) Good
   c) Fair
   d) Poor
   e) Very poor
   f) Don't know

3. Twenty years ago?

   a) Very good
   b) Good
   c) Fair
   d) Poor
   e) Very poor
   f) Don't know

4. Does your evaluation of EPA's use of scientific information differ among programs? How?

5. In what agency decisions has science played a particularly large role? Why?
6. In what EPA decisions has science played little or no role? Why?
7. What are the principal ways in which EPA's use of science has changed during the past decade or more? What accounts for these changes?

INTRODUCTION: Before we (move on to discuss the case studies/conclude the questionnaire), we turn to identifying the most important factors that impede or facilitate the use of science and your suggestions regarding reform of the process.

8. What are the greatest impediments to the use of scientific information in EPA regulatory decisionmaking? Why?
9. What factors facilitate the use of scientific information in the agency's regulatory decisionmaking? Why?
10. Does the current role of science in EPA decisionmaking need to be strengthened or reformed?

   a) Yes. How?
   b) No. How do you interpret the many calls for strengthening and reforming the process?

Additional comments regarding the general role of science at EPA

III. Supplementary Questions on the Case(s)

INTRODUCTION: With which, if any, of the eight case studies (listed above) were you directly involved. How? Follow up by posing Questions 1–3, 9–17 from the Case-Specific Interview Guide.

Appendix K
List of Acronyms

μg/dl	micrograms per deciliter
μg/m³	micrograms per cubic meter
μm	micrometer
AA	Assistant Administrator
ACGIH	American Council of Governmental and Industrial Hygienists
ADI	acceptable daily intake
AET	Alliance for Environmental Technology
AF&PA	American Forest and Paper Association
Ah	aryl hydrocarbon
AHERA	Asbestos Hazard Emergency Response Act of 1986
AIA	Asbestos Industry Association
AIHC	American Industrial Health Council
AIRS	Aerometric Information Retrieval System
ALA	American Lung Association
AMS	Agricultural Marketing Service
AO	Administrator's Office
AOX	adsorbable organic halide indicator
APA	Administrative Procedures Act
API	American Paper Institute; American Petroleum Institute
ARAR	applicable or relevant and appropriate requirements
ARCO	Atlantic Richfield Company
ARS	Agricultural Research Service, USDA
ASARCO	American Smelting and Refining Company
ASTM	American Society for Testing and Materials
ATSDR	Agency for Toxic Substances and Disease Registry, DHHS
AWQC	ambient water quality criteria
AWWA	American Water Works Association
AWWARF	American Water Works Association Research Foundation
BAT	best available technology

BCF	bioconcentration factor
BCT	best conventional (pollutant control) technology
BEST	Board on Environmental Studies and Toxicology
BLM	Bureau of Land Managememt
BMP	best management practice
BoR	Bureau of Reclamation
BOSC	Board of Scientific Counselors
BS	British smokeshade
BSSAF	biota to suspended solids accumulation factor
CAA	Clean Air Act
CAG	Cancer Assessment Group, EPA
CASAC	Clean Air Science Advisory Committee, EPA
CD	criteria document
CDC	Centers for Disease Control and Prevention
CED	Criteria and Evaluation Division, EPA
CEQ	Council on Environmental Quality
CERCLA	Comprehensive Environmental Response, Compensation and Liability Act of 1980
CERCLIS	Comprehensive Environmental Response Compensation and Liability Information System
CFC	Chlorofluorocarbon
CFR	Code of Federal Regulations
CHESS	Community Health and Environmental Surveillance System
CIIT	Chemical Industry Institute of Toxicology
Cl_2	elemental chlorine
ClO_2	chlorine dioxide
CMA	Chemical Manufacturers Association
CO	carbon monoxide
CPSC	Consumer Product Safety Commission
CRAVE	Carcinogen Risk Assessment Verification Endeavor
CRS	Congressional Research Service
CWA	Clean Water Act
DBCP	Dibromochloropropane
DDT	Dichlorodiphenyltrichloroethane
DHHS	Department of Health and Human Services
DOD	Department of Defense
DOE	Department of Energy
DOI	Department of the Interior
DWC	Drinking Water Committee
DWEL	drinking water equivalent level
EAG	Exposure Assessment Group, EPA
EC	European Community
ECAO	Environmental Criteria and Assessment Office, EPA

EDB	ethylene dibromide
EDF	Environmental Defense Fund
EMAP	Environmental Monitoring and Assessment Program
EOP	Executive Office of the President, The White House
EPA	Environmental Protection Agency
EPRI	Electric Power Research Institute
ERC	environmental research center, EPA
ERDDA	Environmental Research, Development and Demonstration Act
ESD	Environmental Services Division, EPA
FACA	Federal Advisory Committee Act
f/ml	fibers per milliliter
FDA	Food and Drug Administration
FFDCA	Federal Food, Drug and Cosmetic Act
FIFRA	Federal Insecticide, Fungicide and Rodenticide Act of 1972
FOIA	Freedom of Information Act
FQPA	Food Quality Protection Act of 1996
FR	*Federal Register*
FWPCA	Federal Water Pollution Control Act
FWS	Fish and Wildlife Service, DOI
FY	fiscal year
GAO	General Accounting Office
GM	General Motors
HAD	health assessment document
HAP(s)	hazardous air pollutant(s)
HEA	health effects assessment
HEAST	Health Effects Assessment Summary Tables
HEED	health and environmental effects document
HEI	Health Effects Insitute
HEI-AR	Health Effects Institute–Asbestos Research
HERL	Health Effects Research Laboratory, EPA
HEW	Department of Health, Education, and Welfare
HHAG	Human Health Assessment Group, EPA
HRS	Hazard Ranking System
HSRC	hazardous substance research center, EPA
HWIR	hazardous waste identification rule
IARC	International Agency for Research on Cancer, WHO
IEHR	Institute for Evaluating Health Risks
IEUBK	Integrated Exposure Uptake Biokenetic (model)
IJC	International Joint Commission
ILZRO	International Lead Zinc Research Organization
IRIS	Integrated Risk Information System
IRLG	Interagency Regulatory Liaison Group

ISI   Institute for Scientific Information
ITER   International Toxicity Estimates for Risk
LC   lethal concentration
LMS   linearized multistage (model)
LOE   level of effort
MACT   maximum available control technology
MCL   maximum contaminant level
MCLG   maximum contaminant level goal
MDI   methylene diphenyl diisocyanate
MIT   Massachusetts Institute of Technology
MLE   maximum likelihood estimate
MMS   Minerals Management Service
MSHA   Mine Safety and Health Administration
MTBE   methyl tertiary butyl ether
MTD   maximum tolerated dose
NAAQS   national ambient air quality standard
NAPA   National Academy of Public Administration
NAPAP   National Acid Precipitation Assessment Program
NAPCA   National Air Pollution Control Administration
NAS   National Academy of Sciences
NASA   National Aeronautics and Space Administration
NASQAN   National Stream Quality Accounting Network
NBS   National Biological Service
NCASI   National Council of the Paper Industry for Air and Stream
           Improvement
NCE   National Commission on the Environment
NCEA   National Center for Environmental Assessment, EPA
NCEH   National Center for Environmental Health
NCERQA   National Center for Environmental Research and Quality
           Assurance, EPA
NCI   National Cancer Institute
NCLAN   National Crop Loss Assessment Network
NCTR   National Center for Toxicological Research, DHHS
NERL   National Exposure Research Laboratory, EPA
NESHAP   National Emission Standards for Hazardous Air Pollutants
NHANES   National Health and Nutrition Examination Survey
NHANES
   II   Second National Health and Nutrition Examination Survey
NHANES
   III   Third National Health and Nutrition Examination Survey
NHATS   National Human Adipose Tissue Survey
NHEERL   National Health and Environmental Effects Research Labora-
           tory, EPA

NHEXAS	National Human Exposure Assessment Survey
NHTSA	National Highway Traffic Safety Administration
NIE	National Institute for the Environment
NIEHS	National Institute of Environmental Health Sciences, DHHS
NIH	National Institutes of Health
NIOSH	National Institute for Occupational Safety and Health, DHHS
NOAA	National Oceanic and Atmospheric Administration
NOAEL	no observed adverse effects level
NOEL	no observed effects level
NO_x	nitrogen oxides
NPDWR	National Primary Drinking Water Regulation
NPL	National Priorities List
NPS	National Park Service
NRC	National Research Council
NRDC	Natural Resources Defense Council
NRMRL	National Risk Management Research Laboratory, EPA
NSF	National Science Foundation
NTP	National Toxicology Program
NWF	National Wildlife Federation
NYU	New York University
OAQPS	Office of Air Quality Planning and Standards, EPA
OAR	Office of Air and Radiation, EPA
OD	oxygen delignification
ODW	Office of Drinking Water, EPA
OER	Office of Exploratory Research
OGC	Office of General Counsel, EPA
OGWDW	Office of Ground Water and Drinking Water, EPA
OHEA	Office of Health and Environmental Assessment, EPA
OMB	Office of Management and Budget, Executive Office of the President
OPA	Office of Policy Analysis, EPA
OPP	Office of Pesticide Programs, EPA
OPPE	Office of Policy, Planning and Evaluation, EPA
OPPTS	Office of Prevention, Pesticides, and Toxic Substances, EPA
OPTS	Office of Pesticides and Toxic Substances, EPA
ORD	Office of Research and Development, EPA
ORI	Office of Research Integrity, DHHS
ORSI	Office of Research and Science Integration, EPA
OSHA	Occupational Safety and Health Administration
OSP	Office of Science Policy, EPA
OSPR	Office of Special Pesticide Reviews, EPA
OST	Office of Science and Technology, EPA
OSTP	Office of Science and Technology Policy, The White House

OSW	Office of Solid Waste, EPA
OSWER	Office of Solid Waste and Emergency Response, EPA
OTA	Office of Technology Assessment, U.S. Congress
OTS	Office of Toxic Substances, EPA
OW	Office of Water, EPA
PbB	blood lead
PCB	polychlorinated biphenyl
PD	position document
PDP	Pesticide Data Program
PHS	Public Health Service
PM	particulate matter
PM-10	particulate matter, 10-micron diameter
PM-2.5	particulate matter, 2.5-micron diameter
PMN	premanufacture notice
ppb	parts per billion
ppm	parts per million
ppq	parts per quadrillion
ppt	parts per trillion
PQL	practical quantitation limit
PRA	Paperwork Reduction Act of 1980
PRG	preliminary remediation goal
PRP	potentially responsible party
PSR	Physicians for Social Responsibility
R&D	research and development
RAC	Risk Assessment Council
RAF	Risk Assessment Forum, EPA
RAGS	Risk Assessment Guidelines for Superfund
RBCA	risk-based corrective action
RCRA	Resource Conservation and Recovery Act of 1976
RFA	request for applications
RfC	reference concentration
RfD	reference dose
RFP	request for proposals
RI	remedial investigation
RIA	regulatory impact analysis
RIHRA	Research to Improve Health Risk Assessment
ROD	record of decision
RPAR	rebuttable presumption against registration
RPM	remediation project manager
RQ	reportable quantity
RSAC	Research Strategies Advisory Committee
RTI	Research Triangle Institute
RTP	Research Triangle Park, North Carolina

S&T	science and technology
SAB	Science Advisory Board, EPA
SAP	Scientific Advisory Panel
SAR	structure activity relationship
SARA	Superfund Amendments and Reauthorization Act of 1986
SBRP	Superfund Basic Research Program
SDWA	Safe Drinking Water Act of 1974
SEGH	Society for Environmental Geochemistry and Health
SEPA	Swedish Environmental Protection Agency
SO_2	sulfur dioxide
SO_x	sulfur oxides
SP	staff paper
SPC	Science Policy Council, EPA
SPRD	Special Pesticide Review Division, EPA
STAR	Science to Achieve Results
TCDD	tetrachlorodibenzo-para-dioxin
TCDF	tetrachlorodibenzofuran
TEF	toxic equivalency factor
TERA	Toxicology Excellence for Risk Assessment
TMDL	total maximum daily load
TRI	Toxic Release Inventory
TSCA	Toxic Substances Control Act of 1976
TSP	total suspended particulates
UCS	Union of Concerned Scientists
USDA	United States Department of Agriculture
USDC	U.S. district court
USGS	United States Geological Survey, DOI
VOC	volatile organic compound
WHO	World Health Organization

Index